Frontiers on Recent Developments in Plant Science (*Volume 1*)

Edited By

Aakash Goyal

Canadian Wheat Breeder
Bayer Crop Science
Canada

&

Priti Maheshwari

Research Scientist Plant Transformation Division
Plantbiosys Ltd, Lethbridge
Canada

CONTENTS

Foreword   *i*

Preface   *ii*

List of Contributors   *iii*

CHAPTERS

1. **Plant Secondary Metabolites: From Diseases to Health**   3
   Rakesh Yadav, Pooja Arora and Ashok Chaudhury

2. **RNAi: New Era of Functional Genomics for Crop Improvement**   24
   Pradeep Kumar, Madhu Kamle and Ashutosh Pandey

3. **Biological Nitrogen Fixation: Host-*Rhizobium* Interaction**   39
   Pooja Arora, Rakesh Yadav, Neeraj Dilbaghi and Ashok Chaudhury

4. **Recent Advances in Sago Palm (*Metroxylon Sagu* Rottboell) Micropropagation**   60
   Annabelle U. Novero

5. **Tools for Generating Male Sterile Plants**   67
   Sudhir P. Singh, Joy K. Roy, Dinesh Kumar and Samir V. Sawant

6. **Anti-Bacterial and Crystallographic Studies of Jatrophone, the Macrocyclic Diterpenoid from the Roots of *Jatropha Gossypifolia* l.**   86
   R. S. Satyan, Ajay Parida and Babu Varghese

7. **Biological Control of Plant Diseases by *Serratia* Species: A Review or a Case Study**   99
   Dipanwita Saha, Gargi Dhar Purkayastha and Aniruddha Saha

8. **Molecular Approaches for Detection of Plant Pathogens**   116
   Dipali Majumder, Thangaswamy Rajesh and Thalhun Lhingkhanthem Kipgen

9. **Role of NACs in Regulation of Abiotic Stress Responses in Plants**   137
   S. Puranik and M. Prasad

10. **Phytoremediation: A New Hope for the Environment**   149
    Sarvjeet Kukreja and Umesh Goutam

    Index   172

FOREWORD

World population has already exceeded seven billion, and further food production is required towards 2060 when the population will be at its peak. Besides this, global climate accompanied with drought across various regions in the world is another big challenge. Also excessive use of ground water and natural resources is drawing towards depletion of these gifts. Currently, food production is unable to meet the increasing demand, and we have failed to compete with the ever increasing food prices. This situation has resulted in a widening disparity between the rich and the poor at both state and personal levels, and consequently has lead to conflicts and political instability.

A combination of wisdoms from different aspects is needed to challenge these global issues, and to avoid periods of tribulations that the humanity may suffer several decades from now. Fortunately, in the last century we could deepen our knowledge of living organisms at the molecular level, which opened a new field of science - biotechnology. This technology expresses great power in medical and agricultural sciences. Many excellent crops have been created by biotechnology and have already appeared in the market. This sharp knife is a key tool to cut out the global issues we are facing.

The eBook edited by Dr. Aakash Goyal and Dr. Priti Maheshwari provides an insight into the current trends in Plant sciences. It gives a broad overview of the implementation of current molecular biology techniques and its increasing role in crop and medicinal plant research. Chapters written by experts in their respective fields will make the reader acquainted with a variety of topics ranging from the plant secondary metabolites, RNAi and its role in crop improvement, biological nitrogen fixation, new trends of micropropagation, medicinal and biological control properties of plants, and the increasing role of plants in phytoremediation. As such, the volume should be particularly useful in getting updates on the given areas of study and will equally help researchers and educators with providing latest advances in Plant Science. The chapters are contributed by up-and-coming young plant researchers, with consideration of plant science applications for agriculture, medical and environmental issues. Great emphases are paid for novel methodologies, use of new crops and undeveloped but promising plant species in the coming decades. I believe that, plant science will strongly contribute to solve the global issues, and hope this century will be named 'The Century of Happiness' when our next generations explore the history.

Hisashi Tsujimoto
Arid Land Research Center
Tottori University
Japan

PREFACE

Nobody knows when and where plants were used for the first time by human beings. Plants were used for food, fire, forage and even for medicinal purposes since prehistoric times. Plants have an entire kingdom of life forms, and are the most essential group of organisms in our world. Without plants, all other forms of life would cease to exist. When humans started building colonies and getting civilized their dependency on plants increased by several folds. Today plants are even more important due to their increased demand of for their different uses.

Research in plant sciences has a significant role to play in many wide areas including agriculture, energy, and environmental stewardship. Today we are dealing with an urgent issue on how to feed the growing world population. With increased and effective plant science research we can develop methods to provide sustainable food supply towards fulfillment of the increased food demand. Second issue is related to the rapidly changing environment leading to deterioration of natural resources and increased global warming. With advancements in plant sciences we can explore how plants can cope with less water, rising temperatures, and other environmental stresses. Plant science research can enable scientists to develop crops that can withstand changing climate conditions alongside of increased productivity. We really need to think today how plants can contribute as environmentally, economically, and socially sustainable sources of energy?

In past decade plant science research has flourished in many different and newer areas *e.g.* plant genetics, genomics, molecular biology, epigenetics and proteomics. This broadened research has augmented our knowledge to understand plants better than before. With the help of several new concepts and powerful strategies like Marker Assisted Selection (MAS), LD mapping, Association mapping, Gene cloning, Radiation hybrid mapping, RNA interference, TILLING, Expression Genetics, and Genome sequencing of model plants, we are now able to improve and think of efficient and different uses of plants.

This multi authored edited eBook "Advances in Plant Science" is an attempt to put forth a compilation of work in various areas of plant science of several scientists and post-doctoral fellows across the world, all well known, distinguished and experts in different frontiers of plant science research. This eBook is an effort to gather the successes achieved in areas of plant genetics, molecular biology and breeding aspects across all the major continents. The chief objective of the eBook hence is to deliver state of the art knowledge, information to comprehend the advancement of plant science research to its readers.

REVIEW PROCESS

Each chapter of the eBook has undergone a double-blinded review process being reviewed by two independent reviewers. The reviewers were selected for their active expertise in the field of the respective chapter. After review, the authors made all probable corrections in the light of reviewers comments after which the chapters were accepted. The constructive comments and critical advice of the reviewers have greatly improved the quality and content of this eBook.

ACKNOWLEDGEMENTS

First of all the Editor's would like to thank all the Author's for their outstanding efforts, and timely work in producing such fine chapters. We also greatly appreciate all the reviewers for their time to review the respective chapters. We would also like to thank Bentham Science Publishers and Salma for her clerical assistance, advice and encouragement during the development of this eBook. Last but not least heartfelt thanks goes to our families, and parents for their love, encouragement and vision that unveiled in us from our earliest years, the desire to thrive on the challenge of always striving to reach the highest mountain in everything we do.

Aakash Goyal
Canadian Wheat Breeder
Bayer Crop Science
Canada

Priti Maheshwari
Research Scientist Plant Transformation Division
Plantbiosys Ltd, Lethbridge
Canada

List of Contributors

Ajay Parida
Bioprospecting & Tissue Culture Lab, M. S. Swaminathan Research Foundation, III Cross Street, Taramani Institutional Area, Taramani, Chennai, (T.N.), India

Aniruddha Saha
Department of Botany, University of North Bengal, Siliguri, (W.B.), India

Annabelle U. Novero
College of Science and Mathematics, University of the Philippines Mindanao, Mintal, Davao City, Philippines

Anu Singh
ITS Paramedical College, Ghaziabad, (U.P.), India

Ashok Chaudhury
Department of Bio and Nano Technology, Guru Jambheshwar University of Science and Technology, Hissar, (Haryana), India

Ashutosh Pandey
Central Institute for Subtropical Horticulture, Rehmankhera, Lucknow, (U.P.), India

Babu Varghese
X-Ray Crystallography Department, Sophisticated Analytical Instrumentation Facility (SAIF), Indian Institute of Technology (IIT), Chennai, (T.N.), India

Dinesh Kumar
University of Lucknow, Lucknow, (U.P.), India

Dipali Majumder
Plant Bacteriology, College of Post Graduate Studies, Central Agricultural University, Umiam, (Meghalaya), India

Dipanwita Saha
Department of Biotechnology, University of North Bengal, Siliguri, (W.B.), India

Gargi D. Purkayastha
Department of Biotechnology, University of North Bengal, Siliguri, (W.B.), India

Joy K. Roy
National Botanical Research Institute, Council of Scientific and Industrial Research, Lucknow, (U.P.), India. Current Address: National Agri-Food Biotechnology Institute, Department of Biotechnology, Mohali, (Punjab), India

Madhu Kamle
Central Institute for Subtropical Horticulture, Rehmankhera, Lucknow, (U.P.), India

Manoj Prasad
National Institute of Plant Genome Research, Aruna Asaf Ali Marg, New Delhi, India

Neeraj Dilbaghi
Department of Bio and Nano Technology, Guru Jambheshwar University of Science and Technology, Hissar, (Haryana), India

Pooja Arora
Department of Bio and Nano Technology, Guru Jambheshwar University of Science and Technology, Hissar, (Haryana), India

Pradeep Kumar
Central Institute for Subtropical Horticulture, Rehmankhera, Lucknow, (U.P.), India

Rakesh Yadav
Department of Bio and Nano Technology, Guru Jambheshwar University of Science and Technology, Hissar, (Haryana), India

R. S. Satyan
Bioprospecting & Tissue Culture Lab, M. S. Swaminathan Research Foundation, III Cross Street, Taramani Institutional Area, Taramani, Chennai, (T.N.), India

Samir V. Sawant
National Botanical Research Institute, Council of Scientific and Industrial Research, Lucknow, (U.P.), India. Current Address: National Agri-Food Biotechnology Institute, Department of Biotechnology, Mohali, (Punjab), India

Sarvjeet Kukreja
Department of Biotechnology, Lovely Professional University, Jalandhar, (Punjab), India

Sudhir P. Singh
National Botanical Research Institute, Council of Scientific and Industrial Research, Lucknow, India-226 001. Current Address: National Agri-Food Biotechnology Institute, Department of Biotechnology, Mohali, (Punjab) India

Swati Puranik
National Institute of Plant Genome Research, Aruna Asaf Ali Marg, New Delhi, India and Department of Biotechnology, Faculty of Science, Jamia Hamdard, New Delhi, India

Thalhun L. Kipgen
Plant Pathology, College of Post Graduate Studies, Central Agricultural University, Umiam, (Meghalaya), India

Thangaswamy Rajesh
Plant Pathology, College of Post Graduate Studies, Central Agricultural University, Umiam, (Meghalaya), India

Umesh Goutam
Department of Biotechnology, Lovely Professional University, Jalandhar, (Punjab), India

2

Frontiers on Recent Developments in Plant Science Vol. 1, 2012, 3-23

CHAPTER 1

Plant Secondary Metabolites: From Diseases to Health

Rakesh Yadav, Pooja Arora and Ashok Chaudhury*

Department of Bio and Nano Technology, Guru Jambheshwar University of Science and Technology, Hisar-125001 (Haryana), India

Abstract: Plant-derived medicines constitute a significant component of today's human healthcare systems in industrialized as well as developing countries. Plants are capable of synthesizing a huge variety of small organic molecules, called secondary metabolites, usually composed of very complex and unique carbon skeleton structures. The abundance and diversity of secondary metabolites are coupled with wide pharmaceutical, therapeutic and medicinal values. Most of the plant derived biomedicines are widely used for recreation and stimulation (the alkaloids nicotine and cocaine; the terpene cannabinol) purposes. Considerable progress has been made in the technology of large-scale plant cell culture for the industrial production of plant-derived fine chemicals. Keeping in view of the huge world market for plant secondary metabolites, it is a challenge for biotechnologists to find techniques to produce these compounds in sufficient quantity and quality. Biotechnological intervention through micropropagation, biotransformation, and metabolic engineering can make the process quite profitable to exploit the productive potential of living cells for the production of bioactive compounds. Therefore, the increasing global demand for biomedicines can only be achieved by application of biotechnological approaches.

Keywords: Secondary metabolites, biomedicines, metabolic engineering, medicinal plants.

INTRODUCTION

Medicinal and aromatic plants have immense worth to human welfare for generation of novel compounds. Over the years, many new disciplines cautious to human health have been pioneered by sighting of the great scientist Louis Pasteur in the year 1822-1895. He gave the first landmark hint to metabolite production with a paradigm of kinetic resolution by selectively destructing d-enantiomer leading to production of dl-ammonium tartarate from *Penicillium glaucun*. Later on, many general and biosynthetic metabolic pathways were discovered *via* biotransformation of intermediates and compounds leading to production of secondary metabolites. Phytochemicals are often classified as primary or secondary metabolites. In plant kingdom, taxonomic-group-specific huge varieties of secondary metabolites are synthesized by specific biosynthetic enzymes from primary metabolites [1]. Secondary metabolites are not essential for plant growth and development but required by them to withstand certain specific conditions such as pathogen defense mechanism. These can be categorized on the basis of chemical structure (having rings, containing a sugar), composition (containing nitrogen or not), their solubility in various solvents, or the pathway by which they are synthesized (phenylpropanoid, which produces tannins). A simple classification includes the alkaloids (vinblastine, vincristine), the terpenes (made from mevalonic acid, composed almost entirely of carbon and hydrogen) and terpenoids (artemisinin, paclitaxel), the phenolics (made from simple sugars, containing benzene rings, hydrogen, and oxygen), and nitrogen-containing compounds (extremely diverse, may also contain sulfur) [2].

Plant secondary products usually act as signal molecules in various environmental associated ecological factors such as pollen attractant, chemical adaptor to stresses, defensive against microbes, predators and other plants. Many of these secondary metabolite products are utilized in pharmaceuticals (steroids and alkaloids), fragrances (scents), dyes (alkanin, shikonin) and pesticides (nicotine, pyrethrins and rotenone) and are often called as fine chemicals. The steroids and alkaloids further include sapogenins, *Digitalis*

*Address correspondence to Ashok Chaudhury:** Professor and Chairman, Department of Bio and Nano Technology, Guru Jambheshwar University of Science and Technology, Hisar-125001 (Haryana), India; Phone: +91-1662-263306, Fax: +91-1662-276240, E-mail: ashokchaudhury@hotmail.com

Aakash Goyal and Priti Maheshwari (Eds)

glycosides, *Catharanthus* anticancer alkaloids, *Belladona* alkaloids, cocaine, colchicines, opium alkaloids, physostigimine, pilocarpine, quinidine glucosinolates and reserpine. Biologically active secondary metabolites serve as a chemical factory for design and synthesis of new drug entities *viz.*, morphine and codeine act as model to design synthetic analgesic drugs meperidine, pentazocine and propoxyphene, while aspirin is generated from simple derivative of naturally present salicylic acid (willow tree-*Salix* spp.).

Biotechnologists need to exercise for finding ways to produce novel bioactive compounds that serve as biomedicines [3]. Traditionally secondary metabolites were produced by growing the respective plants in the field or in greenhouses and to extract the prodrugs from them. For several medicinal plants, new genotypes have been selected with superior quality and yields. Plant cell, tissue and organ culture are significant systems for *in vitro* propagation of medicinal plants. Certainly, metabolic engineering has imposed a direct influence to improve secondary metabolite quality as well as quantity. Key factor in modern medicine novelty relies upon screening of fine chemicals against molecular targets. Therefore, how plants hold tremendous applications in production of useful secondary metabolites for human welfare have been discussed and presented here.

PLANT CELL AS A SOURCE OF BIOMEDICINE

Exploitation of therapeutic values in plants root a new phenomenon and people around the world have been applying preparations and extracts from thousands of medicinal plants for curing several diseases, dating back to prehistoric times. Moreover, in rural areas that previously lacked medical facilities, people utilize extracts from local medicinal plants such as *Artemisia, Azadirachta indica, Withania somnifera, Zingiber officinalis, Ocimum sacnctum, Calotropis procera, Cathranthus roseus* to cope up with various gastrointestinal, urinary, respiratory, endocrinal, dermatological, behavioral, ritual, and reproductive diseases. In the modern era, there has been a considerable development in the plant derived biomedicine consumption in healthcare sector. Plant based secondary metabolites furnish characteristic effects such as growth promotion, immunostimulant, antistress, antibacterial, antifungal, antivirals, appetite stimulators and aphrodisiac biological effects.

The application of certain antibiotics and other synthetic drugs has shown sensitization reaction, other undesirable side effects leading to biomagnifications. In view of serious admonition, FDA and other regulatory authorities have strictly instructed not to use certain antibiotics, such as chloramphenicol, nitrofurans including nitrofurazone, furazolidone, furaltadone, furylfuramide, nitrofurantoin, nifuratel, nifursoxime, nifurprazine and their derivatives, and also neomycin, nalidixic acid, sulphamethoxazole, and preparations thereof, and chlorpromazine, colchicine, dapsone, dimetridazole, metronidazole, ronidazole, ipronidazole and other nitroimidazoles, clenbuterol, diethylstilbestrol (DES), sulonamide (except approved sulfadimethoxine, sulfabromomethazine and sulfaethoxyrpyiadine) and floroquinolones and glycopeptides [3]. Globally, people have understood the direct effect of synthetic medicines and antibiotics, in consequence, they are now shifting over to herbal drugs and natural biomedicines.

MOLECULAR MODES OF ACTION OF BIOMEDICINES

The bioactivity of plant secondary metabolites has been extensively described by several authors [4, 5]. Fortunately, the evolution of bioactive compounds in plants not only provides defence mechanism to them but also promises beneficial confrontation in biotechnology, pharmacy and medicine. The structures of many secondary metabolites have been so shaped that interact with diverse molecular and cellular targets, including enzymes, hormone receptors, neurotransmitter receptors and transmembrane transporters, and can thus mimic a response at the corresponding molecular target as depicted in Fig. **1**. Plant secondary metabolites are extremely able compounds that can modulate any cellular target. Therefore, plants produce a wide range of bioactive compounds, and many of these are already in extensive use in the pharmacological, medical and agricultural industries, while others are under development. In many cases plant secondary metabolites, *e.g.* the terpenoid essential oils, can be more effective than synthetic drugs because they get a complex mixture of bioactive components. Their complexity enables the secondary metabolites to interact with multi-molecular targets and, thus, create a difficulty for target microorganisms or herbivores to acquire any effective resistance or response. Indeed, plant secondary metabolites carry well-established properties that make them active against diverse human and animal diseases.

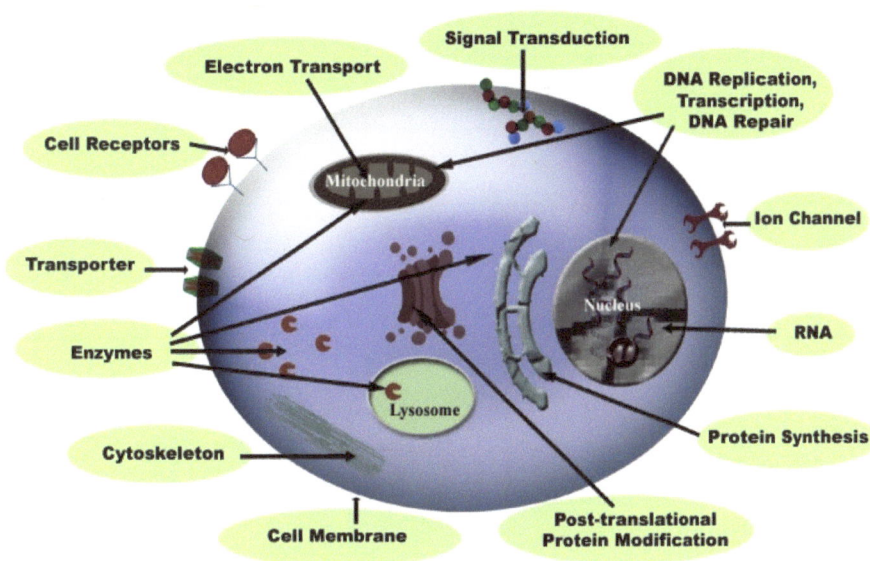

Figure 1: Summary of potential cellular targets of plant secondary metabolites.

BIOACTIVITY OF PLANT SECONDARY METABOLITES

Biomedicine as Antibacterial, Antifungal and Antiviral Agents

Plants produce an enormous variety of secondary metabolites as natural protection against microbial and insect attack. While some of these compounds are toxic to animals, but others may not be toxic. Undeniably, many of these compounds have been used in the form of whole plants or plant extracts for food or medical applications in service to mankind. Some selective examples of antimicrobial properties of medicinal plants that have been progressively reported from different parts of the world have been summarized in Table **1**. The broader potential of plant secondary metabolites illustrated by World Health Organization (WHO) report estimated that 80% of the world's population use plants, their extracts and, active compounds as biomedicine in traditional therapies.

Table 1: Biomedicine obtained from different medicinal plants as antibacterial, antifungal and antiviral agents

Plant Source	Compound	Activity Against Microorganism	Reference(s)
Aegle marmelos	Terpenoid	Fungi	[6]
Allium cepa	Allicin Sulfoxide	Bacteria, *Candida*	[7]
Allium sativum	Allicin, ajoene Sulfoxide	General	[8-10]
Aloe barbadensis, A. vera	Latex	*Corynebacterium, Salmonella, Streptococcus, S. aureus*	[11]
Anacardium pulsatilla	Salicylic acids	*P. acnes*, Bacteria, fungi	[12]
Berberis vulgaris	Berberine	Bacteria, protozoa	[13, 14]
Camellia sinensis	Catechin	General, *Shigella, Vibrio, S. mutans*, Viruses	[15-18]
Capsicum annuum	Capsaicin	Bacteria	[19, 20]
Carum carvi	Coumarins	Bacteria, fungi, viruses	[21-23]
Citrus paradisa	Terpenoid	Fungi	[24]
Citrus sinensis	Terpenoid	Fungi	[24]
Curcuma longa	Curcumin	Bacteria, protozoa	[25]

Table 1: cont....

Galium odoratum		General, Viruses	[20, 22, 26]
Harrisonia abyssinica	Flavonoids	*Salmonella, Klebsella*	[27]
Hydrastis canadensis	Berberine, hydrastine	Bacteria, *Giardia duodenale, trypanosomes*	[28]
Mahonia aquifolia	Berberine	*Plasmodium, Trypansomes*, general	[14, 28]
Matricaria chamomilla	Anthemic acid	*M. tuberculosis, S. typhimurium, S. aureus*, helminths	[20-22]
Melissa officinalis	Tannins	Viruses	[29]
Millettia thonningii	Alpinumisoflavone	*Schistosoma*	[30]
Ocimum basilicum	Terpenoids	Salmonella, bacteria	[31]
Olea europaea	Hexanal	General	[32]
Onobrychis viciifolia	Tannins	Ruminal bacteria	[33]
Petalostemum petalostemumol	Flavonol	Bacteria, fungi	[34]
Piper nigrum	Piperine	Fungi, *Lactobacillus, Micrococcus, E. coli, E. faecalis*	[35]
Podocarpus nagi	Totarol	*P. acnes*, other gram-positive bacteria	[36]
Rabdosia trichocarpa	Trichorabdal A	*Helicobacter pylori*	[37]
Santolina chamaecyparissus		Gram-positive bacteria, *Candida*	[38]
Satureja montana	Carvacrol	General	[39]
Tribulus terrestris	Flavonoids	*Klebsella*	[27]
Vaccinium spp.	Fructose	*E. coli*	[40]
Vaccinium spp.	Fructose	Bacteria	[40, 41]

BIOMEDICINE AS IMMUNOSTIMULANTS

The term immunostimulant describes drugs capable of growing the resistance of an organism against stressors of disparate origin. Drugs achieve this enhancement primarily by imprecise mechanism of actions and stimulate the function and efficiency of immune system in a non-antigen dependant manner in order to reinforce it against microbial attack or immunosuppressive states. Predominantly, they influence the humoral and cellular immune system and are effective prophylactically as well as therapeutically. They are similar to vaccine as must be applied in low doses to obtain optimal results. Their exact mode of action is still not clear. Two major types of plant-derived immunostimulants that are available at drug market include Plant extracts and polysaccharides of fungal origin [42]. They are widely used in Europe and Asia in traditional therapeutic applications. Although numerous plants have been known to produce immunostimulants, metabolites of *Echinacea purpurea, E. angustifolia* and *E. pallida* are most prominent. It comes into view that their immunostimulating activity is associated with lipophylic compounds including alkylamides and polar fraction including cichoric acid and polysaccharides. Most frequently they have been used for chronic and recurrent infection in respiratory and urogenital organs; chronic inflammations; sinusitis; retarted wound healing; ulcuscruris, eczemas, and psoriasis; chronic bronchitis; malignant diseases [42, 43]. Application of *Echinacea* extract blended with *Baptisia* extract during the treatment of leucopenia cancer patients undergoing radiotherapy can significantly increase the number of leucocytes [44]. Among the isolated plant immunostimulants available today, Lentinan (*Lentinus edodes*), Schizophyllan (*Schizophyllum commune*) and Kresin (*Coriolus versicolor*) are highly significant [42]. With the exception of those diseases for which vaccine treatment is obligatory, immunostimulants drugs may be sufficient for preventive measures and adjuvant therapy.

BIOMEDICINE AS ANTI-CANCEROUS DRUG

Cancer is a potent degenerative disease, affecting people at all ages with the risk for most types increasing with age [45]. Cancer caused about 13% of all human deaths in 2007 [46]. Cancer research is the intense scientific effort to understand disease processes and discover possible therapies. Resveratrol, a low molecular weight stilbene-type phytoalexins is induced in grapevine in response to fungal attack [47]. Besides antioxidant and antimutagen activity, it can induce phase-II drug-metabolizing enzyme (anti-initiation activity); and inhibit cyclooxygenase and hydroperoxidase function (anti-promotion activity), therefore, persuade promyelocytic leukemia cell differentiation in human (anti-progression activity) [48]. Also, shikimate pathway derived polyphenols (vitamin P) have been reported to uphold interesting potential to reduce risk of cancer disease [49]. Likewise, neoflavonoids, another polyphenol derived from 4-phenylcoumarine (4-phenyl-1,2-benzopyrone) structure can be used as anti-cancer drug.

BIOMEDICINE AS ANTI-STRESS AGENTS

Stress symptoms usually include a state of alarm and adrenaline production, short-term resistance as a coping mechanism, and exhaustion, as well as irritability, muscular tension, inability to concentrate and a variety of physiological reactions such as headache and elevated heart rate [50]. Antistress plant drugs have been reported earlier [51-54] to produce non-specifically increased resistance in animal or human under stress. They are known to reduce alarm reaction of stress in animals and emotional stress in man. Emotional stress in man causes a variety of intestinal disturbances including hypermotility of the gastrointestinal tract. Geriforte, a plant metabolite and roots of *Panax ginseng* both can be used clinically as anti-stress (adaptogenic) agents and *Ocimum sanctum* and *Rosa damacena* being in process for clinical application showed significant decrease in intestinal transit, while, this effect did not directly relate with their potency as anti-stress agents.

BIOMEDICINE WITH EFFICACY AGAINST HUMAN IMMUNODEFICIENCY VIRUS (HIV)

Far and wide interest has been garnered for effectual therapies against HIV infection both inside as well as outside the laboratories, worldwide. A successful *in vivo* anti-HIV test has been studied through mice infection. Glycyrrhizin, a characteristic metabolite in *Glycyrrhiza* plants extended the life of infected mice up to 14-17 weeks [55]. *In vitro* antiviral effects and cellular toxicity studies of crude extract of *Opuntia streptacantha* was found to be curative and safe in mice, horses, and humans [56]. Since, plant drugs/biomedicines enclose a very broad scope of anti-HIV studies; their contribution has been summarized in Table **2**. Bioactive compounds from plants target various events of virion integration, replication processes or enzymes essential for them including reverse transcriptase, integrase and protease. A few other biomedicines have unidentified effect against HIV.

Table 2: Various plant secondary metabolite exhibiting anti-HIV activity

Target Viral component	Metabolite	Plant Source	Reference(s)
Adsorption	Mannose-specific lectins	Snowdrop (*Galanthus*), daffodil (*Narcissus*), amaryllis (*Amaryllis*), *Gerardia*	[57]
	Prunellin	*Prunella*	[58]
	Schumannificine 1	*Schumanniophyton magnificum*	[59]
Integrase	Curcumin	*Curcuma longa*	[60]
	MAP30, GAP31, DAP 32, DAP 30	*Momordica charantia, Gelonium multiflorum*	[60-64]
	Quercetin	*Quercus rubra*	[65]
Interference with cellular factors	Chrysin	*Chrysanthemum morifolium*	[66]
	Hypericin	*Hypericum*	[67, 68]
	Protein	*Phytolacca*	[67]
	Trichosanthin, momorcharins	Cucurbitaceae family	[69]
Protease	Carnosolic acid	*Rosmarinus officinalis*	[70]
	Ursolic acid Terpenoids	*Geum japonicum*	[71]
Reverse transcriptase	Betulinic acid, platanic acid	*Syzigium claviflorum*	[72]

Table 2: cont….

	Faicalein, quercetin, myricetin, baicalin	*Quercus rubra*, others	[73-75]
	Nigranoic acid	*Schisandra sphaerandra*	[76]
	Psychotrines	Ipecac (*Cephaelis ipecacuanha*)	[77]
	Michellamine B	*Ancistrocladus korupensis*	[78]
	Suksdorfin	*Lomatium suksdorfii*	[79]
	Coriandrin	*Coriandrum sativum*	[68]
	Caffeic acid	*Hyssop officinalis*	[80]
	Cornusin, others	*Cornus officinalis* and others	[81]
	Swertifrancheside	*Swertia franchetiana*	[82]
	Salaspermic acid	*Trypterygium wilfordii*	[83]
	Glycyrrhizin	*Glycyrrhiza rhiza*	[55, 84]
	Methyl nordihydroguaiaretic acid	Various trees	[85]
	Thuja polysaccharide	*Thuja occidentalis*	[86]
	Hydroxymaprounic acid, hydroxybenzoate	*Maprounea africana*	[87]
Syncytium formation	MAR-10	*Hyssop officinalis*	[88]
	Michellamine B	*Ancistrocladus korupensis*	[78]
	Propolis	Various trees	[89, 90]
Unidentified	Chrysin	*Chrysanthemum morifolium*	[91]
	Sulfated polysaccharide	*Prunella vulgaris*	[92]
	Zingibroside R-1	*Panax zingiberensis*	[93]
Viral fusion	Mannose- and *N*-acetylglucosamine-specific lectins	*Cymbidium* hybrid, *Epiactic helleborine*, *Listeria ovata*, *Urtica dioica*	[94]

MAJOR GROUPS OF SECONDARY METABOLITES OBTAINED FROM PLANTS PHENOLICS AND POLYPHENOLS

Phenolics and polyphenols are the group of chemical substances characterized by the presence of one or more phenol unit or building block per molecule. These are generally divided into (a) hydrolyzable tannins (gallic acid esters of glucose and other sugars) and (b) phenylpropanoids, such as lignins, flavonoids, and condensed tannins. Most of the polyphenols including tannins, lignins, and flavonoids are derived from shikimate pathway of plant secondary metabolism.

They were once briefly known as vitamin P because of their well-established health benefits. It is indicated that polyphenols have antioxidant characteristics contributing to potential health welfare and can reduce the risk of degenerative diseases including cancer and cardiovascular ailments [49]. It can bind with nonheme iron (*e.g.* from plant sources) *in vitro* in model systems [95] and may decrease its absorption by the body. Xanthohumol, a prenylated chalcone from hops and beer, has a range of interesting biological properties that have therapeutic utility thus, have received the most attention. [96]. It is entailed that high levels of polyphenols can generally be found in the fruit skins out of which only a fraction of the total content can be estimated, which includes "non-extractable" polyphenols [97].

Simple Phenols and Phenolic Acids

Phenols and phenolic acids, the active constituents of plant cells, are fat-soluble antioxidant and food additives. Salicylic acid is good example that naturally occurs as plant hormone and used for its analgesic, antipyretic, and anti-inflammatory properties, is also a precursor compound to Aspirin.

Quinones and Naphthoquinones

Quinones are phenolic compounds having a fully conjugated cyclic dione structure, such as that of benzoquinones, derived from aromatic compounds by oxidation of an even number of –CH= groups into –C(=O)– groups with any necessary polycyclic and heterocyclic analogues rearrangement [98]. Few are common constituents of biologically relevant molecules (*e.g.* Vitamin K1 is phylloquinone), others serve as electron acceptors in electron transport chains including Photosystems I & II of photosynthesis, and aerobic respiration. Naphthoquinones, an important group of bioactive compounds comprised of two enatiomers alkanin and shikonin. These compounds are major component of the deep red pigments that are easily extracted from the roots of the plants. Alkanine is found in the roots of the plant *Alkanna tinctoria* and known to share medicinal values for healing of injuries from prehistoric era. On the other hand, shikonin and their derivatives have more recently been tested, scientifically, for their therapeutic values including analgesics, antipyretic and wound healing properties. It is naturally obtained from *Lithospermum erythrorhizone* and has multifaceted pharmacological applications as an antibacterial agent, granulation tissue forming activity and anti-ulcer activity [99].

Flavones, Flavonoids, and Flavonols

Flavonoids are synthesized by the phenylpropanoid metabolic pathway in plant where amino acid phenylalanine is employed to produce 4-coumaroyl-CoA [100]. These ketone-containing compounds can be classified into three classes; (a) flavonoids, derived from 2-phenylchromen-4-one (2-phenyl-1,4-benzopyrone) structure (*e.g.* quercetin, rutin), (b) isoflavonoids, derived from 3-phenylchromen-4-one (3-phenyl-1,4-benzopyrone) structure, (c) neoflavonoids, derived from 4-phenylcoumarine (4-phenyl-1,2-benzopyrone) structure. Apart from antioxidant properties, they can be used as anti-cancer (Stauth) and anti-diarrhoea drug [101].

Tannins

Tannins are astringent, bitter polyphenolics that either can bind and precipitate or shrink proteins and several other organic compounds including amino acids and alkaloids. The dry and puckery feeling in the mouth after the consumption of unripened fruit or red wine is caused by astringency from the tannins [102]. Likewise, the destruction or modification of tannins with time plays an important role in the ripening of fruit and the aging of wine. They have molecular weights ranging from 500 to over 3,000 and are large polyphenolic compound containing sufficient hydroxyls and other suitable groups (such as carboxyls) to form strong complexes with proteins and other macromolecules [103]. Tannins are incompatible with alkalis, gelatin, heavy metals, iron, lime water, metallic salts, strong oxidizing agents and zinc sulfate, since they form complexes and precipitate in aqueous solution. The compounds are widely distributed in plant Kingdom, where they play a vital role in protection from predation and perhaps also in regulation of growth. They are natural preservative in wine and have broad medical applications, particularly antidiarrhoeal, hemostatic, and antihemorrhoidal activity.

Coumarins

Coumarin found in many plants such as tonka bean (*Dipteryx odorata*), vanilla grass (*Anthoxanthum odoratum*), woodruff (*Galium odoratum*), mullein (*Verbascum* spp.), and sweet grass (*Hierochloe odorata*), notably in high concentration. It has a sweet odor, readily recognised as the scent of newly-mown hay, and has been used in perfumes. It has clinical and medicinal value as the precursor for several anticoagulants, particularly warfarin, alongwith appetite-suppressing properties. It has been used in the treatment of lymphedema [104]. The biosynthesis of coumarin in plants takes place *via* hydroxylation, glycolysis and cyclization of cinnamic acid. Coumarin can also be prepared in laboratory by Perkin reaction between salicylaldehyde and acetic anhydride or Pechmann condensation reaction.

TERPENOIDS AND ESSENTIAL OILS

Terpenoids, also known as isoprenoids, are a large and varied class of chemically modified hydrocarbons (oxidized or rearranged carbon skeleton), produced primarily by a wide variety of plants, particularly conifers. Terpenes and terpenoids are derived biosynthetically from units of isoprene. Terpenes and terpenoids are the primary constituents of the essential oils of many types of plants and their flowers. Essential oils are used widely as natural flavor additives for food, as fragrances in perfumery, and in traditional and alternative medicines such as aromatherapy (curative effects of specific aromas carried by essential oils). Synthetic variations and derivatives of natural terpenes and terpenoids also greatly expand the variety of aromas used in perfumery and flavors used in food additives. Vitamin A is a good example that belongs to terpenoids group. These are generally extracted by distillation, expression, or solvent extraction method. A multi-thousand tons of essential oil have been produced annually from various plant species such as sweet orange, *Mentha arvensis*, peppermint, cedarwood, lemon, *Eucalyptus globulus*, *Litsea cubeba*, clove, and spearmint worldwide.

ALKALOIDS

Plant alkaloids are one of the largest groups of natural products, provide many pharmacologically active compounds. They are heterocyclic compound containing nitrogen, with an alkaline pH, bitter taste and a marked physiological action on animal physiology. Many alkaloids can be purified from crude extracts by acid-base extraction protocol. Some alkaloids are toxic to other organism, that's why; they often have pharmacological effects and are used in medications, as recreational drugs, or in entheogenic rituals. Few customary examples are the local anesthetic and stimulant cocaine, the stimulant caffeine, nicotine, the analgesic morphine, or the antimalarial drug quinine. Taking into account their molecular structure, chemical features, and biological as well as biogenetic origin, they enjoy tremandous diversity in nature. Major groups of alkaloides are summarized in Table **3** with examples.

Table 3: A summary of major group of alkaloids

Sr. No.	Group	Example (s)
1	Pyridine	Piperine, coniine, trigonelline, arecoline, arecaidine, guvacine, cytisine, lobeline, nicotine, anabasine, sparteine, pelletierine
2	Pyrrolidine	Hygrine, cuscohygrine, nicotine
3	Tropane	Atropine, cocaine, ecgonine, scopolamine, catuabine
4	Indolizidine	Senecionine, swainsonine
5	Quinoline	Quinine, quinidine, dihydroquinine, dihydroquinidine, strychnine, brucine, veratrine, cevadine
6	Isoquinoline	Opium alkaloids (papaverine, narcotine, narceine), pancratistatin, sanguinarine, hydrastine, berberine, emetine, berbamine, oxyacanthine Phenanthrene alkaloids: opium alkaloids (morphine, codeine, thebaine, oripavine)
7	Phenethylamine	Mescaline, ephedrine, dopamine
8	Indole	Tryptamines: serotonin, DMT, 5-MeO-DMT, bufotenine, psilocybin Ergolines (the ergot alkaloids): ergine, ergotamine, lysergic acid Beta-carbolines: harmine, harmaline, tetrahydroharmine Yohimbans: reserpine, yohimbine Vinca alkaloids: vinblastine, vincristine Kratom (Mitragyna speciosa) alkaloids: mitragynine, 7-hydroxymitragynine Tabernanthe iboga alkaloids: ibogaine, voacangine, coronaridine Strychnos nux-vomica alkaloids: strychnine, brucine
9	Purine	Xanthines: caffeine, theobromine, theophylline
10	Terpenoid	Aconitum alkaloids: aconitine Steroid alkaloids (containing a steroid skeleton in a nitrogen containing structure): Solanum (*e.g.* potato and tomato) alkaloids (solanidine, solanine, chaconine) Veratrum alkaloids (veratramine, cyclopamine, cycloposine, jervine, muldamine) Fire Salamander alkaloids (samandarin) Others: conessine
11	Quaternary ammonium compounds	Muscarine, choline, neurine
12	Miscellaneous	Capsaicin, cynarin, phytolaccine, phytolaccotoxin Lectins and Polypeptides

GLYCOSIDES

Chemicaly, glycosides are those biomolecules in which a sugar is bound to a non-carbohydrate moiety (usually a small organic molecule) through its anomeric carbon to another group *via* a glycosidic bond. They play numerous important roles in living systems. Several plants store chemicals in the form of inactive glycosides which can be activated by enzyme hydrolysis so that make them available for use as medications [105]. The diverse group of glycosides can be classified by the glycone, by the type of glycosidic bond, and by the aglycone. If the glucose is as glycone, the molecule is reffered to as a glucoside; if it is fructose, then the molecule is a fructoside; if it is glucuronic acid, then the molecule is a glucuronide. They are classified as α-glycosides or β-glycosides, depending on whether the glycosidic bond lies "below" or "above" the plane of the cyclic sugar molecule. For the purposes of biochemistry and pharmacology, the most useful classification is based on aglycone moiety. Few medicinally important classes of glycosides are reviewed in Table **4**.

Table 4: A summary of medicinally important classes of glycosides

Type of Glycosides	Examples	Occurrence	Medicinal Effects
Alcoholic glycosides	Salicin	*Salix*	Analgesic, antipyretic and antiinflammatory
Anthraquinone glycosides	Antron and anthranol	Senna, Rhubarb and *Aloe*	Laxative
Coumarin glycosides	Apterin	*Psoralea corylifolia*	Vasodilation
Cyanogenic glycosides	Amygdalin	Almonds, Rose family, and Cassava	Toxic
Flavonoid glycosides	Hesperidin, Naringin, Rutin, Quercitrin		Antioxidant effect
Phenolic glycosides	Arbutin	*Arctostaphylos uva-ursi*	Urinary antiseptic effect
Saponins	Saponin	Liquorice	Expectorant
Steroidal glycosides or cardiac glycosides		*Digitalis, Scilla*, and *Strophanthus*	Treatment of heart diseases and arrhythmia
Steviol glycosides	Steviol	*Stevia rebaudiana*	Natural sweeteners
Thioglycosides	Sinigrin and sinalbin	Black mustard and white mustard	

OTHER COMPOUNDS

Many phytochemical not mentioned above have been found in different medicinal plants that include therapeutic properties. Certain polyamines (in particular spermidine) [106], isothiocyanates [107, 108], thiosulfinates [109], and glucosides [110, 111] are also obtained from medicinal plants. Polyacetylene compound from *Bupleurum salicifolium* exhibit inhibitory effect against *Staphyllococcus aureus* and *Bacillus subtilis* [112]. Latex, a milky sap extracted from *Carica papaya* is a complex mixture of chemicals including papain, (a well-known proteolytic enzyme) [113] and carpaine [114]. It contributes a bacteriostatic property against *B. subtilis, Enterobacter cloacae, E. coli, Salmonella typhi, Staphylococcus aureus*, and *Proteus vulgaris* [115].

IN VITRO CULTURE AND PLANT CELL BIOREACTORS

The pharmacologically active metabolites are extracted from plants [116, 117]. In plant systems, they accumulate in leaves (nicotine), roots (ajmalicine), bark (quinine), or in the whole plants (ephedrine). Since plant material may not be available for the whole part of year, an alternative source is required to achieve the industrial demand. In the recent years, it has become increasingly difficult to maintain constant supply of many of the important medicinal plants due to several factors including their hardnosed exploitation, lack of environment upkeeping, rising cost of labor and economical and technical problems associated with the cultivation of medicinal plants that occurring in their native natural environment. Besides this, harvesting of plants from natural habitats is not only difficult, but also makes them into endangered species category. Therefore, the application of plant cell, tissue and organ culture techniques for the synthesis of secondary metabolites, particularly in the plants of pharmaceutical worth, holds tremendous alternative for controlled product of plant origin [118, 119].

In vitro, mass cultivation of plant cells is an operational alternative for production of high value, low yielding secondary metabolites. Considerable attention has been focused to fabricate bioreactors for commercial scale plant cell cultivation during the last three decades [39, 120, 121]. Healthy growing cell suspension cultures have been regarded as the only relevant biotechnological production strategy for biomedicines and phytopharmaceuticals. Plant cell, tissue or organ cultures are grown in a specific bioreactor for commercial scale production of metabolites, either by biogenesis or biotransformation approach. In case of biogenesis, a metabolite is synthesized inside the cell by a set of enzyme-mediated reactions, using precursors from basic metabolic pathways. On the other hand, sometime a compound is chemically modified (generally a functional group) by plant cell using its enzymatic machinery is called biotransformation.

Biotransformation *via* plant cell cultures mediates numerous reactions, such as glycosylation, glucosylesterification, hydroxylation, oxido-reduction, hydrolysis, epoxidation, isomerization, methylation, demethylation and dehydrogenation. Remarkably, exogenously added substances may pertain toxicity to the cell culture, therefore, a glucose conjugate of compounds are preferred as substrate for biotransformation purposes. Certain additives including antioxidants can increase both cell viability as well as productivity in cultures [122]. Sugars act as antioxidants and known to be scavengers of hydroxyl radical produced during biotransformation reaction [123]. Firstly, the cell cultures of *Catharanthus roseus* were used for production of arbutin from hydroquinone. The biotransformation potential of the cells significantly varies with the age and species of culture plants [124]. Almost unlimited number of compounds can undergo modification by plant cell cultures involving phenylpropanoids, alkaloids, and mevalonates.

Genetic manipulation approaches to biotransformation offer great potential to express heterologous genes and to clone and over-express genes for key biosynthetic enzymes. Biotransformation efficiencies can further be improved using molecular techniques involving site-directed mutagenesis and gene manipulation for substrate specificity.

Fast growing hairy root cultures have been grown in bioreactors because of their high production specific potential. The red Shikonin of the cork layer of the roots of *lithospermum erythrorhizon* are derivatives of 1,4- naphthoquinones. This compound has been used for treatment of burns and skin diseases, and also as a dye in silk and lipstics. Earlier, the plants have to be grown for a longer duration of 3-4 years to yield around 1-2 % shikonin [125]. Since 1983, the Mitsui Company, Japan first to obtain patent and started commercial production of shikonin through suspension culture techniques [126]. The roots of *Arnebia hispidissima* are also the most important sources of biosynthesis of shikonin and its derivatives. A simple and efficient protocol for high frequency direct plant regeneration, micropropagation and shikonin induction in *Arnebia hispidissima* has been reported in our laboratory [127-130].

More careful examination of bioreactors for the cultivation of plant cells is needed to deal with certain drawbacks including clustering of cells, their mixing, oxygen transfer, shear insolence and wall growth. Since the cultivation characteristics and the size of the cell aggregates vary from species to species, the selection of bioreactor principally relies upon the plant cell of interest. A suitable bioreactor can be designed for a particular plant cell system by considering several important criteria including optimum aeration conditions, intensity of culture, aggregate size, aseptic conditions, and control of physical conditions such as temperature, pH, nutrient availability [131, 132].

Various types of bioreactors have been used for commercial scale cultivation of plant cells. Stirred tank bioreactors have been largely employed so as to achieve the best process parameters. Although, stirred tank reactors wield high hydrodynamic strain on plant cells, they have enormous potential if used with low agitation rapidity with modified impeller. Stirred tank bioreactor has also been the first bioreactor used to produce shikonin by cell cultures of *Lithospermum erythrorhizon* [133]. Few modified stirred tank bioreactors have been used to culture cells of *Catharanthus roseus*, *Dioscorea deltoidea* [134], *Digitalis lanata* [135], *Panax notoginseng* [136], *Taxus baccata* [137] and *Podophyllum hexandrum* [138]. Other known modified reactors are bubble column reactor, air-lift bioreactor, and rotating drum reactor. The performance and application of the different bioreactors make them suitable for different type of cell cultures [131]. Major drawbacks include the high capital cost of bioreactors and low stability of cell lines, which often lose their capacity to produce target molecules over a passage of time in culture.

OVERVIEW OF DESIGNING STRETEGIES FOR METABOLIC ENGINEERING

Genetic engineering of medicinal plants entails the alteration of endogenous metabolic pathways to intensify flux towards particular molecules of interest. Sometime, the objective is to enhance the production of a natural metabolite, although in others it is to synthesize a novel metabolite. Over the past few decades, considerable advances have been made in metabolic engineering *via* application of genomics and proteomics technologies to enlighten and characterize metabolic pathways in a combinatorial manner, more broadly than on a step-by-step basis [139]. Before designing a metabolic engineering strategy, an understanding of the cell metabolism is a prerequisite. Worldwide, interminable efforts have been made by the researchers to enlighten the complex metabolic pathways underlying the synthesis of medicinal-value-phytochemicals in huge diversity, over the years. Consequently, a number of pathways for secondary metabolites are known to date. Amongst them, some key link pathways connecting the primary metabolism with secondary metabolism are illustrated in Fig. **2** in order to gain knowledge of their genetic control. Notably, the primary metabolic routes provide elementary material and precursors for numerous secondary metabolites, while there are other routes available for supply of skeletal material for biosynthesis of plant metabolites. In essence, all molecular and genetic events associated with metabolite production including primary metabolism, its interaction with secondary metabolism and, post-synthetic actions need to be completely characterized for manipulation or reprogramming of metabolic pathways.

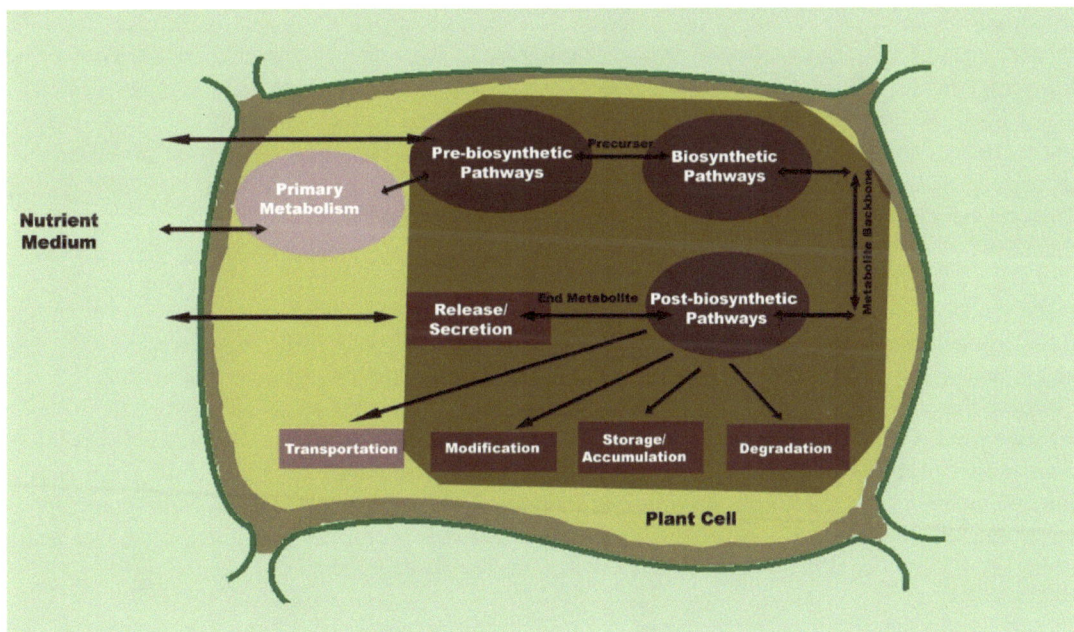

Figure 2: Schematic diagram of major cellular events involved in biosynthesis of plant secondary metabolite and link between primary metabolism and secondary metabolism.

HAIRY ROOT CULTURE FOR IMPROVING SECONDARY METABOLITE PRODUCTION

Improvement of the yield and quality of these natural plant products through conventional breeding is still a challenge. Genetic engineering has been vested as the most effective tool in improving secondary metabolites production in medicinal plant species, if transformation systems are previously established [140], although predictable results were not achieved consistently. Efforts are also being made in this direction of developing a well-established transformation protocols in many medicinal plant species including the *Artemisia annua, Taxus* sp.*, Papaver somniferum, Ginkgo biloba, Camptotheca acuminate, Lavandula intermedia and Withania somnifera* [141-147]. Rather, *in vitro* culture and hairy root induction techniques pledge an eternal solution to certain upcoming problems in the area of plant biotechnology.

Agrobacterium-mediated genetic transformation of medicinal plants to obtain hairy root cultures, for production of transformed cells to be maintained in culture or to regenerate whole plants is gaining importance. Examples of direct DNA transfer methodology to engineer medicinal plants and cultures have also been reported [148, 149]. Although, the establishment of hairy root cultures and transgenic cell lines is relatively easy in some of the species, yet, the production potential for desirable metabolite is still a limitation in most of the species.

The flux rate can be increased towards the downstream reactions by genetic manipulation of medicinal plants using three major ways: (a) engineering the plants with individual structural genes, (b) genetic modification of plants by introducing multiple genes together, and (c) transformation using regulatory genes for complex pathways. All the three approaches require systematic studies of the biosynthesis pathway of particular metabolite and elucidation of its genetic control. Numerous advanced molecular approaches including application of RNA interference technology, redirecting of the common precursor, targeting metabolite to specific cellular compartments, creation of metabolic sinks for storage of over expressed metabolite make the genetic transformation strategies highly efficient than the techniques which were available earlier [5].

There are a few good examples of multi gene engineering, for introducing entirely novel pathways and thus to produce completely new foreign products. In one such experiment two multifunctional cytochrome P450 enzymes and uridine diphosphate glucose (UDPG)-glucosyltransferase were transferred together from Sorghum into *A. thaliana*, leading to the production of the cyanogenic glucoside dhurrin which confers resistance against the flea beetle *Phyllotreta nemorum* [150]. The finely studied PHA, poly (3-hydroxybutyrate) (PHB), can be reprogrammed to accumulate in the plastids of transgenic *A. thaliana* plants at levels beyond 40% of its dry shoot weight. This is obtained *via* introduction of three bacterial genes, encoding the enzymes 3-ketothiolase, acetacetyl-CoA reductase and PHA synthase [151]. Therefore, combinatorial genetic modification of medicinal plants is the rational approach for production of pharmaceutically active compounds.

In case of root-derived pharmaceuticals, the ability of *Agrobacterium rhizogenes* for inducing hairy roots in a diversity of host plants has lead to increase in production of secondary metabolites. Hairy root cultures of *Arnebia hispidissima* furnish high productivity potential for shikonin [127-130]. The interest in hairy roots is largely because of their ability to grow fast devoid of an external supply of auxins and many times, does not require light. Since, they are genetically stable, they have been found to be highly established in metabolite yield. Interestingly, many of the root-derived plant products once not considered feasible for production by cell culture are being reinvestigated for production using the hairy root culture technology. A number of such examples have been summarized in Table **5**. Several hairy root cultures have been put to scale-up studies in bioreactors. However, due to their structural features and metabolite localization characteristics, they need different type of reactors than the ones used for plant cell cultures.

Table 5: Hairy root culture technology for production of bioactive compounds

Sr. No.	Plant species	Product	References
1	*Arnebia hispidissima*	Shikonin	[127- 130]
2	*Artemesia absynthium*	Volatile oil	[152]
3	*Bidens* spp.	Polyacetylenes	[153]
4	*Cassia* spp.	Anthroquinones	[154]

Table 5: cont....

5	*Cicharium intybus*	Esculetin	[155]
6	*Cinchon ledgeriana*	Quinolene alkaloids	[156]
7	*Datura* spp.	Tropane	[157]
8	*Dubaisia leichhardtii*	Tropane alkaloids	[158]
9	*Echinacea purpurea*	Alkaloids	[159]
10	*Glycyrrhiza uralensis*	Isoprenylated flavonoids	[160]
11	*Glabra glabra*	Glycyrrhizin	[161]
12	*Hyoscyamus albus*	Alkaloids	[163]
13	*Hyoscyamus muticus*	Hyoscyamine	[163]
14	*Lithospermum erythrorhizon*	Shikonin	[162]
15	*Panax ginseng*	Saponin	[164]
16	*Panax ginseng*	Ginsenoside	[165]
17	*Rauwolfia serpentina*	Ajmaline, serpentine	[166]
18	*Rubia cordifolia*	Anthroquinones	[167]
19	*Salvia miltorrhiza*	Diterpenes	[168]

CONCLUSIONS AND FUTURE PROSPECTS

Herbal medicines have been increasingly used worldwide for hundreds of decades. WHO reports that the world market for herbal medicine, including herbal products and raw materials, has been estimated to have an annual growth rate between 5 and 15%. The aggregate global herbal drug market is estimated as US $62 billion and is probable to grow to US $5 trillion by the year 2050 [169]. High prices and harmful side effects of synthetic drugs persuade the people more towards herbal medicines. In India, more than 7800 manufacturing units are involved in the production of herbal drugs and pharmaceutical preparations which requires about 2000 tons of medicinal plant raw materials annually [170].

Although, the plant cell for production of secondary metabolite on a large scale is quite feasible for many species, but it is not economically viable for a number of known plant derived drugs. However, it opens remarkable perspectives for screening cell cultures for new biologically active compounds and to exploit this technology for development of novel drugs from rare medicinal plants. In future, certainly a supply crisis will approach for other promising drugs as for taxol, which entails alternative approaches of phytochemical production. Cell cultures technology will, therefore, be an important substitute. Further advancement in various strategies for improving productivity of secondary metabolites will be highly rewarding. The applications of molecular and genetic improvement techniques are quite promising for enhancement of metabolite production as well as biosynthesis of novel compounds.

Recent progress in understanding secondary metabolite pathways and metabolic engineering can be applied to plant cell cultures, for the production of desired pharmaceutical compounds. Comparatively a small number of secondary metabolic pathways have been engineered successfully in medicinal plants [5, 171]. Partially, enormous complexity of secondary metabolism which is regulated at different levels has been responsible for this. Besides, the highly compartmentalized nature of enzymes, substrate precursors, and metabolic intermediates also contribute to these problems. The understanding that ecological interactions between plants and their environment significantly contribute to this complexity has begun to correct our views of how secondary metabolic pathways products interact and operate at the whole plant level. Biotic and abiotic stresses play a large role in these interactions. Now, we are poised at appreciating entire system biology rather than choosing a single gene/pathway manipulation which might be best suited for the understanding of biosynthetic pathways in complex living systems. This understanding eventually requires the advances of mechanistic models that are beginning to depart from the idea that such pathways are a simple collection of sequential enzymes with catalytic, transport or other functions which function in recital with specific transcription factors.

In the past few decades, rare number of plant biosynthetic genes and enzymes has been cloned and identified, adjoin the recent status of knowledge towards biosynthesis pathways involved in production of secondary metabolites. Key biosynthesis cascades of flavonoids, phenolic derivatives, terpenes and alkaloids have been elucidated and numerous enzymes and genes related to the bioactive compounds have been characterized and identified as well. The improvement of plant transformation and gene expression approaches is also a major advantage. Altogether, these advances evidently aid metabolic engineering to have further impact on reprogramming metabolic pathways for efficient production of desired compounds. Furthermore, the upcoming new functional genomic tools including transcriptomics, proteomics and metabolomics as well as computational biology shed new horizon to this discipline. The function of metabolomics in metabolic engineering is understandable. The production of secondary metabolites is closely regulated by the complex networks as one pathway may interlink to several other pathways of primary and secondary metabolism, redirecting a target pathway may cause the production of unintended metabolites. In addition, plant secondary metabolism pathways are normally induced in response to ecological stress and single pathway may produce more than one metabolite [172], thus exploitation of a pathway for a desired compound might change the concentration of other product of the same pathway [173]. Metabolomics is quite competent to diagnose such alterations. So far, transcriptomics and metabolomics data are integrated in order to get insight into the impact of manipulating biosynthesis pathways to the whole plant metabolism. Therefore, the application of these high throughput techniques will accelerate the research on metabolic engineering as the analysis of the changes as a result of modified pathways can be made in holistic way at the level of genome, transcriptome and metabolome.

ABBREVIATIONS USED

DES: Diethylstilbestrol

FDA: Food and drug administration

HIV: Human immunodeficiency virus

WHO: World health organization

ACKNOWLEDGEMENTS

The authors duly acknowledge Department of Biotechnology, Ministry of Science & Technology, Government of India, New Delhi for financial support for EU Patent.

REFERENCES

[1]   Facchini PJ, De Luca V. Opium poppy and Madagascar periwinkle: model non-model systems to investigate alkaloid biosynthesis in plants. Plant J 2008; 54:763–84.

[2]   Wink M. Functions and Biotechnology of Plant Secondary Metabolites. Ann Plant Rev 2009; 39:1–20.

[3]   Sanandakumar S. MPEDA asks aquafarms not to use banned antibiotics, Times news network, 9th April 2002.

[4]   Acamovic T, Brooker JD. Biochemistry of plant secondary metabolites and their effects in animals. Proc Nutr Soc 2005; 64: 403–12.

[5]   Kumar J, Gupta PK. Molecular approaches for improvement of medicinal and aromatic plants. Plant Biotechnol Rep 2008; 2: 93–112

[6]   Rana BK, Singh UP, Taneja V. Antifungal activity and kinetics of inhibition by essential oil isolated from leaves of *Aegle marmelos*. J Ethnopharmacol 1997; 57: 29–34.

[7]   Vohora SB, Rizwan M, Khan JA. Medicinal uses of common Indian vegetables. Planta Med 1973; 23: 381–93.

[8]   Naganawa R, Iwata N, Ishikawa K, Fukuda F, Fujino T, Suzuki A. Inhibition of microbial growth by ajoene, a sulfur-containing compound derived from garlic. Appl Environ Microbiol 1996; 62: 4238–42.

[9]   San-Blas G, Marino L, San-Blas F, Apitz-Castro R. Effect of ajoene on dimorphism of *Paracoccidioides brasiliensis*. J Med Vet Mycol 1993; 31: 133–41.

[10] Yoshida S, Kasuga S, Hayashi N, Ushiroguchi T, Matsuura H, Nakagawa S. Antifungal activity of ajoene derived from garlic. Appl Environ Microbiol 1987; 53: 615–7.

[11] Martinez MJ, Betancourt J, Alonso-Gonzalez N, Jauregui A. Screening of some Cuban medicinal plants for antimicrobial activity. J Ethnopharmacol 1996; 52: 171-4.

[12] Himejima M, Kubo I. Antibacterial agents from the cashew *Anacardium occidentale*. J Agric Food Chem 1991; 39: 418–21.

[13] McDevitt JT, Schneider DM, Katiyar SK, Edlind TD. Berberine: a candidate for the treatment of diarrhea in AIDS patients, abstr. 175. In Program and Abstracts of the 36th Interscience Conference on Antimicrobial Agents and Chemotherapy. American Society for Microbiology, Washington, D.C.USA 1996.

[14] Omulokoli E, Khan B, Chhabra SC. Antiplasmodial activity of four Kenyan medicinal plants. J Ethnopharmacol 1997; 56: 133–7.

[15] Vijaya K, Ananthan S, Nalini R. Antibacterial effect of theaflavin, polyphenon 60 (*Camellia sinensis*) and *Euphorbia hirta* on *Shigella* spp—a cell culture study. J Ethnopharmacol 1995; 49: 115–8.

[16] Toda M, Okubo S, Ikigai H, Suzuki T, Suzuki Y, Hara Y, Shimamura T. The protective activity of tea catechins against experimental infection by *Vibrio cholerae* O1. Microbiol Immunol 1992; 36: 999–1001.

[17] Ooshima H, Kurakake M, Kato J, Harano Y. Enzymatic activity of cellulase adsorbed on cellulose and its change during hydrolysis. Appl biochem biotechnol 1991; 31(3): 253-66.

[18] Keating GJ, O'Kennedy R. The chemistry and occurrence of coumarins, In R. O'Kennedy and R. D. Thornes (ed.), Coumarins: biology, applications and mode of action. John Wiley & Sons, Inc., New York, 1997; pp. 348.

[19] Cichewicz RH, Thorpe PA. The antimicrobial properties of chile peppers (Capsicum species) and their uses in Mayan medicine. J. Ethnopharmacol 1996; 52: 61–70.

[20] Jones, S. B., Jr., Luchsinger AE. Plant systematics. McGraw- Hill Book Co., New York 1986.

[21] Bose PK. On some biochemical properties of natural coumarins. J Indian Chem Soc 1958; 58: 367–75.

[22] Hamburger H, Hostettmann K. The link between phytochemistry and medicine. Phytochem 1991; 30: 3864–74.

[23] Scheel LD. The biological action of the coumarins. Microbiol Toxins 1972; 8: 47–66.

[24] Stange RR Jr Midland SL, Eckert JW, Sims JJ. An antifungal compound produced by grapefruit and Valencia orange after wounding of the peel. J Nat Prod 1993; 56: 1627–29.

[25] Apisariyakul A, Vanittanakom N, Buddhasukh D. Antifungal activity of turmeric oil extracted from *Curcuma longa* (*Zingiberaceae*). J Ethnopharmacol 1995; 49: 163–9.

[26] Berkada B. Preliminary report on warfarin for the treatment of herpes simplex. J Irish Coll Phys Surg 1978; 22(Suppl.): 56.

[27] Mohamed LET, El-Nur EBES, Abdelrahman MEN. The antibacterial, antiviral activities and phytochemical screening of some Sudanese medicinal plants. EurAsia J BioSci 2010; 4:8-16.

[28] Freiburghaus F, Kaminsky R, Nkunya MHH, Brun R. Evaluation of African medicinal plants for their *in vitro* trypanocidal activity. J Ethnopharmacol 1996; 55: 1–11.

[29] Wild R. (ed.). The complete book of natural and medicinal cures. Rodale Press Inc, Emmaus Pa 1994.

[30] Perrett S, Whitfield PJ, Sanderson L, Bartlett A. The plant *molluscicide Millettia thonningii* (Leguminosae) as a topical antischistosomal agent. J Ethnopharmacol 1995; 47: 49–54.

[31] Wan J, Wilcock A, and Coventry M J. The effect of essential oils of basil on the growth of *Aeromonas hydrophila* and *Pseudomonas fluorescens*. J Appl Microbiol 1998; 84: 152–8.

[32] Kubo A, Lunde CS, Kubo I. Antimicrobial activity of the olive oil flavor compounds. J Agric Food Chem 1995; 43: 1629–33.

[33] Jones GA, McAllister TA, Muir AD, Cheng KJ. Effects of sainfoin (*Onobrychis viciifolia* scop.) condensed tannins on growth and proteolysis by four strains of ruminal bacteria. Appl Environ Microbiol 1994; 60: 1374–8.

[34] Hufford CD, Jia Y, Croom EM Jr, Muhammed I, Okunade AL, Clark AM, Rogers RD. Antimicrobial compounds from *Petalostemum purpureum*. J Nat Prod 1993; 56: 1878–89.

[35] Ghoshal S, Krishna Prasad BN, Lakshmi V. Antiamoebic activity of *Piper longum* fruits against *Entamoeba histolytica in vitro* and *in vivo*. J Ethnopharmacol 1996; 50:167–70.

[36] Kubo I, Muroi H, Kubo A. Naturally occurring anti-acne agents. J Nat Prod 1994; 57:9–17.

[37] Kadota S, Basnet P, Ishii E, Tamura T, Namba T. Antibacterial activity of trichorabdal from *Rabdosia trichocarpa* against Helicobacter pylori. Zentbl Bakteriol 1997; 286: 63–7.

[38] Suresh B, Sriram S, Dhanaraj SA, Elango K, Chinnaswamy K. Anticandidal activity of *Santolina chamaecyparissus* volatile oil. J Ethnopharmacol 1997; 55: 151–9.

[39] Ali-Shtayeh MS, Al-Nuri MA, Yaghmour RMR, Faidi YR. Antimicrobial activity of *Micromeria nervosa* from the Palestinian area. J Ethnopharmacol 1997; 58: 143–7.

[40] Ofek I, Goldhar J, Sharon N. Anti-Escherichia coli adhesion activity of cranberry and blueberry juices. Adv Exp Med Biol 1996; 408: 179–83.

[41] Avorn J. The effect of cranberry juice on the presence of bacteria and white blood cells in the urine of elderly women. What is the role of bacterial adhesion? Adv Exp Med Biol 1996; 408: 185–6.

[42] Wagner H. Search for plant derived natural products with immunostimulatory activity (recent results). Pure app chem 1990; 62: 1217-22.

[43] Bauer, R, Wagner H. Echinacea species as potential immunostimulatory drugs In: Economic and medicinal plant resource. Vol.5 (H Wagner, HN Farnsworth. Eds) Academic press, London 1991; pp. 253-321.

[44] Bendel R, Bendel V, Renner K, Carstens V, Stolze K. Zusatzbe handlung mit Esberitox N bei patientinnen mit chemo-strahlen-therapeutischer Behandlung eines fortgeschrittenen Mamma-Karzinoms. Onkologie 1989; 12: 32-8

[45] American Cancer Society (December 2007). Report sees 7.6 million global 2007 cancer deaths. Reuters. http://www.reuters.com/article/healthNews/idUSN1633064920071217. Retrieved 8th july, 2008.

[46] WHO (Feburary 2006) "Cancer". World Health Organization. http://www.who.int/mediacentre/factsheets/fs297/en/. Retrieved 25th june, 2007.

[47] Langcake P, Pryce RJ. A new class of phytoalexins from grapevines, Specialia 1977; 15(2): 151.

[48] Jang M, Cai L, Udeani GO *et al.* Cancer chemopreventive activity of resveratrol, a natural product derived from grapes. Sci 1997; 275(5297): 218-20.

[49] Arts IC, Hollman PC. Polyphenols and disease risk in epidemiologic studies. Am J Clin Nutr 2005; 81(1): 317-25.

[50] Selye H. The Stress of Life. McGraw-Hill, New York 1956.

[51] Brekhman II, Dordymov IV. New substances of plant origin which increase nonspecific resistance. Ann Rev Pharmacol 1969; 9: 419.

[52] Bhargava KP, Singh N. Anti-stress activity in Indian Medicinal Plants. J Res Edu Ind Med 1985; 4: 27.

[53] Singh N. A pharmacoclinical evaluation of some Ayurvedic crude drugs as anti-stress agents and their usefulness in some stress diseases of man. Ann Nat Acad Ind Med 1986; 1: 14.

[54] Singh N, Nath R, Mishra N, Kohli RP. An experimental evaluation of anti-stress effects of 'Geriforte' (Any Ayurvedic Drug). Quart J Crude Drug Res 1978; 16: 125.

[55] Watanbe H, Miyaji C, Makino M, Abo T. Therapeutic effects of glycyrrhizin in mice infected with LP-BM5 murine retrovirus and mechanisms involved in the prevention of disease progression. Biotherapy. 1996; 9: 209–20.

[56] Ahmad A, Davies J, Randall S, Skinner GRB. Antiviral properties of extract of *Opuntia streptacantha*. Antiviral Res 1996; 30: 75–85.

[57] Muller WEG, Renneisen K, Kreuter MH, Schroder HC, Winkler I. The D-mannose-specific lectin from *Gerardia savaglia* blocks binding of human immunodeficiency virus type I to H9 cells and human lymphocytes *in vitro*. J Acquired Immune Defic Syndr 1988; 1: 453–8.

[58] Yao XJ, Wainberg MA, Parniak MA. Mechanism of inhibition of HIV-1 infection *in vitro* by purified extract of *Prunella vulgaris*. Virol 1992; 187: 56–62.

[59] Houghton PJ, Woldemariam TZ, Khan AI, Burke A, Mahmood N. Antiviral activity of natural and semi-synthetic chromosome alkaloids. Antiviral Res 1994; 25: 235–44.

[60] Mazumder A, Cooney D, Agbaria R, Pommier Y. Inhibition of human immunodeficiency virus type 1 integrase by 39-azido-39-deoxythymidylate. Pro. Natl Acad Sci USA 1994; 91: 5771–5.

[61] Huang PL, Chen HC, Kung HF, Huang PL, Huang HI, Lee-Huang S. Anti-HIV plant proteins catalyze topological changes of DNA into inactive forms. Biofactors 1992; 4: 37–41.

[62] Lee-Huang S, Kung HF, Huang PL *et al.* Human immunodeficiency virus type 1 (HIV-1) inhibition, DNA-binding, RNA-binding and ribosome inactivation activities in the N-terminal segments of the plant anti-HIV protein. Proc Natl Acad Sci USA 1994; 91: 12208–12.

[63] Lee-Huang S, Huang PL, Bourinbaiar AS, Chen HC, Kung HF. Inhibition of the integrase of human immunodeficiency virus (HIV) type 1 by anti-HIV plant proteins MAP30 and GAP31. Proc Natl Acad Sci USA 1995; 92: 8818–22.

[64] Lee-Huang S, Huang PL, Chen HC *et al.* Anti-HIV and anti-tumor activities of recombinant MAP30 from bitter melon. Gene (Amsterdam) 1995; 161: 151–6.

[65] Fesen MR., Kohn KW, Leteurtre F, Pommier Y. Inhibitors of human immunodeficiency virus integrase. Proc Natl Acad Sci USA 1993;90:2399–2403.

[66] Critchfield JW, Butera ST, Folks TM. Inhibition of HIV activation in latently infected cells by flavonoid compounds. AIDS Res Hum Retroviruses 1996;12:39.

[67] Chessin M, DeBorde D, Zipf A (ed.). Antiviral proteins in higher plants. CRC Press, Inc., Boca Raton, Fla. Hudson, J. B., E. A. Graham, L. Harris, and M. J. Ashwood-Smith. 1993. The unusual UVA-dependent antiviral properties of the furoisocoumarin, coriandrin. Photochem Photobiol 1995; 57: 491–6.

[68] Hudson JB, Graham EA, Harris L, Ashwood-Smith MJ. The unusual UVA-dependent antiviral properties of the furoisocoumarin, coriandrin. Photochem. Photobiol 1993; 57: 491–6.

[69] McGrath MS, Hwang KM, Caldwell SE *et al.* GLQ223: an inhibitor of human immunodeficiency virus replication in acutely and chronically infected cells of lymphocyte and mononuclear phagocyte lineage Proc Natl Acad Sci USA 1989; 86: 2844–8.

[70] Paris A, Strukelj B, Renko M, Turk V. Inhibitory effect of carnosolic acid on HIV-1 protease in cell-free assays. J Nat Prod 1993; 56: 1426–30.

[71] Xu HX, Zeng FQ, Wan M, Sim KY. Anti-HIV triterpene acids from *Geum japonicum*. J Nat Prod 1996; 59: 643–5.

[72] Fujioka T, Kashiwada Y. Anti-AIDS agents. 11. Betulinic acid and platanic acid as anti-HIV principles from *Syzigium claviflorum*, and the anti-HIV activity of structurally related triterpenoids. J Nat Prod 1994; 57: 243–7.

[73] Ono K, Nakane H, Fukushima M, Chermann JC, Barre-Sinoussi F. Differential inhibitory effects of various flavonoids on the activities of reverse transcriptase and cellular DNA and RNA polymerases. Eur J Biochem 1990; 190: 469–76.

[74] Spedding G, Ratty A, Middleton EJ. Inhibition of reverse transcriptases by flavonoids. Antiviral Res 1989; 12: 99–110.

[75] Tan GT, Miller JF, Kinghorn AD, Hughes SH, Pezzuto JM. HIV-1 and HIV-2 reverse transcriptases: a comparative study of sensitivity to inhibition by selected natural products. Biochem Biophys Res Commun 1992; 185: 370–8.

[76] Sun HD, Qiu SX, Lin LZ *et al.* Nigranoic acid, a triterpenoid from *Schisandra sphaerandra* that inhibits HIV-1 reverse transcriptase. J Nat Prod 1996; 59: 525–7.

[77] Tan GT, Pezzuto JM, Kinghorn AD. Evaluation of natural products as inhibitors of human immunodeficiency virus type 1 (HIV-1) reverse transcriptase. J Nat Prod 1991; 54:143–54.

[78] McMahon JB,Currens MJ, Gulakowski RJ, Buckheit RWJ, Lackman-Smith C, Hallock YF, Boyd MR. Michellamine B, a novel plant alkaloid, inhibits human immunodeficiency virus-induced cell killing by at least two distinct mechanisms. Antimicrob Agents Chemother 1995; 39: 484–8.

[79] Lai PK, Donovan J, Takayama H, Sakagami H, Tanaka A, Konno K, Nonoyama M. Modification of human immunodeficiency viral replication by pine cone extracts. AIDS Res Hum Retroviruses 1990; 6: 205–17.

[80] Kreis W, Kaplan MH, Freeman J, Sun DK, Sarin PS. Inhibition of HIV replication by *Hyssop officinalis* extracts. Antiviral Res 1990; 14: 323–37.

[81] Kaul TN, Middletown E, Jr., Ogra PL. Antiviral effect of flavonoids on human viruses. J Med Virol 1985; 15: 71–9.

[82] Pengsuparp T, Cai L, Constant H *et al.* Mechanistic evaluation of new plant-derived compounds that inhibit HIV-1 reverse transcriptase. J Nat Prod 1995; 58: 1024–31.

[83] Chen K, Shi Q, Kashiwada Y *et al.* Anti-AIDS agents. 6-Salaspermic acid, an anti-HIV principle from Tripterygium wilfordii, and the structure-activity correlation with its related compounds. J Nat Prod 1992; 55: 340–6.

[84] Ito M, Nakashima H, Baba M, Pauwels R, Clercq E De, Shigeta S, Yamamoto N. Inhibitory effect of glycyrrhizin on the *in vitro* infectivity and cytopathic activity of the human immunodeficiency virus [HIV (HTLV-III/LAV)]. Antivir Res 1987; 7: 127–37.

[85] Gnabre JN, Brady JN, Clanton DJ *et al.* Inhibition of human immunodeficiency virus type 1 transcription and replication by DNA sequence-selective plant lignans. Proc Natl Acad Sci USA 1995; 92: 11239–43.

[86] Offergeld R, Reinecker C, Gumz E, Schrum S, Treiber R, Neth RD, Gohla SH. Mitogenic activity of high molecular polysaccharide fractions isolated from the *cuppressaceae Thuja occidentalis* L. enhanced cytokine-production by thyapolysaccharide, g-fraction (TPSg) Leukemia 1992; 6(3): 189S–91S.

[87] Pengsuparp T, Cai L, Fong HHS *et al.* Pentacyclic triterpenes derived from *Maprounea africana* are potent inhibitors of HIV-1 reverse transcriptase. J Nat Prod 1994; 57: 415–418.

[88] Gollapudi S, Sharma HA, Aggarwal S, Byers LD, Ensley HE, Gupta S. Isolation of a previously unidentified polysaccharide (MAR-10) from *Hyssop officinalis* that exhibits strong activity against human immunodeficiency virus type 1. Biochem Biophys Res Commun 1995; 210: 145–51.

[89] Amoros M, Simoes CMO, Girre L. Synergistic effect of flavones and flavonols against herpes simplex virus type 1 in cell culture. Comparison with the antiviral activity of *propolis*. J Nat Prod 1992; 55: 1732–40.

[90]   De Clercq E. New perspectives for the chemotherapy of chemoprophylaxis of AIDS (acquired immunodeficiency syndrome). Verhandelingen 1992 54: 57–89.

[91]   Hu CQ, Chen K, Shi Q, Kilkuskie RE, Cheng YC, Lee KH. Anti-AIDS agents. 10. Acacetin-7-O-beta-D-galactopyranoside, and anti-HIV principle from *Chrysanthemum morifolium* and a structure-activity correlation with some related flavonoids. J Nat Prod 1994; 57: 42–51.

[92]   Tabba HD, Chang RS, Smith KM. Isolation, purification, and partial characterization of prunellin, an anti-HIV component from aqueous extracts of Prunella vulgaris. Antiviral Res 1989; 11:263–73.

[93]   Hasegawa H, Matsumiya S, Uchiyama M *et al.* Inhibitory effect of some triterpenoid saponins on glucose transport in tumor cells and its application to *in vitro* cytotoxic and antiviral activities. Planta Med 1994; 6: 240–3.

[94]   Balzarini J, Schols D, Neyts J, Van Damme E, Peumans W, De Clercq E. α-(1,3)- and α-(1,6)-D-mannose-specific plant lectins are markedly inhibitory to human immunodeficiency virus and cytomegalovirus infections *in vitro*. Antimicrob Agents Chemother 1991; 35: 410–6.

[95]   Matuschek E, Svanberg U. Oxidation of Polyphenols and the Effect on *In vitro* Iron Accessibility in a Model Food System. J Food Sci 2002; 67 (1): 420–4.

[96]   Magalhães PJ, Carvalho DO, Cruz JM, Guido LF, Barros AA. Fundamentals and Health Benefits of Xanthohumol, a Natural Product Derived from Hops and Beer, Nat Prod Commun Vol. 4 (5) 2009.

[97]   Arranz S, Saura-Calixto F, Shaha S, Kroon PA. High Contents of Nonextractable Polyphenols in Fruits Suggest That Polyphenol Contents of Plant Foods Have Been Underestimated. J Agri Food Chem 2009; 57: 7298–303.

[98]   International Union of Pure and Applied Chemistry. "Quinones". Compendium of Chemical Terminology. Internet edition. 1995.

[99]   Papageorgiou VP, Assimopoulou AN, Couladouros EA, Hepworth D, Nicolaou KC. The Chemistry and Biology of Alkannin, Shikonin, and Related Naphthazarin Natural Products. Angew. Chem Int Ed 1999; 38: 270-300.

[100]   Filippos V, Trantas F, Emmanouil, Douglas Carl, Vollmer Guenter, Kretzschmar Georg, Panopoulos Nickolas (October). Biotechnology of flavonoids and other phenylpropanoid- derived natural products. Part I: Chemical diversity, impacts on plant biology and human health. Biotechnol J 2007; 2(10): 1214.

[101]   Schuier M, Sies H, Illek B, Fischer H. Cocoa-related flavonoids inhibit CFTR-mediated chloride transport across T84 human colon epithelia. J Nutr 2005; 135(10): 2320–5.

[102]   McGee H. On food and cooking: the science and lore of the kitchen. New York: Scribner 2006; pp. 714.

[103]   Bate-Smith S. Flavonoid compounds. In : Comparative biochemistry. Florkin M. Mason H.S. Eds. Vol III. Academic Press, New-York 1962; pp.75-809.

[104]   Nicholas F, Neil P. Pharmacogenomics: Its Role in Re-establishing Coumarin as Treatment for Lymphedema. Lymph Res Biol 2005; 3(2): 81–6.

[105]   Brito-Arias, M. Synthesis and Characterization of Glycosides. Springer2007

[106]   Flayeh KA, Sulayman KD. Antimicrobial activity of the amine fraction of cucumber (*Cucumis sativus*) extract. J App Microbiol 1987; 3: 275–9.

[107]   Dornberger K, Bockel V , Heyer J, Schonfeld C, Tonew M, Tonew E. Studies on the isothiocyanates erysolin and sulforaphan from *Cardaria draba*. Pharmazie 1975; 30: 792–6.

[108]   Iwu MM, Unaeze NC, Okunji CO, Corley DG, Sanson DR, Tempesta MS. Antibacterial aromatic isothiocyanates from the essential oil of *Hippocratea welwitschii* roots. Int J Pharmacogn 1991; 29: 154–8.

[109]   Tada M, Hiroe Y, Kiyohara S, Suzuki S. Nematicidal and antimicrobial constituents from *Allium grayi Regel* and *Allium fistulosum* L var caespitosum. Agric Biol Chem 1988; 52: 2383–5.

[110]   Murakami A, Ohigashi H, Tanaka S, Tatematsu A, Koshimizu K. Bitter cyanoglucosides from *Lophira alata*. Phytochem 1993; 32: 1461–6.

[111]   Rucker G, Kehrbaum S, Sakulas H, Lawong B, Goeltenboth F. Acetylenic glucosides from *Microglossa pyrifolia*. Planta Med 1992; 58: 266–9.

[112]   Estevez-Braun A, Estevez-Reyes R, Moujir LM, Ravelo AG, Gonzalez AG. Antibiotic activity and absolute configuration of 8S-heptadeca-2(Z), 9(Z)-diene-4,6-diyne-1,8-diol from *Bupleurum salicifolium*. J Nat Prod 1994; 57: 1178–82.

[113]   Oliver-Bever B. Medicinal plants in tropical West Africa. Cambridge University Press, New York, 1986.

[114]   Burdick EM. Carpaine, an alkaloid of Carica papaya. Its chemistry and pharmacology. Econ Bot 1971; 25: 363–5.

[115]   Osato JA, Santiago LA, Remo GM, Cuadra MS, Mori A. Antimicrobial and antioxidant activities of unripe papaya. Life Sci 1993; 53: 1383–9.

[116]   Balandrin MF, Klocke JA, Wurtele ES, Bollinger WH. Natural plant chemicals: sources of industrial and medicinal materials. Sci 1985; 228: 1154–60.

[117] Parr AJ. Secondary products from plant cell culture. In: Mizrahi A, ed Advances in biotechnology progress, vol 9 Biotechnology in agriculture. New York: Alan R. Liss; 1988; 1–34.

[118] Fransworth NR. The role of medicinal plants in drug development. In: Krogsgaard-Larsen P, Christensen SB and Kofod H (eds) Natural products and drug development. Munksgaard, Copenhagen (Alfred Benzon Symp. 20) 1984; pp 17-30.

[119] Moreno PRH, Heijden R, Verpoorte R. Cell and Tissue Culture of Catha*rantus roseus*: A literature survey II. Updating from 1988 to 1993. Plant Cell Tiss Org Cult 1995; 42(1): 1-25.

[120] Alper J. Effort to combat microbial resistance lags. ASM News 1998; 64: 440–1.

[121] Amoros M, Sauvager F, Girre L, Cormier M. *In vitro* antiviral activity of *propolis*. Apidologie 1992; 23: 231–240.

[122] Yocoyama M, Inomata S, Seto S, Yanagi M.Effects sugar glycoslation of exogenous hydroquinones by *C. roseus* cell in suspension cultures. 1990.

[123] Asada Y, Kiso K. initiation of aerobic oxidation of sulphide by illuminated stomach chloroplasts. Eur J biochem 1973; 33: 253-7.

[124] Tabata, M, Umetani Y, Ooya M, Tanks S, glucosylation of phenolic compoundsby *datura innoxia* suspension cultures. Phytochem 1988; 15: 1225-9.

[125] Fujita Y. Shikonin production by plant (*Lithospermum erythrorhizon*) cell culture. In: Bajaj YPS (ed.) Biotechnology in agriculture and forestry. Medicinal and aromatic plants. Springer, Berlin Heiidelberg 1988; 4: 227-36.

[126] Takahashi S, Fujita Y. Production of shikonin, In: Plant Cell Culure in japan. Progress in the production of useful plant metabolites by Japanese enterprises using plant cell culture technology. Komanine A, Misava, M, DiCosmo, F, (eds.) CMC com, Tokyo 1991; pp 72-8.

[127] Chaudhury A, Pal M. Guru Jambheshwar University of Science & Technology, Hisar, Haryana, India & Department of Biotechnology, Ministry of Science & Technology, Government of India, New Delhi, India. Method of Direct Regeneration, Shikonin Induction in Callus and *Agrobacterium*-mediated genetic transformation of *Arnebia hispidissima*. European Patent # 05256545 10th October, 2008.

[128] Pal M, Chaudhury A. High Frequency Direct Plant Regeneration, Micropropagation and Shikonin Induction in *Arnebia hispidissima*. J Crop Sci Biotech 2010; 13:13-20.

[129] Pal M. *In vitro* culture and induction of Shikonin production in *Arnebia hispidissima*. Ph. D. Thesis submitted to Department of Biotechnology, Guru Jambheshwar University of Science & Technology, Hisar, Haryana, India. 2005.

[130] Chaudhury A., Pal M. Method for enhanced Shikonin production in *Arnebia* species through induction of hairy root cultures. Patent Application # 2483/DEL/2004 (Indian Patent granted).

[131] Panda AK, Mishra S, Bisaria VS, Bhojwani SS. Plant cell reactors: A perspective. Enzyme Microb Technol 1989; 11: 386-97.

[132] Tanaka H. Technological problems in cultivation of plant cells at high density. Biotechnol Bioeng 2000; 67: 1203-18.

[133] Payne GF, Shuler ML, Brodelius P. Plant cell culture. In: B. K. Lydensen (ed). Large Scale Cell Culture Technology. Hanser Publishers, New York, USA 1987; pp. 193-229.

[134] Drapeau D, Blanch HW, Wilke CR. Growth kinetics of *Dioscorea deltoidea* and *Catharanthus roseus* in batch culture. Biotechnol Bioeng 1986; 28: 1555-63.

[135] Fulzele D, Kreis W, Reinhard E. Cardenolide biotransformation by cultured *Digitalis lanata* cells: Semi-continuous cell growth and production of deacetyllanatoside-C in a 40-L stirred tank bioreactor. Planta Med 1992; 58: A601-2.

[136] Zhong JJ, Chen F, Hu WW. High density cultivation of *Panax notoginseng* cells in stirred bioreactors for the production of ginseng biomass and ginseng. saponin. Process Biochem 2000; 35: 491-6.

[137] Srinivasan V, Pestchanker L, Moser S, Hirasuma TJ, Taticek RA, Shuler ML. Taxol production in bioreactors: kinetics of biomass accumulation, nutrient uptake, and taxol production by cell suspensions of *Taxus baccata*. Biotechnol. Bioeng 1995; 47: 666-76.

[138] Chattopadhyay S, Farkya S, Srivastava AK, Bisaria VS. Bioprocess Considerations for Production of Secondary Metabolites by Plant Cell Suspension Cultures. Biotechnol Bioprocess Eng 2002 ;7: 138-49

[139] Jacobs DI, van der Heijden R, Verpoorte R. Proteomics in plant biotechnology and secondary metabolism research. Phytochem Anal 2000; 11: 277-87.

[140] Verpoorte R, Van der Heijden R, Ten Hoopen HJG, Memelink J. Metabolic engineering of plant secondary metabolic pathways for the production of fine chemicals. Biotechnol Lett 1999; 21: 467–79.

[141] Nisha KK, Seetha K, Rajmohan K, Purushothama MG. *Agrobacterium tumefaciens*-mediated transformation of Brahmi [*Bacopamonniera* (L.) Wettst.], a popular medicinal herb of India. Curr Sci 2003; 85: 85–9.

[142] Fu CX, Xu YJ, Zhao DX, Ma FS. A comparison between hairy root cultures and wild plants of *Saussurea involucrata* in phenylpropanoids production. Plant Cell Rep 2006; 24: 750–4.

[143] Dronne S, Colson M, Moja S, Faure O. Plant regeneration and transient GUS expression in a range of lavandin (*Lavandula intermedia* Emeric ex Loiseleur) cultivars. Plant Cell Tiss Org Cult 1999; 55: 193–8.

[144] Nebauer SG, Arrillaga I, Del Castillo-Agudo L, Segura J. *Agrobacterium tumefaciens*-mediated transformation of the aromatic shrub *Lavandula latifolia*. Mol Breed 2000; 6: 539–52.

[145] Park SU, Facchini PJ. *Agrobacterium*-mediated genetic transformation of California poppy, *Eschscholzia californica* Cham, *via* somatic embryogenesis. Plant Cell Rep 2000; 19: 1006–12.

[146] Gomez-Galera S, Pelacho AM, Gene A, Capell T, Christou P. The genetic manipulation of medicinal and aromatic plants. Plant Cell Rep 2007; 26: 1689–715.

[147] Pandey V, Misra P, Chaturvedi P, Mishra MK, Trivedi PK, Tuli R. *Agrobacterium* tumefaciens-mediated transformation of *Withania somnifera* (L.) Dunal: an important medicinal plant. Plant Cell Rep. 2009; 29(2): 133-41.

[148] Tor M, Ainsworth C, Mantell SH. Stable transformation of the food yam *Dioscorea alata* L. by particle bombardment. Plant Cell Rep 1993; 12(7–8): 468–73.

[149] Hosokawa K, Matsuki R, Oikawa Y, Yamamura S. Production of transgenic gentian plants by particle bombardment of suspension-culture cells. Plant Cell Rep 2000; 19(5): 454–8.

[150] Tattersall DB, Bak S, Jones PR, Olsen CE, Nielsen JK, Hansen ML, Hoj PB, Moller BL: Resistance to an herbivore through engineered cyanogenic glucoside synthesis. Sci 2001; 293: 1826-8.

[151] Bohmert K, Balbo I, Kopka J, Mittendorf V, Nawrath C, Poirer Y, Tischendorf G, Trethewey RN, Willmitzer L. Transgenic *Arabidopsis* plants can accumulate polyhydroxybutyrate at up to 4% of their fresh weight. Planta 2000; 211:841-5.

[152] Kennedy AI, Deans SG, Svoboda KP, Waterman PG. Volatile oils from normal and transformed root of *Artemisia absynthium*. Phytochem 1993; 32: 1449-51.

[153] McKinely TC, Michaels PJ, Flores HE. Is lipoxygenase involved in polyacetylene biosynthesis in *Asteraceae*. Plant Physiol Biochem 1993; 31: 835-53.

[154] Ko KS, Ebizyka Y, Noguchi H, Sanakawa U (1988). Secondary metabolite production by Hairy Roots and regenerated plants transformed by Ri plasmids. Chem. Pharm. Bull. 36: 417-20.

[155] Bais HP, George J, Ravishankar GA. Production of esculin by Hairy Root Culture of *Cinchorium intybus* L. CV. Lucknow Local. Indian Exp Biol 1991; 37: 269-73.

[156] Hamill JD, Parr AJ, Robins RJ, Rhodes MJC, Robins RJ, Walton NJ. New routes to plant secondary products. Biotechnol 1987; 5: 800-4.

[157] Rhodes MJC. In: Kurz WGW, editor. Primary and secondary metabolism of plant cell cultures. Springer-Verlag, Berlin 1989; pp. 58-72.

[158] Mano Y, Ohkawa H, Yamada Y. Production of tropane alkaloids by hairy root cultures of *Duboisia leichhardtii* transformed by *Agrobacterium rhizogenes*. Plant Sci 1989; 59: 191-201.

[159] Trypsteen M, Van Lijsebettens M, Van Severen R, Van Montagu M. *Agrobacterium rhizogenes* mediated transformation of *Echinacea purpurea*. Plant Cell Rep 1991; 10: 85-9.

[160] Asada Y, Li W, Yoshikawa T. Isoprenylated flavonoids from Hairy Root Cultures of *Glycyrrhiza glabra*. Phytochem 1998; 47: 389-92.

[161] Ko KS, Noguchi H, Ebizyka Y, Sanakawa U. Oligoside production by Hairy Root cultures transformed by Ri plasmids. Chem Pharm Bull 1989; 37: 245-58.

[162] Shimomura K, Sudo H, Saga H, Komada H. Shikonin production and secretion by Hairy Root Cultures of *Lithospermum erythrorhizon*. Plant cell Reports 1991; 10: 282-5.

[163] Sevon N, Hiltunen R, Caldentey KMO. Somoclonal variation in transformed roots and protoplast derived hairy root cultures of *Hyoscyamus muticus*. Planta Med 1998; 64: 37-41.

[164] Yoshikawa T, Furuya T. Saponin production by cultures of *Panax ginseng* transformed with *Agrobacterium rhizogenes*. Plant Cell Rep 1987; 6: 449-52.

[165] Kunshi M, Shimomura K, Takida M, Kitanaka S. Growth and ginsenoside production of adventitious and hairy root cultures in an interspecific hybrid ginseg (*Panax ginseng x P. quinquefolium*). Nat Med 1998; 52: 1-4.

[166] Benjamin BD, Roja G, Heble MR. Alkaloid synthesis by root cultures of *Rauwolfia serpentine* transformed by *Agrobacterium rhizogenes*. Phytochem 1994; 35: 381-3.

[167] Shin SW, Kim YS. Production of anthroquinones derivatives by hairy root culture of *Rubia cordifolia* var pratensis. Korean J Pharmacogn 1996; 27: 301-8.

[168] Hu BZ, Alfermann AW. Diterpenoid production in Hairy Root Cultures of *Saliva miltiorrhiza*. Phytochem 1993; 32: 699-703.

[169] Maggon K. Best Selling Human Medicines 2002-2004. Drug Discovery Today 2005; 10: 738-42.

[170] Kochhar SL. Tropical Crops: A Textbook of Economic Botany (Paperback). MacMillan Pub Ltd, London 1981; pp. 467.

[171] Go´mez-Gale´ra S, Pelacho AM, Gene A. The genetic manipulation of medicinal and aromatic plants. Plant Cell Rep 2007; 26:1689–715.

[172] Fischbach MA, Clardy J. One pathway, many products. Nat Chem Biol 2007; 3(7): 353-5.

[173] Mazid M., Khan TA. , Mohammad F. Role of secondary metabolites in defense mechanisms of plants. Biol Medc.2011 3(2): 232-49.

CHAPTER 2

RNAi: New Era of Functional Genomics for Crop Improvement

Pradeep Kumar[*], Madhu Kamle and Ashutosh Pandey

Central Institute for Subtropical Horticulture, Rehmankhera, Lucknow, (U.P.), India

Abstract: RNA interference mediated gene silencing is a knock down technology that brings revolution in functional genomics and has great promises for new applications of commercial value in crop improvement. Although the technology is still in its infancy and many mysteries of RNAi have yet to unfold, but the new field of RNAi based on genomics is increasingly being qualified as a fundamental paradigm shift for biotechnology and future genomics. The principle behind RNAi is that it starts with introduction of a dsDNA in cell, followed by activation of DICER gene and RISC complex, which eventually leads to loss of gene expression. An important aspect of silencing in plants is that it can be triggered locally and then spread *via* a mobile silencing signal. RNAi technology may be used for generating improved crop varieties in terms of disease resistance, insect resistance, enhancing nutritional qualities, shelf life and abiotic stress tolerance as a biological tool has been employed to investigate gene function *in vitro* and *in vivo*. It is also being used for regulating genes in crop plants in a very specific manner without affecting the expression of other genes, thus increasing their productivity.

Keywords: RNA interference, gene silencing, crop improvement, DICER, RISC complex and disease resistance.

INTRODUCTION

RNA silencing is a novel gene regulatory mechanism that limits the transcript level by either suppressing transcription (TGS) or by activating a sequence-specific RNA degradation process (PTGS/RNA interference [RNAi]) [1]. The term RNA interference, or "RNAi," was initially coined by Fire and coworkers 1998 for unraveling the mechanism of gene silencing by double stranded RNA, based on their studies in nematode worm *Caenorhabditis elegans*. During the 1990s, a number of gene silencing phenomena that occur at the post-transcriptional level were discovered in plants, fungi, animals and ciliates [2, 3]. The silencing effect was first observed in plants in 1990, when the Jorgensen laboratory introduced exogenous transgenes into petunias in an attempt to up-regulate the activity of a gene for chalcone synthase, an enzyme involved in the production of specific pigments [1, 4]. Unexpectedly, flower pigmentation did not deepen, but rather showed variegation with complete loss of color in some cases. This indicated that not only were the introduced trangenes themselves inactive, but that the added DNA sequences also affected expression of the endogenous loci [5]. This phenomenon was referred to as "co-suppression" [4, 6]. The molecular mechanism of RNAi is most elegant and efficient biochemical pathway that occurs in nature. RNAi based gene silencing was first discovered in plants where it was termed co-suppression or post-transcriptional gene silencing (PTGS). RNA silencing is well suited to the systematic analysis of gene function.

The improvement of RNA silencing approach and other related methods, will firmly entrench as a potent and handy tool in the study of plant functional genomics for crop improvement. In plants, RNA silencing, is an efficient part of gene silencing, not only serves as an essential component of the defense system being targeted against transposable elements and viral infection, but also plays important roles in the regulation of endogenous gene expression [7]. These wider applications in reference prompt us to explore on RNAi mediated gene silencing in crops [8], a rather simple but highly effective way to control gene expression in living organisms. In present chapter we focuses on the landmarks in RNAi (RNA-interference) discovery, its mechanism of action and the promises and pitfalls it offers in post translational gene silencing for the crop improvements through biotechnological applications.

*Address correspondence to Pradeep Kumar: Central Institute for Subtropical Horticulture, Rehmankhera, Lucknow, (U.P.), India; E-mail: pradeepgkp17@yahoo.co.in

Aakash Goyal and Priti Maheshwari (Eds)

LAND MARKS IN RNAi DISCOVERY

RNAi was firstly discovered and observe in transcriptional inhibition by antisense RNA expressed in transgenic plants and more directly by reports of unexpected outcomes in experiments performed in 1990s. In an attempt to produce more intense purple coloured *Petunias*, researchers introduced additional copies of a transgene encoding chalcone synthase (a key enzyme for flower pigmentation). But were surprised at the result that instead of a darker flower, the *Petunias* were variegated (Fig. **1**). This phenomenon was called co-suppression of gene expression [4], since both the expression of the existing gene (the initial purple colour) and the introduced gene /transgene (to deepen the purple) were suppressed. It was subsequently shown that suppression of gene activity could take place at the transcriptional level (transcriptional gene silencing, TGS) or at the post-transcriptional level (post-transcriptional gene silencing, PTGS).

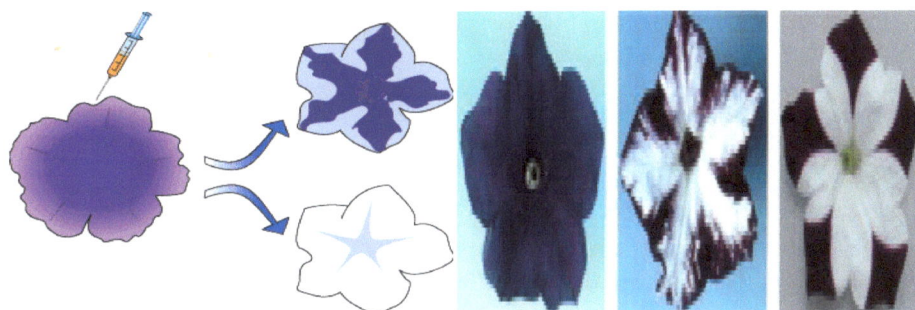

Figure 1: Upon injection of the transgene responsible for purple colorings in *Petunias,* the flowers became variegated or white rather than deeper purple as was expected.

The mystery of molecular mechanism responsible for gene silencing now known as RNA interference (RNAi) exploded in 1998. It was discovered that PTGS was triggered by double-stranded RNA (dsRNA), provided most unexpected explanation with many profound consequences. Thus, it was concluded that co-suppression in plants, quelling in fungi and RNAi in nematodes all shared a common mechanism called as gene silencing. Discovery and landmarks in RNAi advancement given in Table **1**.

Table 1: A brief history and developments in RNAi

Discovery	References
Double-stranded RNA (dsRNA) works better than anti-sense RNA	[9]
Feeding *C. elegans* with bacteria expressing dsRNA	[10]
Micro-RNAs	[11]
RNA interference (RNAi) in mammalian cell culture	[12]
RNAi delivered by plasmids (short-hairpin RNA, shRNA)	[13-17]
Viral delivered shRNA	[14]
Viral shRNA-induced animal knockdowns	[18]
C. elegans genome screen using RNAi	[19]
Human genome screen using RNAi	[20, 21]
Enzyme-mediated RNAi library synthesis	[22]
RNAi-mediated suppression of DET1 expression under fruit-specific promoters to improve carotenoid and flavonoid levels in tomato	[23]
Reduction of lysine catabolism specifically during seed development produced white-flowered transgenic gentians by suppressing the chalcone synthase (*CHS*)	[24, 25]

COMPONENTS OF RNAi

Among the components of gene silencing process, some serve as initiators and others serve as effectors, amplifiers and transmitters.

DICER

RNAi mechanism involves dsRNA processing, mRNA degradation where the Dicer acts as an essential component in the process. Dicer is a ribonuclease in RNase III family enzyme whose functions is the processing of dsRNA to short double-stranded RNA fragments called small interfering RNA (siRNA) [26]. DsRNA continuously cleaves by dicer at 21-25 bp distance and produce siRNA with 2-nt 3' overhangs and 5' phosphorylated ends. Dicer contains helicase domain, dual RNase III motifs and a region homologous to the protein of RDE1 or QDE2 or Argonaute family [27]. Dicer works in the first step of RNAi pathway as a catalyst starting production of RNA-induced silencing complex (RISC). Argonaute, a catalytic component of dicer, have the capability to degrade mRNA complementary to that of the siRNA guide strand [28]. Dicer digests dsRNA to siRNA of uniformed size.

RNA INDUCED SILENCING COMPLEX (RISC)

RNA-induced silencing complex or RISC is a siRNA directed endonuclease contains proteins and siRNA. It targets and destroys mRNAs in the cell complementary to the siRNA strand. When RISC finds the mRNA complementary to siRNA, it activates RNase enzyme resulting the cleavage of targeted RNA. About 20-23 bp siRNA are able to associate with RISC and guide the complex to the target mRNA and combined together and degrades them, resulting in decreased levels of protein translation and knockdown the gene function [27, 29]. RISC acts as catalyst to cleave single phosphodiester bond of mRNA [30]. Although the composition of RISC is not completely known, it includes members of the Argonaute family [27] that have been implicated in processes directing post-transcriptional silencing [31].

RNA DEPENDENT RNA POLYMERASE (RDRPS)

RNA-dependent RNA polymerases play an important role in the silencing effect in RNAi and Post Transcriptional Gene Silencing (PTGS) mechanism. RNAi is more powerful technique than the antisense approach of gene silencing in plants [32] and *C. elegans* [33] due to the activity of RdRPs and this activity was first observed in RNA viruses. Transcription and replication of viral genome is the result of the activity of viral RdRP. In some plants, RdRP activity can also be found in healthy tissue. RdRP was isolated and purified from the leaves of tomato by [34] and characterized its catalytic properties by [35]. RdRPs activity was normally demonstrated by performing RNA transcription reactions in the presence of DNA dependent RNA polymerase inhibitors.

TRANSLATION INITIATION FACTORS

Initiation and effector steps are involved in RNAi mechanism [36]. In the initiation step, dsRNA is digested into 21-23 nucleotides siRNAs known as "guide RNAs" [27, 36]. It was evident that siRNAs are produced by the activity of dicer on dsRNA strand. Initiation step of RNAi involves the activity of some genes. In the effector step, double stranded siRNA associate with RISC. The RISC activation is linked with ATP dependent siRNA unwinding. The activated RISC then break down the mRNA ~12 nucleotides from the 3' terminus of the siRNA [36, 37].

MOLECULAR MECHANISM BEHIND RNAi MEDIATED GENE SILENCING

RNA interference is an RNA dependent gene silencing (RdGS) mechanism that includes the endogenously induced gene silencing effects of miRNA as well as silencing triggered by foreign dsRNA. The dsRNA binds with a protein complex DICER which is an endonuclease that cleaves it into short fragments with 20 to 25 base pairs with a few unpaired overhung bases at both ends. The short dsRNA fragments produced by DICER, is called small interfering RNAs (siRNAs). Thus, miRNA and siRNA share same cellular machinery as well as functional analogy. These fragments (SiRNA or miRNA) integrate with another active protein complex RISC (RNA induced silencing complex). The catalytic active component of the RISC complex is known as a protein which acts as endonuclease and mediates SiRNA induced cleavage of the target mRNA strand. Consequently, one of the RNA strands (antiguide strand or passenger strand) is degraded while the other is selected as a guide strand which remains bound to RISC complex. When a complementary mRNA is located by an RISC bound guide strand it binds to it and is cleaved and degraded. The expression of the gene corresponding to the mRNA is silenced. RNAi has been particularly well studied in certain organisms such as *C. elegans*, *Drosophila*, and in plants where the effect can be spread from cell to cell within the organism. The protein complex like RISC and the enzyme DICER have been found to be conserved throughout eukaryotes. The RNA interference pathway, as well, is conserved across all eukaryotes. It has been established that RNAi mechanism provides genomic stability, especially in plants, keeping the transposons production under control. RNAi pathway can be divided into three major steps.

1. Initiator Step: Cleavage of dsRNA

This is the first step in which, dsRNA is converted into 21-23bp small fragments by the enzyme Dicer. Dicer is the enzyme involved in the initiation of RNAi. It is a member of Rnase III family of dsRNA specific endonuclease that cleaves dsRNA in ATP dependent, processive manner to generate siRNA duplexes of length 21-23 bp with characteristic 2 nucleotide overhang at 3'- OH termini and 5' PO_4 (Fig. **2**).

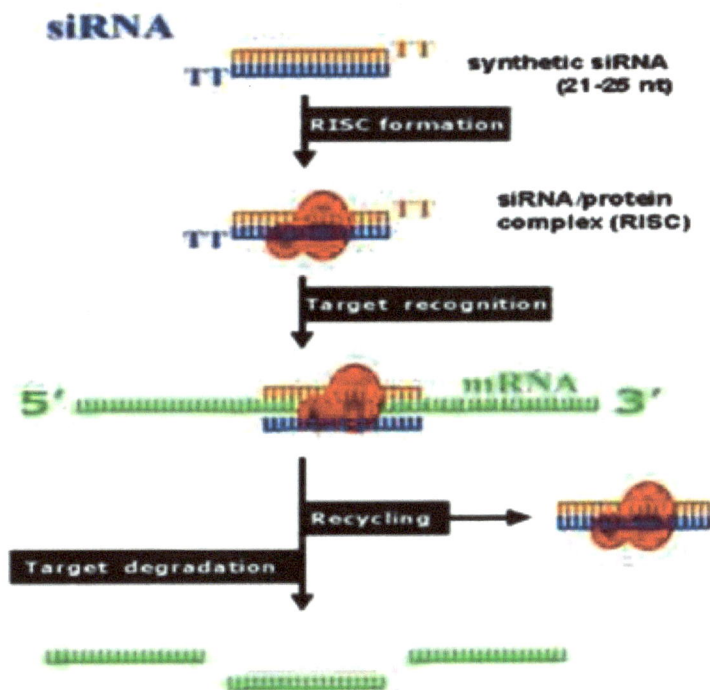

Figure 2: RNAi mechanism: The mRNA degrading pathway [45].

2. Effector Step: Entry of siRNA into Risc

The siRNAs generated in the initiator step now join a multinucleate effector complex RISC that mediates unwinding of the siRNA duplex. RISC is a ribonucleoprotien complex and its two signature components are the single-stranded siRNA and Argonaute family protein. The active components of an RISC are endonucleases called argonaute proteins, which cleave the target mRNA strand complementary to their bound siRNA, therefore argonaute contributes "Slicer" activity to RISC. As the fragments produced by dicer are double-stranded, they could each in theory produce a functional siRNA. However, only one of the two strands, which is known as the guide strand, binds the argonaute protein and directs gene silencing. The other anti-guide strand or passenger strand is degraded during RISC activation. The process is actually ATP-independent and performed directly by the protein components of RISC. Although it was first believed that an ATP-dependent helicase separated these two strands, the process is actually ATP-independent and performed directly by the protein components of RISC (Fig. **3**).

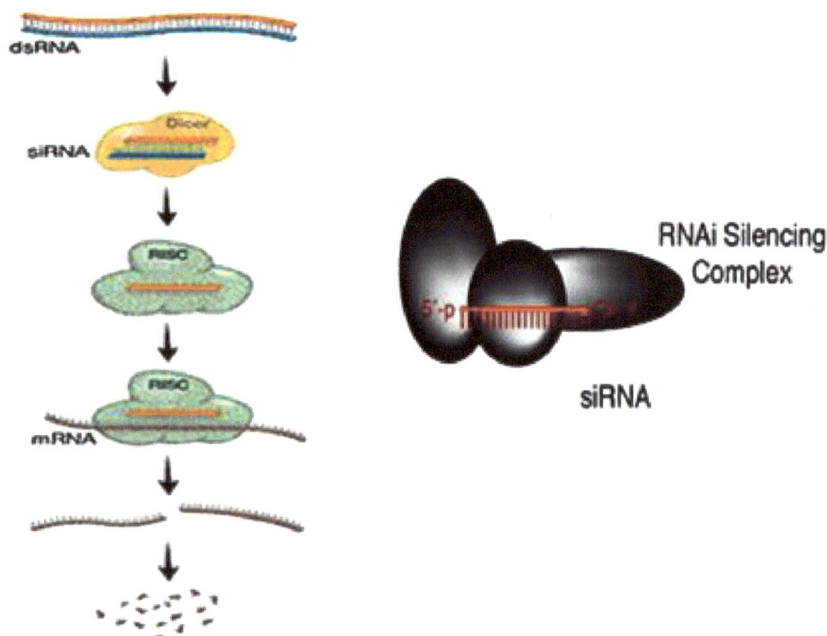

Figure 3: The RNA interference process and the biochemical machinery involved. Double-stranded RNA is cut into short pieces (siRNA) by the endonuclease Dicer. The antisense strand is loaded into the RISC complex and links the complex to the mRNA strand by base-pairing. The RISC complex cuts the mRNA strand, and the mRNA is subsequently degraded.

3. Sequence-Specific Cleavage of Targeted mRNA

The active RISC further promotes unwinding of siRNA through an ATP dependent process and the unwound antisense strand guides active RISC to the complementary mRNA. The targeted mRNA is cleaved by RISC at a single site that is defined with regard to where the 5' end of the antisense strand is bound to mRNA target sequence. The RISC cleaves the complimentary mRNA in the middle, ten nucleotides upstream of the nucleotide paired with the 5' end of the guide siRNA. This cleavage reaction is independent of ATP. The target RNA hydrolysis reaction requires Mg^{2+} ions. Cleavage is catalyzed by the PIWI Domain of a subclass of Argonaute proteins. This domain is a structural homolog of RNase H, Mg^{2+} dependant endoribonuclease that cleaves the RNA strand of RNA-DNA hybrids. But each cleavage-competent RISC can break only one phosphodiester bond in its RNA target. The siRNA guide delivers RISC to the target region, the target is cleaved, and then siRNA departs intact with the RISC. Thus the two important conditions to be fulfilled for the success of silencing by RNAi are established as: 5' phosphorylation of the antisense strand and the double helix of the antisense target mRNA duplex to be in the A form. The A-form helix is required for the stabilization of the heteroduplex formation between the siRNA antisense strand and its target mRNA.

STRATEGIES FOR CROP IMPROVEMENT

The utilization of the RNAi technology to enhance specifically the nutritional value of plant organs that are consumed as foods is progressively increasing with several published successful efforts, suggesting that this approach has tremendous potential [25].

NUTRITIONAL VALUE

RNAi technology has also been used in several other plants to improve their nutritional quality. For example, caffeine content in coffee plants has been markedly reduced by RNAi-mediated suppression of the caffeine synthase gene [38]. In another study, RNAi has been successfully used to generate a dominant high-lysine maize variant by knocking out the expression of the 22-kD maize zein storage protein, a protein that is poor in lysine content [39]. Traditional breeding has been successful only for the screening of a recessive lysine-rich mutant called opaque 2 (O2). The O2 gene encodes a maize basic leucine zipper transcriptional factor that controls the expression of a subset of storage proteins, including the 22-kDa zein storage protein. Although it is rich in lysine, the opaque 2 mutant is not very useful in agriculture because of its adverse effects on seed quality and yield. By contrast, down regulation of the maize lysine-poor 22-kDa zein gene *via* RNAi does not alter the general functions of O2, but generates quality and normal maize seeds with high levels of lysine-rich proteins [40].

RNAi technology has also been successful in genetic modification of the fatty acid composition of oil. RNAi mediated by a hairpin RNA has been used in cotton to down regulate two key fatty acid desaturase genes encoding stearoyl-acyl-carrier protein D9-desaturase and oleoylphosphatidylcholine u6-desaturase [41]. Knockdown of these two genes in cotton leads to an increase in nutritionally improved high-oleic and high-stearic cottonseed oils, which are essential fatty acids for health of the human heart.

Two general approaches are commonly used to reduce the levels of undesirable gene products: recessive gene disruption and dominant gene silencing. In gene disruption approaches the target sequence is mutated to eliminate either expression or function, whereas in dominant gene silencing either destruction of the gene transcript or inhibition of transcription is induced.

The advantages of the dominant gene silencing methodologies over the gene disruption approach are two fold. First, dominant gene silencing is easier to bring about genetically and screening of the resultant transgenic plants is also more straightforward. Second, in contrast to the gene disruption approach, dominant gene silencing can be done in a spatial and temporal manner by using specific promoters of the dominant gene silencing approaches, dsRNA triggered RNAi is apparently the most powerful: it is the most efficient in terms of the extent of gene silencing, and the resulting silencing is almost as complete as that achieved in a gene knockout approach. It seems that dsRNA-triggered RNAi directly bypasses the requirement for dsRNA synthesis *via* RdRPs, which probably is the rate limiting step in the plant RNAi pathway.

RNAi FOR THE GENETIC IMPROVEMENT OF CROP PLANTS

Double stranded RNA-based gene silencing is fairly common in plants; actual technological improvements in RNAi, especially for the comprehensive analysis of the whole genome, in plant cells are still limited. The RNAi can be an excellent mechanism to control viral diseases in plants, animals and humans [42]. RNAi techniques have been working with mRNA degradation, gene silencing, gene expression regulation, and resistance to virus infection, regulation of chromatin structure and genome integrity. The immersing role of RNAi technology may be used for gene silencing and generating improved crop varieties in terms of disease, insect resistance, enhancing nutritional qualities and Trait stability [40].

INCREASING THE LEVEL OF LYSINE IN PLANTS

Synthesis of Lysine was regulated by the feedback inhibition loop in which lysine inhibits the activity of dihydrodipicolinate synthase (DHPS), the first enzyme on the pathway specifically committed to lysine biosynthesis [40]. This high lysine levels in seeds are beneficial, increases in the level of this amino acid in vegetative tissues are undesirable, because high levels of lysine cause abnormal vegetative growth and flower development that, in turn, reduces seed yield [43, 44]. Because lysine accumulation in plants is negatively affected by its catabolism (degradation), constitutive knockout of lysine catabolism using a gene insertion knockout approach accelerates lysine accumulation in seeds when combined with the seed-specific expression of a feedback-insensitive DHPS [45]. Seed germination has been improved by RNAi approach because of reduction of lysine catabolism [24, 25]. Another effort done by [39], in that RNAi has been successfully used to generate a dominant high-lysine maize variant by knocking out the expression of the 22-kD maize zein storage protein,a protein that is poor in Lysine content and RNAi generates quality and normal maize seeds with high levels of lysine-rich proteins [40, 46].

INCREASING GRAIN AMYLOSE CONTENT

Starch is the major content of plant carbohydrates, which is formed by the amylopectin and amylose polysaccharides, synthesized by two competitive pathways [47]. When cooked food was undergoing cooling process before eating, amylose molecules tend to efficiently form digestion-resistant complexes that are part of healthy dietary fiber [48]. The aim to increase the relative content of amylose in wheat grains, a RNAi construct designed to silence the genes encoding the two starch-branching isozymes of amylopectin synthesis, were expressed under a seed-specific promoter in wheat [49]. This resulted in increased grain amylose content to over 70% of the total starch content [25].

COLOR MODIFICATION OF PLANTS BY RNAi-MEDIATED GENE SILENCING GENTIAN

Gentians, *Gentiana triflora*, *Gentiana scabra* and their interspecific hybrid, are one of the most popular floricultural plants in Japan, and more than half of gentian production is from the Iwate prefecture. Gentians come into bloom from early summer to late autumn in Japan, and are often used as ornamental cut flowers. Genetic engineering approaches are being applied to several ornamental plants [50-52]. Industry like Florigene Ltd. and Suntory Ltd. have developed blue-flowered carnations using genetic engineering, and now they are commercialized in many countries *viz.*, North America, Australia and Japan [52]. Workers developed the white-flowered transgenic gentians by using gene silencing for the suppression of the chalcone synthase [53]. Other workers [54-58] developed the different colored flower of many plant species including petunia, torenia and tobacco plants.

COLOUR MODIFICATION OF ROSE BY USING RNAi

Anthocyanins, which are major precursors for all pigments, among them the cyanidin gene is responsible for a synthethic pathway that leads to formation of red pigment and a correspondent Delphinidin gene is the key gene for formation of blue color. Scientists at Florigene (Australia) and Suntory (Japan) have been sucessful in knock -downing the cyanidin genes in rose and carnation by RNAi technology and introduce delphinidin genes, which, in natural condition, are absent in these two important cut flowers. The result has been spectacular to the flower industry, Horticultural Societies of Britain and Belgium has offered a prize of 500 000 francs to the first person to produce a blue rose [59].

RNAi AND FLOWERING TIME

Flowering time is one of the important aspects of crop production. As a examples, if flowers of creal crop appear too early, it may have not yet made sufficient energy stores to fuel its maximum grain production. Similarly, in case of late flowering there may be chances of low yield. In *Arabidopsis*, there is a gene called FLC which represses flowering and we have used RNAi to switch it off and bring on flowering. This clearly shows that the technology has the potential to regulate flowering time in crops [60].

PRODUCTION OF HEALTHIER OIL

An important application of RNAi in plants that is much closer to agricultural use is the silencing of genes involved with seed-oil production. Few seeds-oils are soun good for human health than others and some oils are more stable at high temperatures than others. It all depends on the fatty acid composition of the oil. As a example palm oils is very high in palmitic acid which makes it stable at high temperatures but also unhealthy for human consumption, as it raises LDL cholesterol levels. Olive oil, on the other hand, is high in linoleic acid which is much healthier for human consumption, but it is not stable at high temperatures and therefore not good for frying. The best oil for heat stability and with no negative effects on cholesterol levels is one which is high in oleic acid. RNAi have been used to silence the gene in cotton which codes for the enzyme that converts oleic acid into a different fatty acid. This has altered the seed-oil from being around 10% to an impressive 75% oleic acid. If these plants were used in agriculture it would produce two crops, fibre and seed-oil, for the price of one [60].

PEST CONTROL BY USING RNAi

All the transgenic resistance to insect and pest now a day developed through using *Bacillus thuringiensis* (Bt) a natural, environmentally friendly manner. Bt are effective in controlling insects in the larva stage only. The larva stage in an insects' life cycle is the stage during which most of the feeding occurs. Since, Bt must be ingested to work, the insect must be controlled during the larval stage. Insect resistance to Bt has not been a significant problem to date, but risk of development of resistance to these proteins can be further reduced with the use of non-Bt-crop or natural refuges of non-resistant plant hosts. Combining Bt technology with a second, independent mode of insect control *via* RNAi would both enhance product performance and further guard against the development of resistance to Bt proteins [61].

IMPROVEMENT OF WOOD AND FRUIT QUALITY

During chemical pulping of wood, one of the most expensive and environmentally hazardous processes is to separate lignin from cellulose and hemicellulose. The production of plant material with lower contents of lignin would mean a significant reduction of cost and pollution to the paper industry. One of the approaches to obtain reduced lignin forest trees has been the down regulation of lignin biosynthesis pathways [62]. The main genes involved with genetic transformation targeting lignin reduction are 4-coumarate: coenzyme A ligase (*Pt4CL1*) [62], cynnamyl alcohol deshydrogenase (CAD - the final enzyme in the biosynthesis of lignin monomers) [63] and caffeate/5-hydroxyferulate O-methyltransferase (COMT - enzyme involved in syringyl lignin synthesis) [64].

Many efforts were tried by workers to improve the fruit quality through gene silencing for improve the self-life [65], reducing the allergence amount of apple [66] and sorbitol distribution affects fruit quality such as starch accumulation and sugar-acid balance which has been suggested by [67].

DNA DIRECTED RNAi

DNA directed RNAi makes use of dsRNA-expressing vectors introduced into plants *via Agrobacterium*-mediated transformation. This approach has been shown to be effective in reducing the amount of a specific gene product. One of the first studies testing DNA-directed RNAi in plants was done to compare the ability of sense, antisense, or dsRNA at generating RNA-mediated virus resistance *via* PTGS in tobacco and silencing of an endogenous GUS reporter gene in rice [60]. In both cases it was shown that duplex RNA was more effective than either sense or antisense RNA at silencing the target gene.

CO-SUPPRESSION

Co-suppression is a phenomenon in plants in which transgene causes silencing of a homologous endogenous gene. In this case, host plants have the endogenous gene which is homologous to the foreign gene [68]. It was found that these suppressive transgenes will generate aberrant RNAs. Although it is hypothesized that the aberrant RNAs can bind to their target mRNA and act as primers for the synthesis of dsRNA, this has never been directly demonstrated. It is likely that the transgene mRNA gets cleaved into siRNAs, which lead to the co-suppressive silencing effect on the endogenous gene [69]. Plants are the only organism that shows sense-mediated silencing (silencing in the apparent absence of dsRNA). Thus co-suppression is unique to plants.

VIRUS-INDUCED GENE SILENCING

Virus-induced gene silencing (VIGS) is another approach often used to analyse gene function in plants. RNA viruses generate dsRNA during their life cycle by the action of virus-encoded RdRP. If the virus genome contains a host plant gene, inoculation of the virus can trigger RNAi against the plant gene. Because this approach does not involve a transformation process, it might be suitable for the functional analysis of essential genes.

TISSUE-SPECIFIC OR STIMULI-RESPONSIVE SUPPRESSION

A seed-specific promoter has been shown to be effective for suppressing constitutively expressed genes in the seeds. Lgc1 acts as a Mendelian factor in F2 seeds on a single F1 plant, suggesting that there is no transmission of the silencing signal among developing seeds. Absence of plasmodesmata between the seed and its surrounding tissues might affect the efficiency of spread of the silencing signal. Alternatively, the signal might be excluded from seeds, as it is excluded from the shoot apex. By such mechanisms, hpRNA-induced RNAi driven by a seed-specific promoter might confer seed-specific suppression; however, when other tissues, particularly where the PTGS signal travels easily, are specific targets of hpRNA-induced RNAi this specificity might be lost. In fact, systemic spread was observed in the chemically regulated RNAi system. This potential problem could be overcome by the use of a virus protein that suppresses the systemic spread of the PTGS signal or through knockout of a gene involved in the spread of the RNAi signal.

TRANSCRIPTIONAL GENE SILENCING (TGS)

First discovered by [70] in tobacco, this form of gene silencing, termed RNA-directed DNA Methylation (RdDM) takes place when the promoter region of a gene is methylated. Either dsRNA or siRNA directs methylation of cytosine residues in both the promoter region and the open reading frame (ORF) [71]. Interestingly, methylation in the ORF region alone is not sufficient to initiate TGS. On the other hand, methylation of the promoter region alone is sufficient for producing stable and heritable silencing of a transgene. Further research into RdDM has illustrated that DNA methylation may be helpful in preventing the incidence of potato spindle tuber viroid [70], plant potyviruses [71], cereal yellow dwarf virus [72], and others.

TRAIT STABILITY

The rice mutant line LGC-1 (Low Glutelin Content-1) was the first commercially useful cultivar produced by RNAi. It is low-protein rice and is useful for patients with kidney disease whose protein intake is restricted. This dominant mutation produces hairpin RNA (hpRNA) from an inverted repeat for glutelin, the gene for the major storage protein glutelin, leading to lower glutelin content in the rice through RNAi. Interestingly, this mutant was isolated in the 1970s, and the mutant trait appears to have been stable for over 20 generations. These examples suggest that the suppression of gene expression by hpRNA induced RNAi would be inherited stably.

DEGREE OF SUPPRESSION

RNAi poses some advantages among systems and its applicability to multigene families and polyploids, as it is not straightforward to knockout a multigene family by the accumulation of mutations for each member of the family by conventional breeding, particularly if members of the family are tightly linked. Another advantage of RNAi lies in the ability to regulate the degree of suppression. Agronomic traits are often quantitative, and a particular degree of suppression of target genes may be required. Control of the level of expression of dsRNA through the choice of promoters with various strengths is thought to be useful in regulating the degree of suppression. Because RNAi is a very efficient knockdown technology in plants it is thought to be useful for genetic improvement, even in plants with low transformation efficiencies.

DOWN REGULATION

Down regulating genes in crop plants have a very specific manner without affecting the expression of other genes, thus increasing their productivity. Down regulation of a particular gene can be achieved by mutation-based reverse genetics, but its use is more limited than that of RNAi. Although the basic concept of the application of transgene-based RNAi to the genetic improvement of crop plants has been established, further feasibility studies are needed for its wider application. For instance, the transient satellite-virus-based SVISS technology developed by Bayer Crop Science allows the production of high levels of dsRNA in plants, which triggers efficiently transient RNAi. The SVISS technology has been implemented as a research tool to discover and validate gene functions of candidate herbicide target genes and genes involved in abiotic stress response. For the abiotic stress-related PARP pathway in canola and corn and the enzymatic pathway underlying seed shattering in oilseed rape, it has been demonstrated that stable transformation of crops with RNAi constructs results in stable modification of biochemical pathways which can result in improved productivity and quality of crops in the field.

SYSTEMIC SILENCING IN PLANTS

The mechanism of PTGS is initiated from dsRNA that result from replicative intermediates of viral RNAs or aberrant transgene coded RNAs. PTGS that starts locally in plants by a transgene or virus can spread systematically to the rest of the plant. DsRNA or siRNAs when introduced locally in the plants can trigger systemic silencing. SiRNAs act as "mobile trigger elements" for systemic silencing. A remarkable feature of RNA silencing is its ability to act beyond the cells in which it is initiated. In plants carrying an expressed GFP transgene, RNA silencing can be initiated by localized introduction of an additional ectopic GFP transgene. The RNA silencing is initially manifested in the tissues containing the ectopic DNA but eventually becomes systemic showing that a silencing signal moves between cells and in the vascular system of the plants.

ADVANTAGES AND DRAWBACKS OF RNAi IN FUNCTIONAL GENOMICS

RNAi has advantages and limitations when used in plant functional genomics. RNAi has many advantages over the functional genomics strategies based on insertional mutagenesis. The first and foremost advantage is that RNAi gives us the ability to specifically target a gene. If the target sequence is carefully chosen, a specific gene or genes can be silenced. RNAi can also be used to achieve varying levels of gene silencing, using the same ihpRNA construct in different lines. This allows for selection of lines with varying degrees of gene silencing. In addition to this, the timing and extent of the gene silencing can be controlled, so that genes that are essential will only be silenced at chosen stages of growth or in chosen plant tissues. So, RNAi provides us with a great degree of flexibility in the field of functional genomics.

There are also limitations however to RNAi. Unlike in insertional mutagenesis, for the use of RNAi the exact sequence of the target gene is required. Once this sequence information is available, the rest of the process is however relatively fast. Secondly, delivery methods for the dsRNA are a limiting step for the number of species for which RNAi based approaches can be used easily. Due to this, improvement and further research into the kinds of vectors that can be used safely and reliably is needed. There have also been some reports that it has been difficult to detect a mutant in which there has been subtle changes in gene expression [73].

VECTORS USED IN RNAi BASED FUNCTIONAL GENOMICS

There are currently many different vectors in use for performing RNAi for the use of functional genomics. These include such vectors as binary vectors used for expression of GUS and GFP proteins, the pHELLSGATE high-throughput gene silencing vector and a high throughput tobacco rattle virus (TRV) based Virus-induced gene silencing (VIGS) vector. One feature that is common to all of these vectors is the inclusion of the Gateway recombination-based technology for cloning that was developed by Invitrogen. The Gateway system is used to replace conventional cloning steps that took up valuable time. This is being exploited by several projects including the AGRIKOLA project [73].

Gateway is a cloning system developed by Invitrogen that is universal, and has speed up the process of plant functional genomics. It is based on the phage lambda system of recombination. It enables segments of DNA to be transferred between different vectors while orientation and reading frame are maintained. It can also be used for transfer of PCR products. It saves valuable time, because once the DNA has been cloned into a Gateway vector, it can be used as many genome function analysis systems as is required. In this way, the use of vectors in the process of plant functional genomics has been made much easier, while the process has also been made faster. This allows for higher throughput analysis to occur.

DELIVERY OF VECTORS

There are many different ways in which vectors can be delivered. Firstly, there is micro particle bombardment with vectors that express intron-containing hairpin RNA (ihpRNA) or dsRNA. The second method of delivery is through the use of *Agrobacterium* carrying a T-DNA that expresses an ihpRNA transgene. Then there is virus induced gene silencing (VIGS), where the target sequence is integrated into the virus' sequence which is then used to infect the plant. These can also be expressed from transgenes introduced by *Agrobacterium*, or by stable transformation by ihpRNAs that express transgenes [73]. Each of these methods of delivery has their advantages and disadvantages. Microparticle bombardment is a transient method of vector delivery. Its advantages are that it is rapid, has a wide range of species on which it can work, and is a valuable tool for work on single cells. The disadvantages are that this limits gene silencing to the cells on the surface of the leaf, and silencing is only temporary. The *Agrobacterium* method is also a transient system for vector delivery. The advantages of this method of vector delivery are that it is rapid and provides a high throughput, it is relatively easy to use and it has a low cost. The disadvantage is that it has not really been tested on most species, so we do not know the scope for use on different species.

Virus induced gene silencing (VIGS) is another method of vector delivery that is transient. It has many advantages. It is rapid and provides a high throughput, and it is easy to use. It can be applied to plants that are mature, and is considered to be good for use on species that difficult to transform. With these many advantages also come many disadvantages because of the limitations in its host range. It might have restricted regions of silencing, and their may be size restrictions on the inserts. It is dependent on the availability of infectious clones. Viral symptoms could be superimposed onto the silenced phenotype. The ihpRNA method of delivery also has many advantages. It has no restrictions on host range, and provides heritable gene silencing. It has a high throughput, and one can control the degree of gene silencing that occurs. In addition to this, one can control the tissue specificity of the gene silencing. This method of vector delivery has one disadvantage: an efficient technique for transformation is needed [74].

FUTURE PROSPECTS AND RNAi

RNAi is an important area of molecular research all over the world. Clear knowledge and understanding of RNAi mechanisms is necessary to know the functions of gene and establishment of this technology. It is now in an advanced state but still did not come out from its infancy. The application of this field is ranging from molecular biology to gene therapy. The RNAi field is linked with RNA interference, transgene silencing and transgene mobilization [75, 76]. DsRNA activates in RNAi as a normal cellular process leading to specific RNA degradation of high specificity and a cell-to-cell dissemination and transfer of this gene silencing effect in several RNAi mechanisms [77]. Practically, RNAi is a very effective and powerful technique to block particular gene expression which can be used to know the function of gene in plants or humans or organisms [78, 79]. Technologically reliable and high-throughput methods of this technique are being developed by understanding of the core RNA silencing mechanism. Viral attack is a major problem existing worldwide for plants and vertebrates. Thus, we can say that RNAi can be potential tool by exploring future in functional genomics by silencing viral genes [42], development of siRNA based drugs [80], systemic silencing and co-suppression *etc.* It is evident that the RNAi gradually become an impressive field for improvement of crops and which could be the major outcome in the near future using functional genomics.

ABBREVIATIONS USED

miRNA: MicroRNA

RNAi: RNA interference

siRNA: Small interfering RNA

TGS: Transcriptional gene silencing

PTGS: Post-transcriptional gene silencing

RISC: RNA induced silencing complex

VIGS: Iiral-induced gene silencing

RdRP: RNA-dependent RNA polymerase

dsRNA: Double-stranded RNA

REFERENCES

[1] Agrawal N, Dasaradhi PVN, Mohommed A, Malhotra P, Bhatnagar RK, Mukherjee SK. RNA Interference: Biology, Mechanism, and Applications. Microb Mol Biol Rev 2003; 67: 657-85.

[2] Baulcombe D. Unwinding RNA silencing. Science 2000; 290: 1108-1109.

[3] Matzke M, Matzke A, Pruss G, Vance V. RNA-based silencing strategies in plants. Curr Opin Genet Dev 2001; 11: 221-7.

[4] Napoli C, Lemieux C, Jorgensen R. Introduction of chimeric chalcone synthase gene into *Petunia* results in reversible cosuppression of homologous genes in trans. Plant Cell 1990; 2: 279-289.

[5] Hannon GJ. RNA interference. Nature 2002; 418: 244-251.

[6] Campbell TN. Choy FYM, RNA Interference: Past, Present and Future. Curr Issues Mol Biol 2005; 7: 1-6.

[7] Voinnet O, Lederer C, Baulcomb DC. A viral movement protein prevents spread of the gene silencing signal in *Nicotiana benthamiana*. Cell 2002; 103 (1) 157-67.

[8] Mittal P, Yadav R, Devi R, Tiwari A, Upadhey SP, Ghoshal SS Woundorous RNAi- Gene Silencing. Biotechnol 2011; 10: 41-50.

[9] Fire A, Xu S, Montgomery MK, Kostas SA, Driver SE, Mello CC. Potent and specific genetic interference by double-stranded RNA in *Caenorhabditis elegans*. Nature 1998; 391: 806-11.

[10] Timmons L, Fire A. Specific interference by ingested dsRNA. Nature 1998; 395: 854.

[11] Reinhart BJ, Slack FJ, Basson M, Pasquinelli AE, Bettinger JC, Rougvie AE, Horvitz HR, Ruvkun G. The 21-nucleotide let-7 RNA regulates developmental timing in *Caenorhabditis elegans*. Nature 2000; 403: 901-6.

[12] Elbashir SM, Harborth J, Lendeckel W, Yalcin A, Weber K, Tuschl T. Duplexes of 21- nucleotide RNAs mediate RNA interference in cultured mammalian cells. Nature 2001; 411: 494-498.

[13] Paddison PJ, Caudy AA, Hannon G. Stable suppression of gene expression by RNAi in mammalian cells. Proc Nat Acad Sci USA 2002; 99: 1443-48.

[14] Blummelkamp TR, Bernards R, Agami R. Stable suppression of tumorigenicity by virus-mediated RNA interference. Cancer Cell 2002; 2: 243-7.

[15] Miyagishi M, Taira K. U6 promoter-driven siRNA with four uridine 3' overhangs efficiently suppress targeted gene suppression in mammalian cells. Nat Biotech 2002; 20: 497-500.

[16] Lee NS, Dohjima T, Bauer G, Li H, Li MJ, Ehsani A, Salvaterra P, Rossi J. Expression of small interfering RNAs targeted against HIV-1 rev transcripts in human cells. Nat Biotech 2002; 20: 500-5.

[17] Paul CP, Good PD, Winer I, Engelke DR. Effective expression of small interfering RNA in human cells. Nat Biotech 2002; 20: 505-8.

[18] Rubinson DA, Dillon CP, Kwiatkowski AV, Sievers C, Yang L, Kopinja J. A lentivirusbased system to functionally silence genes in primary mammalian cells, stem cells and transgenic mice by RNA interference. Nat Genet 2003; 33: 401-7.

[19] Kamath RS, Fraser AG, Dong Y, Poulin G, Durbin R, Gotta M. Systematic functional analysis of the *Caenorhabditis elegans* genome using RNAi. Nature 2003; 421: 231–237.

[20]   Kiger AA, Baum B, Jones S, Jones MR, Coulson A, Echeverri C, Perrimon N. A functional genomic analysis of cell morphology using RNA interference. *J Biol 2003;* 2: 27-41.

[21]   Berns K, Hijmans EM, Mullenders J, Brummelkamp TR, Velds A, Heimerikx M, Kerkhoven RM, Madiredjo M, Nijkamp W, Weigelt B, Agami R, Ge W, Cavet G, Linsley PS, Beijersbergen RL, Bernards R. A large-scale RNAi screen in human cells identifies new components of the p53 pathway. Nature 2004; 428: 431-7.

[22]   Sen G, Wehrman TS, Myers JW. Restriction enzyme generated siRNA (REGS) vector and libraries. Nat Genet 2004; 36: 183-9.

[23]   Davuluri GR, van Tuinen A, Fraser PD, Manfredonia A, Newman R, Burgess D, Brummell DA, King SR, Palys J, Uhlig J, *et al.* Fruit-specific RNAi-mediated suppression of DET1 enhances carotenoid and flavonoid content in tomatoes. Nature Biotechnol 2005; 23: 890–895.

[24]   Zhu X, Galili G. Lysine metabolism is concurrently regulated by synthesis and catabolism in both reproductive and vegetative tissues. Plant Physiol 2004; 135: 129–36.

[25]   Tang G, Galili G, Zhuang X. RNAi and microRNA: breakthrough technologies for the improvement of plant nutritional value and metabolic engineering. Metabol 2007; 3:357–69.

[26]   Bernstein E, Caudy A, Hammond SM, Hannon GJ. Role for a bidentate ribonuclease in the initiation step of RNA interference. Nature 2001; 409(6818): 295-6.

[27]   Hammond SM, Caudy AA, Hannon GJ. Post-transcriptional Gene Silencing by Double-stranded RNA. Nat Rev Gen 2001; 2: 110-9.

[28]   Jaronczyk K, Carmichael J, Hobman T. Exploring the functions of RNA interference pathway proteins: some functions are more RISCy than others? Biochem J 2005; 387(3): 561-71.

[29]   Hammond SM, Bernestein E, Beach D, Hannon GJ. An RNA directed nuclease mediates posttranscriptional gene silencing in Drosophila cells. Nature 2000; 404: 293-6.

[30]   Schwarz D, Tomari Y, Zamore P. The RNA-Induced Silencing Complex Is a Mg-Dependent Endonuclease. Curr Biol 2004; 14(9): 787-91.

[31]   Stevenson M. Therapeutic Potential of RNA Interference. The New Engl Jour med 2004; 351:1772-77.

[32]   Dalmay T, Hamilton A, Rudd S, Angell S, Baulcombe DC. An RNA-dependent RNA polymerase gene in *Arabidopsis* is required for posttranscriptional gene silencing mediated by a transgene but not by a virus. Cell 2000; 101: 543-53.

[33]   Smardon A, Spoerke J. M, Stacey SC, Klein M. E, Mackin N, Maine EM. EGO-1 is related to RNA-directed RNA polymerase and functions in germ-line development and RNA interference in *C. elegans.* Curr Biol 2000; 10: 169-178.

[34]   Schiebel W, Haer B, Marinkovic S, Kianner A, Sanger HL. RNA-directed RNA polymerase from tomato leaves, purification and physical properties. J Biol Chem 1993a; 268: 11851-57.

[35]   Schiebel W, Haer B, Marinkovic S, Kianner A, Sanger HL. RNA directed RNA polymerase from tomato leaves II catalytic *in vitro* properties. J Biol Chem 1993b; 268: 11858-67.

[36]   Hutvagner G, Zamore PD. A micro RNA in a multiple turn –over RNAi enzyme complex. Science 2002; 297:2056-60.

[37]   Naykanen A, Haley B, Zamore PD. ATP requirements and small interfering RNA structure in the RNA interference pathway. Cell 2001; 107: 309-21.

[38]   Ogita S, Uefuji H, Yamaguchi Y, Koizumi N, Sano H. Producing decaffeinated coffee plants. Nature 2003; 423: 823.

[39]   Segal G, Song R, Messing J. A new opaque variant of maize by a single dominant RNA-interference-inducing transgene. Genetics 2003; 165: 387–97.

[40]   Angaji SA, Hedayati SS, Poor RH, Poor SS, Shiravi S, Madani S. Application of RNA interference in plants. Plant Omics Journal 2010; 3: 77-84.

[41]   Liu, Q, Singh S, Green A. High-oleic and high-stearic cottonseed oils: nutritionally improved cooking oils developed using gene silencing. J Am Coll Nutr 2002; 21 (3): 205–11.

[42]   Rahman M, Ali TH, Riazuddin S. RNA interference: The story of gene silencing in plants and humans. Biotech Adv 2008; 26(3): 202- 9.

[43]   Negrutiu I, Cattoir-Reynearts A, Verbruggen I, Jacobs M. Lysine overproducer mutants with an altered dihydrodipicolinate synthase from protoplast culture of Nicotiana sylvestris (Spegazzini and Comes). Theor Appl Gen 1984; 68: 11–20.

[44]   Frankard V, Ghislain M, Jacobs M. Two feedbackinsensitive enzymes of the aspartate pathway in Nicotiana sylvestris. Plant Physiology 1992; 99: 1285–1293.

[45]   Zhu X, Galili G. Increased lysine synthesis coupled with a knockout of its catabolism synergistically boosts lysine content and also transregulates the metabolism of other amino acids in Arabidopsis seeds. Plant Cell 2003; 15: 845–53.

[46]   Tang G, Galili G. Using RNAi to improve plant nutritional value: from mechanism to application. TRENDS in Biotech 2004; 22(9): 463-9.

[47]   Williams CL. Importance of dietary fiber in childhood. Jour Amer Diet Assoc 1995; 95: 1140–6.

[48]   Crowe TC, Seligman SA, Copeland L. Inhibition of enzymic digestion of amylose by free fatty acids *in vitro* contributes to resistant starch formation. Journal of Nutrition 2000; 130: 2006–2008.

[49]   Regina A, Bird A, Topping D, Bowden S, Freeman J, Barsby T, Kosar-Hashemi B, Li Z, Rahman S, & Morell, M. High-amylose wheat generated by RNA interference improves indices of large-bowel health in rats. Proc of the Nat Acad of Sci of the USA 2006; 103, 3546–3551.

[50]   Forkmann G, Martens S. Metabolic engineering and applications of flavonoids. Curr Opin Biotechnol 2001; 12: 155–160.

[51]   Tanaka Y, Katsumoto Y, Brugliera F, Mason J. Genetic engineering in floriculture. Plant Cell Tiss Org Cult 2005; 80: 1–24.

[52]   Tanaka Y, Tsuda S, Kusumi T. Metabolic Engineering to Modify Flower Color. Plant Cell Physiol 1998; 39: 1119–26.

[53]   Nishihara M, Nakatsuka T, Hosokawa K, Yokoi T, Abe Y, Mishiba K, Yamamura S. Dominant inheritance of white-flowered and herbicide-resistant traits in transgenic gentian plants. Plant Biotechol 2006; 23: 25–31.

[54]   Nishihara M, Nakatsuka T, Yamamura S. Flavonoid components and flower color change in transgenic tobacco plants by suppression of chalcone isomerase gene. FEBS Lett 2005; 579: 6074–78.

[55]   Tsuda S, Fukui Y, Nakamura N, Katsumoto Y, Yonekura-Sakakibara K, Fukuchi-Mizutani M, *et al.* Flower color modification of *Petunia 37ybrid* commercial varieties by metabolic engineering. Plant Biotechnol 2004; 21: 377–86.

[56]   Nakamura N, Fukuchi-Mizutani M, Miyazaki K, Suzuki K, Tanaka Y. RNAi suppression of the anthocyanidin synthase gene in *Torenia hybrida* yields white flowers with higher frequency and better stability than antisense and sense suppression. Plant Biotechnol 2006; 23: 13–18.

[57]   Nakatsuka T, Abe Y, Kakizaki Y, Yamamura S, Nishihara M. Production of red-flowered plants by genetic engineering of multiple flavonoid biosynthetic genes. Plant Cell Rep 2007; 26: 1951–9.

[58]   Nakatsuka T, Mishibaa KI, Abe Y, Kubota A, Kakizaki Y, Yamamura S, Nishihara M. Plant Biotechnology, Flower color modification of gentian plants by RNAi-mediated gene silencing. Plant Biotechnology 2008; 25:61-68.

[59]   Van Uyen N. Novel approaches in plants breeding RNAi technology. In: Proceedings of International Workshop on Biotechnology in Agriculture 2006; pp12-16.

[60]   Waterhouse PM, Graham MW, Wang MB. Virus resistance and gene silencing in plants can be induced by simultaneous expression of sense and antisense RNA. Proc Natl Acad Sci USA 1998; 95: 13959–13964.

[61]   Goldstein DA, Songstad D, Sachs E, Petrick J (2009). RNA interference in plants. Monsanto. http://www.monsanto.com.

[62]   Hu W, Harding SA, Lung J, Popko JL, Ralph J, Stokke DD, Tsai C, Chiang VL Repression of lignin biosynthesis promotes cellulose accumulation and growth in transgenic trees. Nature Biotechnology 1999; 17: 808-812.

[63]   Baucher M, Chabber B, Pilate G, Vandoorsselaere J, Tollier MT, Petit-Conil M, *et al.* Red xylem and higher lignin extractability by down-regulating a cinnamyl alcohol dehydrogenase in Poplar. Plant Physiology 1996; 112: 1479-1490.

[64]   Lapierre C, Pollet B, Petit-Conil M, Toval G, Romero J, Pilate G, Leple J, Boerjan W, Ferret V, De Nadai V, Jouanin L. Structural alterations of lignins in transgenic poplars with depressed cinnamyl alcohol dehydrogenase or caffeic acid *O*-methyltransferase activity have an opposite impact on the efficiency of industrial kraft pulping. Plant Physiol 1999; 119: 153-164.

[65]   Dandekar AM, Teo G, Defilippi BG, Uratsu SL, Passey AJ, Kader AA, Stow JR, Colgan RJ, James DJ. Effect of down-regulation of ethylene biosynthesis on fruit flavor complex in apple fruit. Transgenic Research 2004; 13: 373-384.

[66]   Gilissen LJWJ, Bolhaar STH, Matos CI, Rouwendal GJA, Boone MJ, Krens FA, Zuidmeer L, Van Leeuwen A, Akkerdass J, Hoffmann-Sommergruber K, Knulst AC, Bosch D, Van De Weg WE, Van Ree R. Silencing of the major apple allergen Mal 1 by using RNA interference approach. J Allergy Clini Immunol 2005; 115: 364-369.

[67]   Teo G, Suziki Y, Uratsu SL, Lampien B, Ormonde N, Hu WK, Dejong TM, Dandekar AM. Silencing leaf sorbitol synthesis alters long-distance portioning and apple fruit quality. Proc Nat Acad Sci USA 2006; 103 :18842-7.

[68]   Hamilton AJ, Baulcoumbe DC. A species of small antisense RNA in posttranscriptional gene silencing in plants. Science 1999; 286(5441):950–952.

[69]   Tijsterman M, Plasterk RH. Dicers at RISC; the mechanism of RNAi. Cell 2004; 117(1): 13.

[70]   Wassenegger M, Heimes S, Riedel L, Sanger H. RNA-directed *de novo* methylation of genomic sequences in plants. Cell 1994; 76: 567-76.

[71]   Jones AL, Thomas CL, Maule AJ. *De novo* methylation and co-suppression induced by a cytoplasmically replicating plant RNA virus. EMBO J 1998; 17: 6385-93.

[72]   Wang M, Abbott D, Waterhouse PM. A single copy of a virus derived transgene encoding hairpin RNA gives immunity to barley yellow dwarf virus. Mol Plant Pathol 2000; 1: 401-10.

[73]   Matthew L. RNAi for plant functional genomics, Comp Funct Genom2004; 5, 240-244

[74]   Waterhouse PM , Helliwell CA. Exploring Plant Genomes by RNA-induced Gene Silencing, Nature Reviews: Genetics 2003; 29-38

[75]   Agami R. RNAi and related mechanisms and their potential use for therapy. Curr Opin Chem Biol 2002; 6: 829-4.

[76]   Couzin J. RNA interference. Mini RNA molecules shield mouse liver from hepatitis. Science 2003; 299 (5609): 995.

[77]   Shuey D, McCallus D, Tony G. RNAi: gene-silencing in therapeutic intervention. Drug Discov Tod 2002; 7(20): 1040-46.

[78]   Schwarz DS, Hutvagner G, Haley B, Zamore PD. Evidence that siRNAs function as guides, not primers, in the Drosophila and human RNAi pathways. Mol Cell 2002; 10: 537-48.

[79]   Ge Q, McManus MT, Nguyen T, Shen CH, Sharp PA, Eisen HN. RNA interference of influenza virus production by directly targeting mRNA for degradation and indirectly inhibiting all viral RNA transcription. Proc Natl. Acad. Sci. USA 2003; 100: 2718-23.

[80]   Judge A, Maclachlan. Overcoming the Innate Immune Response to Small Interferring RNA. Hum Gene Ther 2008; 19: 111-24.

CHAPTER 3

Biological Nitrogen Fixation: Host-*Rhizobium* Interaction

Pooja Arora, Rakesh Yadav, Neeraj Dilbaghi and Ashok Chaudhury[*]

Department of Bio and Nano Technology, Guru Jambheshwar University of Science and Technology, Hisar-125001 (Haryana), India

Abstract: *Rhizobium*-legume symbiosis represents a classical example of mutualism wherein, *Rhizobium* fixes inert nitrogen in the form of ammonia and equips it to plant; in return, plants provide shelter, mandatory source of carbon (dicarboxylic acids) and energy to the bacterium. Although nodulation is not a pre-requisite for leguminous plants but it has a selective advantage with respect to richness in protein (25%) enhancement as compared to wheat (10%). Perhaps, in symbiosis studies extra requirement for nitrogen is the key factor for selection of nitrogen fixation in legumes. *Rhizobium* infects leguminous host plants by way of growing root hairs to form nodules. Once specifically recognized, bacteria enter the cells of host plants through infection thread and infect its central tissue where they differentiate morphologically into nitrogen-fixing bacteroids. Important features of *Rhizobium* such as host-bacterium range, infection process, nodulation genes and its signaling have been elaborated. Studies on *Rhizobium*-legume symbiosis over the last 20 years, is gigantic, therefore, present chapter is restricted to plant-microbe interaction and its future prospects using biotechnological tools.

Keywords: Symbiosis, *Rhizobium*, nitrogen, legumes, infection.

INTRODUCTION

Optimum plant growth requires specific physical and chemical conditions along with various microbiological processes for maintaining fertility of soil. These microbiological processes are a part of nitrogen, phosphorus and carbon cycles. In recent past, there was substantial increase in food grain production due to various crop improvement methods (applications of biofertilizers, biopesticides) and integrated management practices (IPM). Biofertilizers are preparations of efficient microbes (nitrogen fixing, phosphate solubilising or cellulolytic) used for soil or composting areas with the objective of accelerating certain microbial processes for augmenting the availability of nutrients in a form easily assimilated by the plants. The substances of biological origin (Green manure and biofertilizers) used as fertilizers supply almost all the nutrients required by the crop plants resulting in increased crop productivity. Most easily available resource of biofertilizers is farmyard manure and composite manure. Farmyard manure is a mixture of cattle dung and remaining unused parts of straw and plant stalks fed to cattle and Composite manure is a mixture of decayed or decomposed and useless parts of plants and animals. To check soil erosion and leaching, at times remains of herbaceous crop and soil are mixed together to enrich the soil texture that forms a protective soil cover. It includes leguminous plants that contribute a huge amount to the total nitrogen fixed into the biosphere.

Leguminous plants can be classified into three major botanical subfamilies of the family *Leguminoseae*; *Ceasalpinoideae*, *Mimosoideae* and *Papilionoideae*. Out of 700 genera and 14,000 species of leguminous plants 500 genera and approximately 10,000 species, belong to the subfamily *Papilionoideae*. Beforehand considered fact elucidates not all leguminous plants have nodules on their root system, infact certain free forms do not possess them at all. Hardly, 10-12% of *Leguminoseae* have so far, been examined for nodulation of which 10% of *Mimosoideae*, 6.5% of *Cesalpinoideae* and 6% of *Papilionoideae* do not possess root nodules. According to eighth Edition of Bergey's Manual of Determinative Bacteriology, Buchmann and Gibbons in 1974 described that the family *Rhizobiaceae* consists of two genera: *Rhizobium* and *Agrobacterium*. *Rhizobium*

*Address correspondence to Ashok Chaudhury:** Professor and Chairman, Department of Bio and Nano Technology, Guru Jambheshwar University of Science and Technology, Hisar-125001 (Haryana), India Phone: +91-1662-263306, Fax: +91-1662-276240, E-mail: ashokchaudhury@hotmail.com

Aakash Goyal and Priti Maheshwari (Eds)

survives in the soil and in the rhizospheres of legumes as well as non-legumes. In a larger sense, biofertilizers includes all organic resources for plant growth, rendering available form for plant adsorption through microorganism-plant interactions [1].

About 200 years ago, Antonie Lavoisier called element nitrogen as **"azote"** meaning "without life" as it is a component of food, poisons, fertilizers, and explosives [2]. Total amount of nitrogen in atmosphere is about 10^{15} tonnes and on global basis nitrogen cycle transform about 3×10^9 tonnes of nitrogen [3, 4] annually. Although, fertilizers offer 25% chemically fixed nitrogen but large amount of nitrogen fixed by biological processes, which contribute 60% to the total amount of nitrogen fixed. Chemical fertilizers cause negative environmental effects, so; symbiotic nitrogen fixation takes hold of cultivated crops. From 1973 to 1988, the consumption of nitrogen fertilizers had increased globally from 8 to 17 kg ha^{-1} of total agricultural lands in both developed and developing countries [5, 6]. Unambiguously for food and crop production huge amount of nitrogen, fertilizers commercialized in market that had degraded the importance of biologically fixed nitrogen (BNF) as a primary sole source of nitrogen in agriculture [6]. Atmospheric fixation of biologically inert nitrogen can be helpful in reducing the use of fossil fuels, reforestation and restoration of lands [7, 4]. A dramatically increased concentration of toxic nitrates in drinking water supplies due to nitrogenous fertilizers has resulted in high eutrophication and water pollution in lakes and rivers [4, 8, 9]. This environmental degradation has increased the need and demand of research for biologically fixed nitrogen. In biologically fixed nitrogen (BNF) nitrogenase, enzyme complex irreversibly inactivates oxygen and reduces dinitrogen to ammonia by using large amount of energy [3, 9]. About 87 species in 2 genera of archea, 38 genera of bacteria, and 20 genera of cyanobacteria have been identified as diazotrophs (organisms that can fix nitrogen) [4, 9, 10].

As illustrated in equation:

$$N_2 + 8H^+ + 8 e^- + 16ATP \implies 2NH_4^+ + H_2 + 16ADP + 16P_i$$

Biological nitrogen fixation (BNF) is the second most important biological process on earth after photosynthesis. Here, atmospheric N_2 converts into ammonia, with the help of enzyme nitrogenase. BNF fulfills the demand of global nitrogen by supplying 175 million tons of nitrogen per year out of which 65% acquires by agriculture [11].

RHIZOBIA

Rhizobium, which is a soil-inhabiting bacteria form specific root structures, called nodules. In rhizobia, nod genes mediate root nodulation on legumes. In effective nodules, the bacteria fix inert nitrogen from the atmosphere and convert it into ammonia [12], which supports plant growth predominantly in nutrient deficient soils. In return, the *Rhizobium* gets accommodation inside the nodule structure [13] and required nutrient supply in form of dicarboxylic acids [14]. Nitrogen-fixing symbiotic microbes can biologically benefit the agriculture by enhancing crop and pasture growth without the addition of surplus chemical nitrogen fertilizers. However, in case of the ineffective nodules, rhizobia are parasitic, as nitrogen does not fix symbiotically, while they still supply it with nutrients [15]. This is the key rationale for majority of research, which focused on cross inoculating herbaceous crop, and forage legumes of agricultural significance (Table **1**).

Table 1: *Rhizobial* legume-cross-inoculation groups

Rhizobium Species	Legume-Hosts
R. leguminosarum	Lens, Pisum, Vicia
R. trifolii	Trifolium
R. phaseoli	Phaseolus
R. meliloti	Medicago, Melilotus, Trigonella
R. lupine	Lupinus, Ornithopus
R. japonicum	Glycine
R. mongolense	Alfalfa

Table 1: cont….

Rhizobium	*Vigna, Arachis*
Rhizobium	*Lupins*
Rhizobium	*Chickpeas*
Rhizobium	*Sanfoin*
Rhizobium	*Crownvetch*
Rhizobium	*Fababean*
Mesorhizobium loti	*Birdsfoot trefoil*
M. huakuii , M. mediterraneum	*Cicer milkvetch*
M. ciceri	*Chickpeas*
Sinorhizobium meliloti	*Alfalfa, Sweetclover*
S. fredii	*Soybean*
S. medicae	*Annual medics*
B. japonicum, B. elkanii, B. liaonginense	*Soybean*
Bradyrhizobium	*Lupins*

In our laboratory *Rhizobium* strains have been identified among exotic (*A. nilotica, Dalbergia and Prosopis*) and endemic (Cowpea, Pigeonpea and Chickpea) cross nodulating legumes and their evolutionary significance based on 16s rRNA has been compared [16]. In contrast, few studies have focused on association of rhizobia with non-crop legumes, which may have ecological importance in the natural landscape processes [17]. There are about 17000–19000 legume species estimated worldwide [18]. Among the identified twelve genera; in total 55 rhizobial species are classified (Table **2**) of which majority of the species comes under *Rhizobium*, *Bradyrhizobium, Mesorhizobium* and *Ensifer* (*Sinorhizobium*). *Rhizobium* belongs to phylum Proteobacteria and class Alphaproteobacteria, which contains six rhizobial families and one order Rhizobiales [19].

Table 2: Classified list of rhizobial species

Genus	Host	Species	Author
Rhizobium	-	*Rhizobium daejeonense*	[20]
	Phaseolus vulgaris	*Rhizobium etli*	[21]
	Galega	*Rhizobium galegae*	[22]
	P. vulgaris	*Rhizobium gallicum*	[23]
	P. vulgaris	*Rhizobium giardinii*	[23]
	Centrosema, Desmodium, Stylosanthes, Tephrosia	*Rhizobium hainanense*	[24]
	Sesbania herbacea	*Rhizobium huautlense*	[25]
	Indigofera spp.	*Rhizobium indigoferae*	[26]
	Trifolium, Vicia	*Rhizobium leguminosarum*	[27]
	Astragalus, Lespedeza	*Rhizobium loessense*	[28]
	Medicago, Ruthenica	*Rhizobium mongolense*	[29]
	Hedysarum, Hedysari	*Rhizobium sullae*	[30]
	Leucaena, P. vulgaris	*Rhizobium tropici*	[31]
	Neptunia natans	*Rhizobium undicola*	[32]
	Amphicarpaea, Trisperma, Corollina varia, Gueldenstaedtia multiflora	*Rhizobium yanglingense*	[33]
	Sesbania rostrata, Abrus precatorius	*Ensifer (Sinorhizobium) abri*	[34]
		Ensifer adhaerens	[35]

Sinorhizobium	Acacia spp.	Ensifer americanum	[36]
	Acacia Senegal,Prosopis chilensis	Ensifer arboris	[37]
	G. max	Ensifer fredii	[38]
	Sesbania rostrata, Abrus precatorius	Ensifer indiaense	[34]
	A. senegal, P. chilensis	Ensifer kostiensis	[37]
	Kummerowia stipulacea	Ensifer kummerowiae	[26]
	Medicago spp.	Ensifer medicae	[39]
	Medicago sativa	Ensifer meliloti	[40]
	Sesbania	Ensifer saheli	[41]
	Acacia, Sesbania	Ensifer terangae	[41]
	G. max	Ensifer xinjiangense	[42]
Mesorhizobium	Amorpha fruticosa	Mesorhizobium amorphae	[43]
	Prosopis alba	Mesorhizobium chacoense	[44]
	Cicer arietinum	Mesorhizobium ciceri	[45]
	Astragalus	Mesorhizobium huakuii	[46]
	Loti	Mesorhizobium loti	[47]
	Cicer arietinum	Mesorhizobium mediterraneum	[48]
	Acacia, Leucaena	Mesorhizobium plurifarium	[49]
	Astragalus adsurgens	Mesorhizobium septentrionale	[50]
	Astragalus adsurgens	Mesorhizobium temperatum	[50]
	Glycyrrhiza, Sophora, Glycine	Mesorhizobium tianshanense	[51]
Bradyrhizobium	Papilionoideae: Genisteae	Bradyrhizobium canariense	[52]
	Glycine max	Bradyrhizobium elkanii	[53]
	G. max	Bradyrhizobium japonicum	[54]
	G. max	Bradyrhizobium liaoningense	[55]
	Lespedeza spp.	Bradyrhizobium yuanmingense	[56]
Burkholderia	M. pudica, M. diplotricha	Burkholderia caribensis	[57]
	Oryza sativa	Burkholderia cepacia	[57]
	Machaerium lunatum, Mimosa invisa	Burkholderia phymatum	[57]
	Aspalathus carnosa	Burkholderia tuberum	[57]
Azorhizobium	Sesbania rostrata	Azorhizobium caulinodans	[58]
	Sesbania virgata	Azorhizobium doebereinerae	[59]
Cupriavidus	Mimosa pudica, M. Diplotricha, M. pigra	Cupriavidus taiwanensis	[60]
Devosia	Neptunia natans	Devosia neptuniae	[61]
Herbaspirillum	Phaseolus vulgaris	Herbaspirillum lusitanum	[62]
		Phyllobacterium trifolii	[62]
Methylobacterium	Crotalaria, Pedocarpa	Methylobacterium nodulans	[63]
Ochrobactrum	Lupinus albus	Ochrobactrum lupini	[64]

A number of species of rhizobial genera do not form nodules, and therefore do not fit the functional definition of rhizobia. These include some species of *Agrobacterium* like *R. larrymoorei*, *R. rubi*, and *R. vitis* [65, 66].

ASSOCIATION BETWEEN PLANT HOST AND RHIZOBIA

Mutual relationship and the symbiotic association are mandatory for the developmental process of infection. In the first step of this process, when plant (host) and bacterial (rhizobia) signals are mutually exchanged, cell signaling in the form of cell-to-cell contact and recognition occurs.

ADHESION OF RHIZOBIA TO ROOT HAIRS

How *Rhizobium* binds tightly to host root hairs, reported in two-step protocol of *R. leguminosarum* [67-69]. The first one is a weak Calcium-dependent binding step where protein rhicadhesin-mediated binding occurs in root hair, reflected in most rhizobia [67, 68]. Consequently, a tight binding step initiates by synthesis of cellulose fibrils in the bacteria [68, 69]. These fibrils further form biofilm-like caps on the tips of pea root hairs in *R. leguminosarum*. Caps are absent in mutants that did not form the fibrils, but they are able to form nitrogen-fixing nodules, which indicate that for a successful symbiosis capping and cellulose-mediated tip binding are not much essential [68]. However, in natural conditions for colonization of root hairs, binding and capping may be a need for rhizobia. Host lectins (carbohydrate containing protein) form determinate and indeterminate nodules, which play key role in rhizobial adhesion to plants. Determinate nodules (tropical legumes: soybean, bean, *Lotus japonicus*) are round in shape, nodule primordia forms in outer cortex, meristem is not persistant. In contrast, indeterminate nodules (temperate legumes: pea, alfalfa, vetch, *Medicago truncatula*) are oval in shape, nodule primordia forms in outer cortex, persistant meristem. These lectins bind simultaneously to the plant cell wall and to saccharide moieties and are thought to convey host-symbiotic specificity [70-72]. Recently, it has identified that lectins and surface polysaccharides of rhizobia play an important role in recognition process. Lectins bind to specific receptors of carbohydrate structure in a manner analogous to the binding of antibody-antigen. Such binding is of high affinity, non-covalent and reversible which totally relies on the structure and conformation of carbohydrate receptors and proteins. The bacterium binds to the host by the mechanistic role of host plant lectins located on most surfaces which recognizes carbohydrate receptors on compatible host cell surfaces. There is also an evidence to suggest the presence of receptors for lectins in legumes root. It is been suggested that lectin first binds to glycosylated receptors present on the root surface. This receptor bears immunological resemblance to the saccharide receptors on *Rhizobium*.

ROOT HAIR DEFORMATION AND CURLING

Root hairs of host plants like alfalfa, *Medicago truncatula*, pea, clover, and vetch undergo Nod factor-induced deformation and branching from their compatible rhizobial species [73-75]. Nod factors alone are not sufficient to cause the entry of bacteria into the plants, as they cannot induce the formation of tightly curled root hairs called 'Shepherd's Crooks'. In Recent experiments, purified Nod factor from *Sinorhizobium meliloti* induced in root hairs of *M. truncatula*; have shown that a resource from Nod factor can grow root hairs into structures resembling shepherd's crooks [76]. Root hairs show a wide range of deformation and morphological changes in a single plant. Some show swelled wavy root hairs, and other regions show tightly curled root hairs that are able to sustain the development of infection threads [77]. Plant hormone ethylene can modulate Nod factor induced root hair deformation, which not only inhibits Nod factor based signal transduction but can also influence the incidence of productive variability in infections of root hair [78]. In addition to root hairs in beans, Nod factor causes the fragmentation of the actin microfilament network [79]. In legume root hairs, addition of compatible Nod factors induces membrane depolarization, calcium influx at the root hair tip, and calcium spiking in the perinuclear region within seconds [80]. In the root hair cytoskeleton, like actin component, the helical cortical array shaped microtubule component also varies by subsequent exposure of compatible rhizobia. The microtubules rearranged to form endoplasmic networks around nuclei in root hair cells and a cortical network gets aligned with long axes of the root hairs leaving helical shape by induction of *S. meliloti*. As curling begins, the endoplasmic microtubule network moves towards the center of the curl. If the curl developed an infection thread, the microtubular array connects the tip of the infection thread to the nucleus [81].

INFECTION

Nitrogen-fixing root nodules form after the infection of bacteria in the roots from the primordium of differentiated root cells. Nodule formation involves differential changes in the root layers of epidermis, cortex, and pericycle. Rhizobia induce morphological changes in the epidermis by colonizing the root surface of their host. *Rhizobium* shows changed growth direction and root hair deformation by reinitiating of tip growth in cells [74, 82]. The first reaction of the root system is the curling and deformation of root hairs, which confirms the presence of rhizobia as evidenced by studies on clover and Lucerne. In some root hairs, adhesion follows deformations and root hair curling that resembles a so-called shepherd's crook [83]. Gradual and constant reorientation of the growth direction leads to root hair curling by curl with a turn effect [84-86]. The formation of tightly curled root hairs signals the formation of a thread-like structure inside the root hair called "infection thread". Production of indole acetic acid (IAA) in the root region by rhizobia confirms the induction of curling effect. Several microorganisms other than rhizobia can also synthesize IAA. In this regard, a specific root hair curling factor, believed to be a water-soluble polysaccharide produced by rhizobia, occupy typical curling of root hairs where infection threads forms.

At this stage, deep interaction occurs between the infection thread, which originates from the tip of the curled portion of the root hair and the nucleus of root hair cell. The path of infection process depicts that organization of nucleus is directly proportional to the growth of the thread. Thus, in the root hair nucleus is the key determinative factor. Initially, the nucleus migrates to the distal end of the hair and then slowly walks towards the proximal end near the cortex, and then the infection thread traverses up and down before signal transfers by the host nucleus to the contents of the infection thread.

Studies on root hair infection in clover seedlings had also revealed many interesting facts. In clover, infection of root hairs take place at primary infection sites and these primary infection sites give rise to "Zones of Infection". Soon after infection of root hairs, number of infected root hairs amplifies exponentially until the first nodule is forms, following a drop in the number of infections. Rhizobia grow and divide inside a tubule called an infection thread and thus, populate cells in the incipient nodule during symbiosis. A deformed root hair forms a sharp bend or curl, and bacteria bound to the root hair become trapped between appressed cell walls which leads to formation of a new infection thread [87]. Plant cell membrane invaginates in the curl by degradation of the wall followed by initiation of an infection thread. Tip starts growing from the invagination of the thread downwards in the root hair, towards the epidermal cell linings. Following this, infection thread fuses with the distal cell wall and exits the epidermal cell and thus, bacterium enters the intercellular space between the epidermal cell and the underlying cell layer. After which, a thread filled with bacteria invaginates and propagates towards the root interior of the underlying cell membrane [88-90]. For the colonization of nodule cells, infection thread branches through the root and enters the nodule primordium that increases the number of sites from which bacteria can exit and enter nodule cells. Inside the nodule cells, the bacterium differentiates and synthesizes proteins required for nitrogen fixation and maintenance for the mutualistic corporation of host and the bacterium.

GROWTH OF ROOT HAIRS

Plant cells that elongate through the process of tip growth are root hairs and pollen tubes [91-93]. Infection threads, a kind of tip-growing structure also develop from growing root hairs. Thus, understanding the processes should shack much needed light that contribute to tip growth in root hairs and pollen tubes on processes involved in infection thread growth.

CYTOLOGY AND DEVELOPMENT OF ROOT HAIRS

The process of root hair development in *M. truncatula* begins with the nucleus of an epidermal cell which starts moving towards the center of the inner cell wall; leading to the development of an embryonic root hair knot [75] (Fig. 1). Growing tip from the root hair accommodates filled mass of cytoplasm, a big vacuole in its shaft, and a thin sheath of cytoplasm connecting the vacuole and the plasma membrane. As growth proceeds, the epidermal cell nucleus starts moving away from the inner periclinal wall and takes a position in the cytoplasmic dense region. As the root hair approaches its mature length in fully-grown cells, the vacuole extends nearly to the tip in the cytoplasmic dense region of the root hair, and the nucleus migrates anywhere in the cell [75]. At the tip region polarized secretion of vesicles are present from where root hairs and pollen tubes starts elongating under the influence of internal turgor pressure (reviewed in references [92, 93]. Vesicles, emerging structures from Golgi bodies are found at a shorter distance from the growing tip blended within root hairs.

During tip growth, vesicles and other organelles comes at the apical region of the cell by the action mechanism of actin-dependent cytoplasm torrent. The cytoplasm first moves outside the cell toward the growing tip of the root hair and then enter back toward the center of the cell through basal regions. This archetype that commonly found in pollen tubes, referred to as reverse fountain streaming [92-95].

MODE OF ENTRY OF RHIZOBIA

Rhizobia can enter the root system by two modes either by entry of small coccoid swarmers through the gaps in cellulose microfibrils or through direct invagination of the root hair cell known as the "Crack Entry" (Fig. **1**). The invagination hypothesis rests on the ground that there are some localized soft regions on the root hair, produced by interaction of auxins and pectic enzymes on the root surface facilitating the inward growth of the root hair cell wall against the hydrostatic pressure of the cell contents [1]. A combination of these two hypothesis explains the entry of *Rhizobium* into root hairs *viz.*, the entry of swarmers through microfibrils of the root hair cell wall and the invagination of the root hair in cell wall. The coccoid swarmers slowly migrate between the gaps in the microfibrils of the root hair cell wall [1]. At this step of infection, the incorporation of the bacteria into the cell wall triggers by superimposed primary or secondary wall material. Subsequently, to allow invagination of the inner layers the outer wall of the root hair strengthens at a specified point (Fig. **2**). Inspite of these elucidations and speculations, the *modus operandi* of infection is still not very clear. Fine structure studies support the invagination hypothesis of infected root hairs showing the continuation of the wall of the infection thread with the cell wall of the root hair [1].

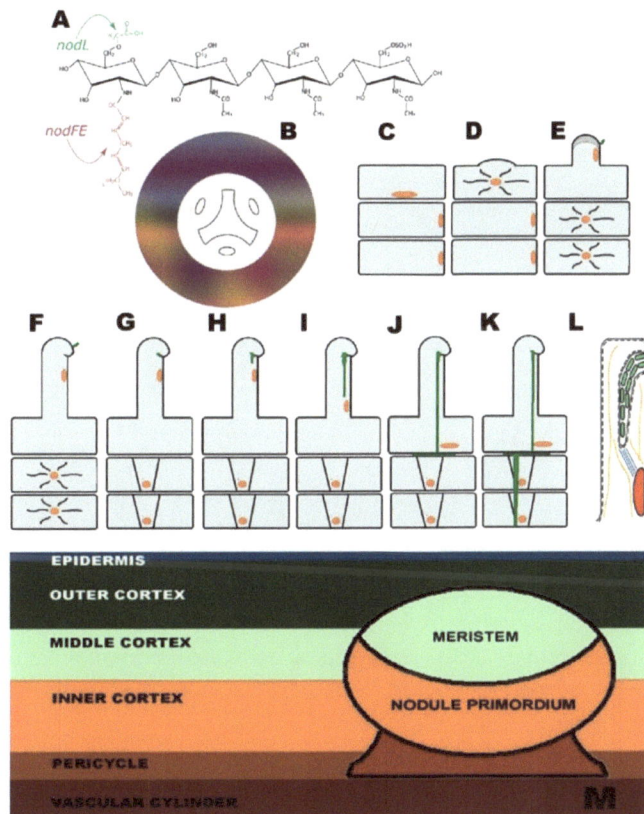

Figure 1: Ontology of events occurring in indeterminate nodules formation-A brief synopsis: A) Nod factor, *NodL* arrow shows acetyl group addition while *NodF* and *NodE* enrolls the lipid moiety, B) Cross section of root showing multiple activating and inhabiting factors(multiple colors) at proto-xylem and phloem poles, C) Pre-initiating step of root hair formation from a nucleus(orange) aligned to the epidermal cell with two underlying cortical cells, D) Initiation of root hair in the epidermal cell, E), F) Initiation of root hair binding with *Rhizobium* with concomitant activation of Nod factor influenced cortical cells, G) Nod factor induced root hair curling and growth of rhizobia inside cytoplasm of polarized cortical cells, H) formation of infection thread, I) Geotropic movement of nucleus and thread in the root hair, J) Fusion of epidermal cell wall with the thread and growth of rhizobia between the epidermal and cortical cells, K) Elongation of infection thread towards outer region of cortical cells, L) Dissected analysis of root hair shows that infection thread is still outside the root hair, thread covered with cell wall and cell membrane connects microtubules (blue) with the nucleus. During the progression of infection threads, cytoplasmic streaming forms actin cables (orange) in the root hairs. Topologically, *Rhizobium* remains outside the root until the tip forms from the infection thread, M). (Adapted source: Gage DJ. 2004 [96]).

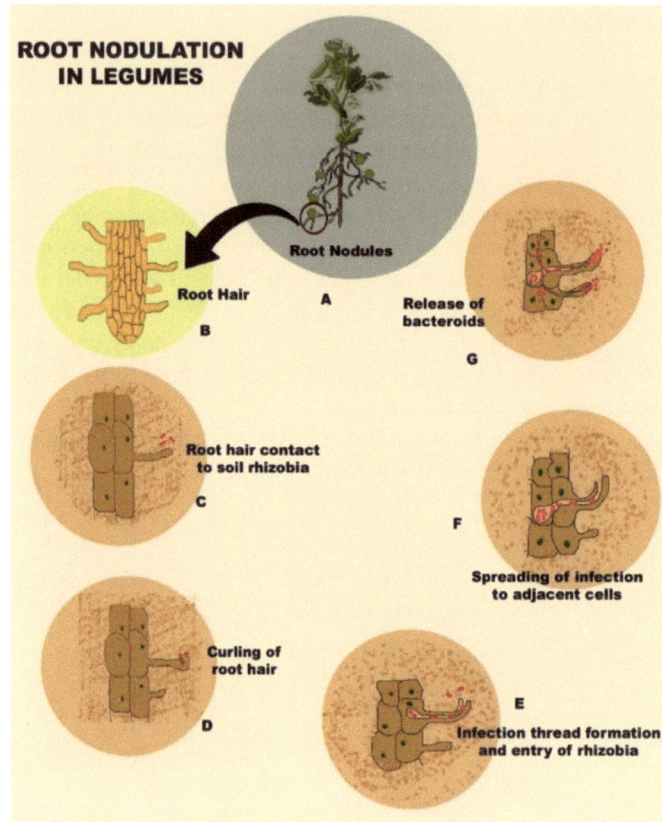

Figure 2: Dissected study of root nodule formation in legumes (Inferred source: *Rhizobium*, Root Nodules and Nitrogen Fixation. 2002 Society for General Microbiology [97]).

NODULE DEVELOPMENT

Root hair contains more than one infection thread, which appears to be originating from the same host. The infection thread then enters into the cortical cells of the root where it branches and then transverses intracellularly [1]. It so happens that the contents of an infection thread (bacteria) liberates into a tetraploid cell of the root cortex stimulating the cell to perform intense meristematic activity (Fig. **3**). The differentiating cellular mass or the initial nodular tissue constitutes a meristem. Later on, the edge demarcations well differentiate between diploid nodule cortex and central tetraploid bacteroid zone, precluding vascular connections with the parent root system [1].

The mechanism of infection in plants where infection takes place directly through the cells of the root cortex does not yet clearly supported by research studies. Rhizobia enter roots by mechanical injury caused to the roots or by some unexplained enzymatic processes is still an assumption [1]. However, **flow diagram-1** summarizes studies based on Cloves and Lucerns, the physiological events of infection.

STRUCTURE OF THE NODULE

The "**Bacteroid Zone**" surrounding several layers of cortical cells is present in the central core of the mature nodule. The relative volume of bacteroid tissue (16 to 50% of the dry weight of the nodules) is much greater in effective nodules than in ineffective ones [1]. The volume of bacteroid tissue in effective nodules is directly proportional to the amount of nitrogen fixed. Ineffective nodules with structural abnormalities produce ineffective strains, are generally small and contain poorly developed bacteroid tissue. In all ineffective associations, starch accumulates in the uninfected cell and dextran in the infected cell with glycogen in the bacteroid. Effective nodules are big slimy mucoid and pink in color (leghaemoglobin) with well-developed and organized bacteroid tissue [1].

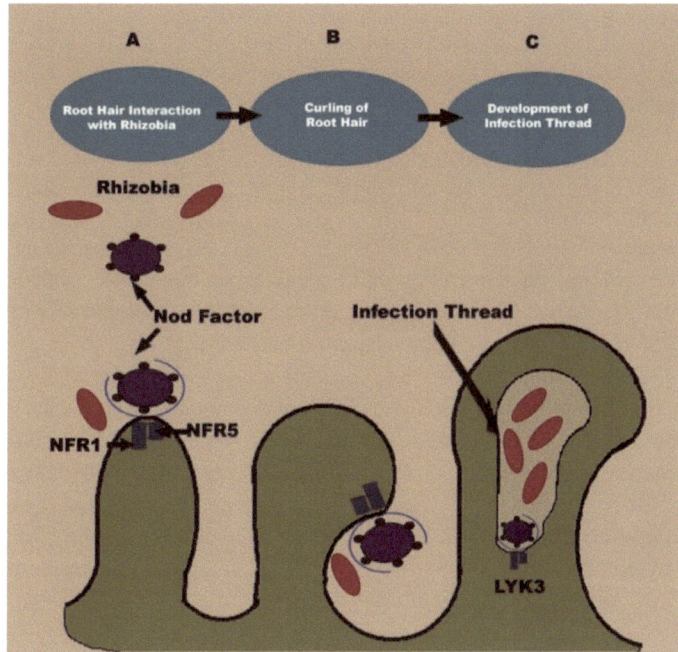

Figure 3: Receptors in nodulation (Adapted source: Parniske M, Downie JA. (2003) Plant biology: Locks, keys and symbioses. *Nature* 425:569-70 [98]).

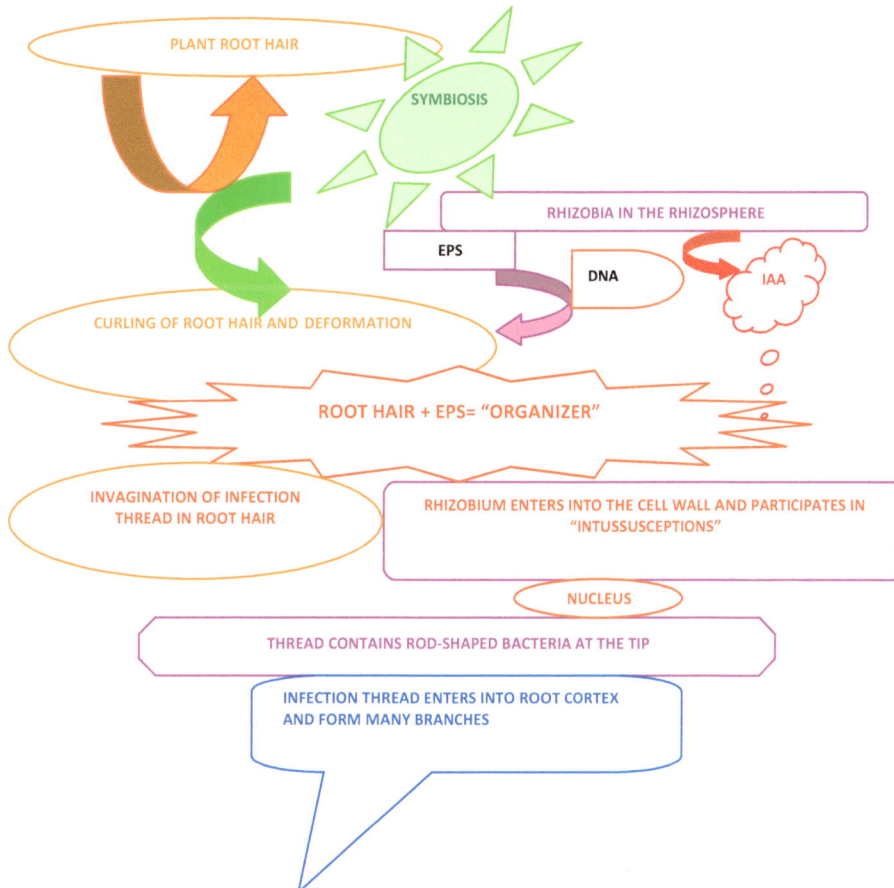

Flow Diagram-1: The Physiological Events of Infection-It is intresting to note that from normal root hairs, organic exudates creep out owing *Rhizobium* accumulation in the rhizosphere. Conversion of tryptophane to IAA confirms bacterial binding with its host. Root hair forms curl with a turn, rhizobial polysaccharide (EPS) and the DNA grows to form infection thread. Simultaneously, "organizer" forms (polysaccharides + root hair cell) and induces the production of polygalacturonase (PG) followed by depolymerisation of cell wall pectin. In such process, rhizobia incorporates into cell wall which further participates in 'intussuusception' (taking in of rhizobia by root hair and its conversion into organic tissues). In the central portions of the nodule, the infection tube or thread intitates branching and the bacteria released into their symbiont's cytoplasm to multiply. The nucleus of the root hair cell further guides the rhizobial pathway.

A fully developed nodule contains bacteroid with no flagella, encircled by three unit membranes. In subterranean clover root system, there exists an intra-cytoplasmic membrane system in the bacteroids of nodules [1]. The nuclear region of bacteroids coupled with granular cytoplasm appears fragmented and distorted. Bacteroids grow *in vitro* on a synthetic medium (yeast extract- 3.5% as an essential component) for nutrition and growth [1]. Various other alkaloids like Caffeine also encourage bacteroid production on artificial media. Depending on the kind of legume and the different techniques used in the fine structure studies, individuality and uniqueness of each bacteroid production or groups of bacteroids surrounded by membrane envelopes has been variously comprehended [1]. The three hypotheses concerning them are- they are formed *de novo* after the release of the bacteria from the infected thread, they are extensions of the endoplasmic reticulum of the host cells and they are derived from plasmalemma by a process of phagocytosis. The number of bacteroids enclosed in membrane envelope appears to vary from one to many depending upon the species of legume in question [1].

GENES AND SIGNALS IN THE *RHIZOBIUM*

Rhizobium shows symbiosis by recognizing and responding to host plant roots. In the rhizosphere of a host plant, rhizobia sense secreted compounds such as flavonoids, betaines (stachydrine and triggonelline) and the adonic acids (erythronic and tetronic acid) by the host root. At higher concentration, they get active and react to induce *nod* genes [99-101]. Nod factor is composed of *nod* genes, which encode approximately 25 proteins for bacterial synthesis. It sends a lipooligosaccharide signal consisting of a chitin backbone. Nod factors are composed of four to five *N*-acetylglucosamine units, with a lipid attached to its nonreducing end (Fig. 1). Nod factor initiates nodulation process and developmental changes in the host plant, which includes root hair deformation, membrane depolarization, intracellular calcium oscillations, and initiation of cell division in root cortex, which further begins the formation of a meristem and nodule primordium.

Two free-living organisms commence *Rhizobium*-legume symbiosis with a friendly cellular co-existence. *Rhizobium* bacteria specifically recognize the plants; aggravate growth of a root nodule by induction in plant tissue. Finally, the *Rhizobium*-host cell relation forms and by nutriotional exchange bacteria brings fixed nitrogen to the plant and in return receives shelter and sugars that the plant cell can provide under symbiotic association [102-105].

Rhizobium Genes

Rhizobium-legume symbiosis had started gaining serious research interest after Beijerinck's demonstration of nodule formation in bacteria [103]. Although, prior to 1975 mechanism and biochemistry of nitrogen fixation was already reported. However, for the mechanistic study of nodule formation, the critical first step was the identification of *Rhizobium* genes. In early 1980s, Gary Ruvkun and Sharon R. Long, USA cloned the first *Rhizobium* gene for nitrogen fixation (*nif*) and nodulation (*nod*) with the help of their colleagues in Fred Ausubel's Laboratory, USA [106], and later on many more *nif*, *nod*, and *fix* (symbiotic fixation) genes were identified in other laboratories worldwide.

Since the cloning of *nif* and *nod* genes during 1980-1981, over 30 different research groups have contributed in understanding of *Rhizobium* symbiosis genes through physical cloning, chromosomal walking, megaplasmid identification, site-directed mutagenesis, and many other genotypic and phenotypic studies on diverse plant host groups [104, 107, 108]. Direct and indirect screens [106, 108] have found many bacterial genes for invasion and differentiation. A novel Bacterial signal, **Lumichrome** manipulates host respiration in early stages of symbiosis [109]. In the nitrogen-fixing bacteria, carbon and nitrogen metabolism shows a new adaptation for the production of **"Rhizopines"** which supplies specialized nutrition to sibling bacteria in the environment [105, 106].

Genes responsible for nodulation, symbiosis resides on large plasmids (pSym) and genomic islands (SI) in cluster form [110, 111]. *nif* genes synthesize fully functional nitrogenase enzyme protein. The *fix* genes are must for rhizobial nitrogen fixation in the nodule [112]. *NifHDK* genes encode two structural proteins DK ($\alpha_2\beta_2$ dinitrogenase) and H (dinitrogenase reductase) of nitrogenase. Each α subunit of dinitrogenase contains FeMo cofactor for synthesis and construction of the pathway. FeMo cofactor synthesis differentiates protein specific genes into three classes *viz.*, *nifU, nifB* and *nifEN* (molecular scaffolds), *nifX* and *nifY* (precursor carrying proteins), and *nifS, nifQ* and *nifV* (enzymes that provide substrate for cofactor synthesis) [113, 114]. Until now, function of *nifW* and *nifT* are unknown [114]. A two component regulatory system of FixL/FixJ is required for expression of nitrogen fixation by *S. meliloti* genes. FixL autophosphorylates and transmits phosphate to the FixJ. FixJ regulates the response and activates transcription of the *nifA* and *fixk* genes [115].

Under micro-aerobic conditions, FixK activates 97 genes to express in both free living and symbiotic bacteria. NifA (enhancer binding protein) binds with sigma factors (σ^{54}) and induce expression of in total 19 genes in bacteroids including *nifHDKEX*, fixABCX operons, *nifB* and *nifN* [115]. NifA also regulates response of hup genes (hydrogenase uptake), which further allows the hydrogen to reassimilate, lost in nitrogenase reaction [116]. In *Bradyrhizobium japonicum*, RegSR-NifA and FixLJ-FixK2 (regulatory cascades) controls the expression of nitrogen fixation genes [117, 118]

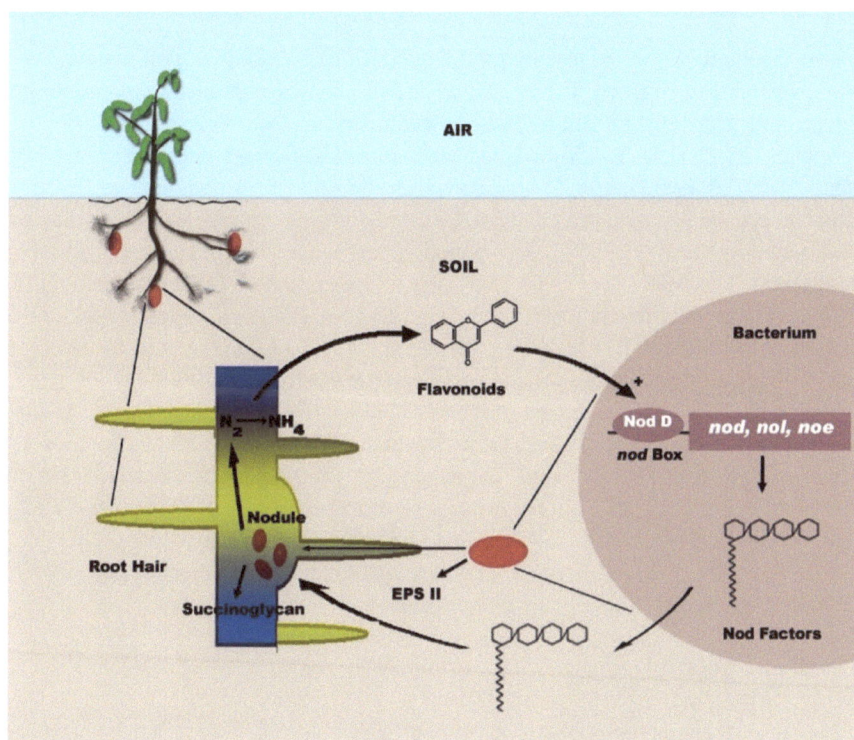

Figure 4: Plant roots produce phenolic signals in form of flavonoids (Inducers for the *NodD* transcriptional activator gene) which triggers nodule invasion and binding of conserved promoter elements (*nod* boxes) to the upstream of the nodulation genes *nod, nol* and *noe*. It resulted in production of species-specific Nod factor molecules. These newly formed Nod factors finally invade nodule formation in root hairs with the help of active exopolysaccharides-EPS (Adapted from: Juan E. Gonza'lez and Melanie M. Marketon 2003 [119]).

PLANT FLAVONOIDS: IN A NEW SIGNALING ROLE

The *Rhizobium* nodulation genes and expression studies have showed that *nod* genes usually do not express in free-living cells. The *nod* genes requires natural compounds like luteolin, flavones or flavanones, daidzein, an isoflavone (soybean) called plant inducers for transcription. Therefore, it was concluded that each legume produces a diverse blend of flavonoids and that the magnitude and spectrum of compounds may vary with the age and physiological state of the plant [106, 107, 111]. *Rhizobium* further uses the flavonoid families of compounds as a positive signal for activation and the events have been illustrated (Fig. **4**).

NOD FACTORS: BACTERIAL CARBOHYDRATES

In order to explore the new category of Nod factor signal, from 1986-1990 genetics, cell biology, and biochemistry researchers performed hand-in-hand studies [120-122]. These approaches to fruition in 1990 with the work in Toulouse by the groups of J. De´narie´, G. Truchet, and J. C. Prome´ they have observed that *Rhizobium meliloti* displayed nodule-like behavior in alfalfa plants when induced and pretreated with flavonoids, they fractionated the *Rhizobium* to identify specific active fractions or "Nod factors" (Fig. **4**). The outcome of this work showed that the active component is a novel hydrophobic oligosaccharide molecule based on a chitin oligomer backbone, which carries sulfate at its reducing end. Plant host bioassay demonstrated that the side groups provide host specificity among plant groups. In the past decade, research groups have identified that most Nod proteins have enzymatic activities and these are consistent with the synthesis of the lipochitooligosaccharide such as polymerases and *N*-acyl transferases (encoded by common *nod* genes), and *O*-sulfonyl, *O*-acetyl, *N*-methyl, and exotic glycosyl transferases (encoded by host-specific *nod* genes) (For details, see [123]).

BEYOND SIGNALS: CELL RESPONSES

Bacterial nod genes and the revelation of signals control symbiosis genes. For bottleneck study of bacterial and plant transcription, cellular organization, exchange and assimilation of nutrients many symbiotic researchers have used genetics, molecular biology, cell biology, biochemistry, and physiology to contribute in expanding worth-plus information. Prof. Desh Pal Verma and colleagues [124] have first cloned the leghaemoglobin genes, subsequently Dr. Ton Bisseling and colleagues identified the early nodulins, or ENODs, as Pro-rich sequences located in cell wall. Nodulins genes induced in rhizobia during nodulation are early nodulins (*ENOD*) and late nodulins (*LNOD*) [125]. During nodule initiation, the infection thread develops ENOD proteins (hydroxyl-proline rich glycoproteins) and targets the remodeling of cell wall [126]. *ENOD12*, *ENOD40* and *RIP1* activates after few hours of infection [127-132]. *ENOD11*, *ENOD12* and *RIP1* activates in the epidermis of the root hair zone [128, 131]. Conversely, *ENOD20* and *ENOD40* expressed in nodule primordium and in mitotically active pericycle cells [133]. Nod Factor signal transduction pathway activates directly *ENOD12* and *RIP1*. *ENOD12* and *RIP1* also show non-symbiotic expression and regulation [131, 134]. Higher concentration of Nod factor activates *ENOD40* for inducing cortical cell division. Cytokinin and auxin transport inhibitors activate *ENOD40* to induce formation of nodule like structures [135-137]. Cellular and tissue rearrangements by microscopic, immunochemical, and biochemical analysis revealed changes in cytoskeletal architecture, cell wall biochemistry, and oxidative metabolism during infection [123, 138]. Stougaard and colleagues have cloned the first plant nodulation gene, *Lotus japonicus NIN-1*, encoding a probable transcription factor required for nodule morphogenesis [139].

NODULATION BY RHIZOBIA

Rhizobium nodulation in leguminous plants is a complex and enthralling developmental phenomenon that requires a series of biochemical interactions between the bacterium-host relationship [140-145] (Fig. **4**). During this association, the bacterium undergoes curling by altering the growth of the epidermal hairs on the surface of the roots and thus shows chemotaxis toward the plant roots. Consequently, the bacterium establishes nodule meristem by inducing cell division in the quiescent cells of the inner cortex of the root. The bacterium entrapped in the curled root hair induces the formation of an infection thread (tube of plant origin) which penetrates the outer plant cells wall by the development of a nodule [146]. The bacteria once released into the cytoplasm of the cell, starts differentiating into morphologically altered forms termed bacteroids and then begin to produce nitrogenase and the other enzymes required for nitrogen fixation such as nodule-specific protein termed nodulins (leghaemoglobin). During nitrogen fixation, host plant accommodates symbiotic interaction that results in the reduction of atmospheric dinitrogen to ammonia by the bacteroids.

NODULATION (NOD) FACTORS

Plant host-*Rhizobium* research is mainly focused on the *nod* genes [147-149], rhizobial mutants of which do not induce root hair curling or nodule formation [150]. *NodD* gene of LysR family (Fig. **4**) regulates transcription by controlling *nod* gene expression in variable rhizobial strains. NodD binds to the *nod* box which is a conserved 47-bp region found upstream of the variable nodulation genes (*nod*, *nol*, and *noe*) [151]. Plant-produced flavonoids stimulate and activate the *nod* gene expression and subsequent Nod factor production [152-154]. The Nod factors has the same generic lipochitooligosaccharide structure with the terminal nonreducing sugar N-acylated by a fatty acid usually of 16 or 18 carbon side chains [155]. The *nodABC* genes are required for the biosynthesis of the Nod factors. The NodC protein (chitin oligosaccharide synthase) which links the UDP-*N*-acetyl glucosamine monomers is a β-glucosaminyl transferase. NodB (chitin oligosaccharide deacetylase) removes an acetyl group from the chitin-like backbone of terminal residue units, while on the free amino group NodA (acyl transferase) catalyzes the transfer of a fatty acyl chain by using acyl-ACP from fatty acid metabolism [156]. Nodulation genes determine the nature of the substitutions at the terminal residues and the structure of the acyl chain, which determines host specificity of the bacterium [157-162]. The structure of the acyl moiety contains acyl groups that occur commonly as moieties of the phospholipids attached to lipochitooligosaccharide [159, 160, 162-165]. Although, biochemical and genetic approaches have led to the characterization of various putative high-affinity binding sites whereas, until now, identity of the Nod factor receptors in legumes is unknown. The analysis of nodulation-deficient plant mutants is another interesting approach for the study of Nod factor signaling in the near future [166].

EXOPOLYSACCHARIDES IN NODULE INVASION

Rhizobium produces a variety of polysaccharides [167-170], and it was assumed that rhizobial exopolysaccharides (EPS) play roles in bacterium-plant interactions. EPS consists of large hetropolymers formed from repeating unit structures. Until now, the best-characterized exopolysaccharides is of *S. meliloti*. It is capable of synthesizing two different exopolysaccharides, succinoglycan and EPS II, which are required by *S. meliloti* Rm1021 for the development of normal nitrogen- fixing nodules [158, 159, 165, 171, 172]. Mutants that lack bacteroids form empty nodules as they were unable to synthesize these exopolysaccharides [173-176]. The empty nodules gets arrested at intermediatory transition state of nodule development and thus, express only one or two of the seventeen nodule-specific plant proteins called nodulins that are synthesized in nodules surrounding wild-type bacteria [176].

CONCUSIONS AND FUTURE PROSPECTS

Nitrogen-fixing rhizobia infect and benefits plant hosts growing in nitrogen-deficient environments. Symbiosis of nitrogen-fixing bacteria and leguminous plants have evolved complex exchange of cell signaling between rhizobia and its specific host plant that forms invasion structures through which the bacteria can enter the plant root. Once the bacterium invades and gets accommodation in the root cells it differentiates and converts atmospheric nitrogen into ammonia by biological nitrogen fixation. Mutually in return, the plant receives nitrogen from the bacteria, which allows it to grow in the absence of external nitrogen source-chemical fertilizers. Symbiotic investigational studies of cellular and molecular aspects of infection thread initiation and growth has been difficult because plant-microbe are the two partners involved and their individual contributions are highly integrated with each other, often find harder to apply traditional reductionistic approaches. However, transgenic plant research and image analysis related studies for nodule formation are the overture of new avenues for investigating symbiotic infection. In addition, the augmentation of significant genetic models by genetic analysis in host plant biology by model host plants-*Lotus japonicus* and *Medicago truncatula* are providing a wealth of opportunities to this new era. Sequencing of plant-microbe partners (plant-rhizobia) will offer vital data that could allow investigators to link mutant phenotypes to genes at a much faster pace. In plant signal transduction, plant response to *Nod* factor and its receptors have opened avenues for symbiotic bottleneck research in rhizobial–plant symbioses, which in future will help to unravel and devise strains of *Rhizobium*, highly efficient in nitrogen fixation. After 100 years of the earliest published elucidations and deliberations of infection threads several important areas of infection thread biology are still opaque [177, 178] which includes nature of infection thread building machinery in plant root hairs, expression and growth rate of bacteria in infection threads and nodule cells. For rapid development of the N_2 fixing system, studies on plant-rhizobial genes involved in nodulation and N_2 fixation is mandatory. In the last two decades, rhizobium-legume symbiosis has produced scant results in its process improvement due to underestimation and ignorance about the regulatory mechanisms involved. Nitrogen-fixing rhizobia benefits plant hosts growing in nitrogen-deficient environments even though, how the bacteria benefits from this association is less clear. Exopolysaccharides are required for successful nodule invasion in bacteria but their exact function is still unknown. Once the bacterium enters the nodule, its regulation of bacterial differentiation and host-bacterium signal exchange is still uncertain. Nodule invasion follows clustering of bacteria around the root hairs; it is therefore possible that the rise in cell density around the root hairs alters the expression of some genes. Bacterial population communicates *via* cell-cell chemical signals and secreted proteins to avoid strong plant defence responses. In future knowledge attained in these fields will allow the fabrication of specific strategies to create plants resistance from phytopathogens and rhizobial strains with improved natural symbiotic properties. To unravel the molecular dialog between rhizobia and leguminous plants several technical advances will also give many new assets. This is an overview of a limited but fascinating part of rhizobial-host relation to avoid Herculean task for readers. Speedy progress in many of the above areas, which have remained unanswered until now, will be put to task in the near future, given the new techniques and tools in the hands of molecular biologists and biotechnologists. For all of these ins and outs of "what" happens during nodulation, we are still in search mode.

ACKNOWLEDGEMENTS

We are grateful to Dr. Aakash Goyal, Visiting Professor Lethbridge Research Centre Agriculture and Agri-Food, Alberta, Canada and the peer reviewers for their helpful comments and suggestions. The authors acknowledge financial support for "part of the work" mentioned in this article from the University Grants Commission, New Delhi, India.

ABBREVIATIONS:

IPM: Iintegrated management practices

BNF: Biological nitrogen fixation

IAA: Indole acetic acid

PG: Polygalacturonase

ENOD: Early nodulins

LNOD: Late nodulins

EPS: Exopolysaccharides

REFERENCES

[1]   (http://microbiologyprocedure.com/index.htm)
[2]   Schoot Uiterkamp AJM. Nitrogen cycling and human intervention, Gresshoff PM, Roth LE, Stacey G, and Newton WE. (ed.), Nitrogen fixation: achievements and objectives. Chapman and Hall, New York, N.Y. 1990; p. 55–66.
[3]   Postgate JR. The fundamentals of nitrogen fixation. Cambridge University Press, Cambridge, United Kingdom. 1982.
[4]   Sprent JI, Sprent P. Nitrogen fixing organisms. Pure and applied aspects. Chapman and Hall, London, United Kingdom. FAO. 1990. Fertilizer yearbook, vol. 39. FAO, Rome, Italy.
[5]   FAO. 1990. Fertilizer yearbook, vol. 39. FAO, Rome, Italy.
[6]   Peoples MB, Herridge DF, Ladha JK. Biological nitrogen fixation: an efficient source of nitrogen for sustainable agricultural production. Plant Soil 1995; 174: 3–28.
[7]   Burris RH. Biological nitrogen fixation—past and future, *In* N. A. Hegazi, M. Fayez, and M. Monib (ed.), Nitrogen fixation with nonlegumes. The American University in Cairo Press, Cairo, Egypt. 1994; p. 1–11.

[8]   Al-Sherif EM. Ecological studies on the flora of some aquatic systems in Beni-Suef district. M.Sc. thesis. Cairo University (Beni-Suef Branch), Beni-Suef, Egypt. 1998.

[9]   Dixon ROD, Wheeler CT. Nitrogen fixation in plants. Blackie, Glasgow, United Kingdom. 1986.

[10]   Zahran HH, Ahmed MS, Afkar EA. Isolation and characterization of nitrogen-fixing moderate halophilic bacteria from saline soils of Egypt J Basic Microbiol 1995; 35: 269–75.

[11]   Graham PH. Biological Dinitrogen Fixation: Symbiotic. In Principles and Applications of Soil Microbiology, eds. D. Sylvia *et al.* Upper Saddle River, NJ: Prentice-Hall, 1998.O'Gara F and Shanmugam KT. Regulation of nitrogen fixation by Rhizobia. Export of fixed N_2 as NH4 [+]. Biochim et Biophys Act 1976; 437(2): 313–21.

[12]   O'Gara F and Shanmugam KT. Regulation of nitrogen fixation by Rhizobia. Export of fixed N_2 as NH4 [+]. Biochim et Biophys Act 1976; 437(2): 313–21.

[13]   Van Rhijn P, Vanderleyden J. The *Rhizobium*–plant symbiosis. Microbiol Rev 1995; 59(1): 124–42.

[14]   Lodwig EM, Poole PS. Metabolism of *Rhizobium* bacteroids." Crit Rev Plant Sci 2003; 22(1): 37–38.

[15]   Denison RF, Kiers ET. Why are most rhizobia beneficial to their plant hosts, rather than parasitic? Microb Inf 2004; 6(13): 1235–39.

[16]   Arora P, Yadav R, Chaudhury A, Dilbaghi N. Genotypic characterization of North Indian isolates of rhizobia nodulating exotic and endemic legumes. Published in Proceedings "49th Annual Conference of the Association of Microbiologists of India" Delhi University, Delhi: India 2008

[17]   Boring LR, Swank WT, Waide JB, Henderson GS. Sources, fates, and impacts of nitrogen inputs to terrestrial ecosystems: review and synthesis. Biogeochem 1988; 6(2): 119–59.

[18]   Martínez-Romero E, Caballero-Mellado J. *Rhizobium* phylogenies and bacterial genetic diversity. Crit Rev Plant Sci 1996; 15(2): 113–40.

[19]   Garrity GM, Bell JA, Lilburn TG. Taxonomic outline of the prokaryotes, Bergey's Manual of Systematic Bacteriology, second edition. Bergey's Manual Trust. 2004 (http://dx.doi.org/10.1007/bergeysoutline).

[20]   Quan Z-X, Bae H-S, Baek J-H, Im W-T, Lee S-T, Chen W-F. *Rhizobium daejeonense* sp. nov. isolated from a cyanide treatment bioreactor. Intern J Sys Evol Microbiol 2005; 55(6): 2543–49.

[21]   Segovia L, Young JPW, Martínez-Romero E. Reclassification of American *Rhizobium leguminosarum* biovar phaseoli type I strains as *Rhizobium etli* sp. nov. Intern J Sys Bacteriol 1993; 43(2): 374–77.

[22]   Lindström K. *Rhizobium galegae*, a new species of legume root nodule bacteria. Intern J Sys Bacteriol 1989; 39(3): 365–67.

[23]   Amarger N, Macheret V, Laguerre G, Amarger N. *Rhizobium gallicum* sp. nov. and *Rhizobium giardinii* sp. nov., from *Phaseolus vulgaris* nodules. Intern J Sys Bacteriol 1997; 47(4): 996–1006.

[24]   Chen, W. X., Tan, Z. Y., Gao, J. L., Li, Y., Wang, E. T. *Rhizobium hainanense* sp. nov, isolated from tropical legumes. Intern J Sys Bacteriol 1997; 47(3): 870–73.

[25]   Wang ET, van Berkum P, Beyene D, Sui XH, Dorado O, Chen WX, *et al. Rhizobium huautlense* sp. nov., a symbiont of *Sesbania herbacea* that has a close phylogenetic relationship with *Rhizobium galegae*. Intern J Sys Bacteriol 1998; 48(3): 687–99.

[26]   Wei GH, Wang ET, Tan ZY, Zhu ME, Chen WX. *Rhizobium indigoferae* sp. nov. and *Sinorhizobium kummerowiae* sp. nov., respectively isolated from *Indigofera* spp. and *Kummerowia stipulacea*. Intern J Sys Evol Microbiol 2002; 52(6): 2231–39.

[27]   Frank B. Ueber die Parasiten in den Wurzelanschwillungen der Papilionaceen. Bot Zeit 1879; 37: 376–87, 394–99.

[28]   Wei GH, Tan ZY, Zhu ME, Wang ET, Han SZ, Chen WX. Characterization of rhizobia isolated from legume species within the genera *Astragalus* and *Lespedeza* grown in the Loess Plateau of China and description of *Rhizobium loessense* sp. nov." Intern J Sys Evol Microbiol 2003; 53(5): 1575–83.

[29]   van Berkum P, Beyene D, Campbell TA, Bao G, and Eardly BD. *Rhizobium mongolense* sp. nov. is one of three rhizobial genotypes identified which nodulate and form nitrogen-fixing symbioses with *Medicago ruthenica* [(L.) Ledebour]. Intern J Sys Bacteriol 1998; 48(1): 13–22.

[30]   Squartini A, Struffi P, Döring H, Selenska-Pobell S, Tola E, Giacomini A, *et al. Rhizobium sullae* sp. nov. (formerly '*Rhizobium hedysari*'), the root-nodule microsymbiont of *Hedysarum coronarium* L. Intern J Sys Evol Microbiol 2002; 52(4): 1267–76.

[31]   Martínez-Romero E, Segovia L, Mercante FM, Franco AA, Graham P, Pardo MA. *Rhizobium tropici*, a novel species nodulating *Phaseolus vulgaris* L. beans and *Leucaena* sp. trees. Intern J Sys Bacteriol 1991; 41(3): 417–26.

[32]   de Lajudie P, Laurent-Fulele E, Willems A, Torck U, Coopman, R, Collins MD, *et al. Allorhizobium undicola* gen. nov., sp. nov., nitrogen-fixing bacteria that efficiently nodulate *Neptunia natans* in Senegal. Intern J Sys Bacteriol 1998a; 48(4): 1277–90.

[33] Tan ZY, Kan FL, Peng GX, Wang ET, Reinhold-Hurek B, Chen WX. *Rhizobium yanglingense* sp. nov., isolated from arid and semi-arid regions in China. Intern J Sys Evol Microbiol 2001; 51(3): 909–14.

[34] Ogasawara M, Suzuki T, Mutoh I, Annapurna K, Arora NK, Nishimura Y, *et al. Sinorhizobium indiaense* sp. nov. and *Sinorhizobium abri* sp. nov. isolated from tropical legumes, *Sesbania rostrata* and *Abrus precatorius*, respectively. Symb 2003; 34(1): 53–68.

[35] Wang ET, Tan ZY, Willems A, Fernández-López M, Reinhold-Hurek B, Martínez-Romero E. *Sinorhizobium morelense* sp. nov., a *Leucaena leucocephala*-associated bacterium that is highly resistant to multiple antibiotics. Intern J Sys Evol Microbiol 2002; 52(5): 1687–93.

[36] Toledo I, Lloret L, Martínez-Romero E. *Sinorhizobium americanus* sp. nov., a new *Sinorhizobium* species nodulating native *Acacia* spp. in Mexico. Sys App Microbiol 2003; 26(1): 54–64.

[37] Nick G, de Lajudie P, Eardly BD, Suomalainen S, Paulin L, Zhang XP, *et al. Sinorhizobium arboris* sp. nov. and *Sinorhizobium kostiense* sp. nov., isolated from leguminous trees in Sudan and Kenya. Intern J Sys Bacteriol 1999; 49(4): 1359–68.

[38] Scholla MH, Moorefield JA, and Elkan GH. Deoxyribonucleic acid homology between fast-growing soybean nodulating bacteria and other rhizobia. Intern J Sys Bacteriol 1984; 34(3): 283–86.

[39] Rome S, Brunel B, Cleyet-Marel J-C, Rome S, Fernandez MP, Normand P. *Sinorhizobium medicae* sp. nov., isolated from annual *Medicago* spp. Intern J Sys Bacteriol 1996; 46(4): 972–80.

[40] Dangeard PA. *Recherches sur les tubercules radicaux des légumineuses*. Series 16. Le Botaniste, Paris1926.

[41] de Lajudie P, Willems A, Pot B, Dewettinck D, Maestrojuan G, Neyra, M, *et al* Polyphasic taxonomy of rhizobia: Emendation of the genus *Sinorhizobium* and description of *Sinorhizobium meliloti* comb. nov., *Sinorhizobium saheli* sp. nov., and *Sinorhizobium teranga* sp. nov. Intern J Sys Bacteriol 1994; 715–33.

[42] Chen WX, Yan GH, Li JL. Numerical taxonomic study of fast-growing soybean rhizobia and a proposal that *Rhizobium fredii* be assigned to *Sinorhizobium* gen. nov." Intern J Sys Bacteriol 1988; 38(4): 392–97.

[43] Wang ET, van Berkum P, Sui XH, Beyene D, Chen WX, Martínez-Romero E. (Diversity of rhizobia associated with *Amorpha fruticosa* isolated from Chinese soils and description of *Mesorhizobium amorphae* sp. nov. Intern J Sys Bacteriol 1999; 49(1): 51–65.

[44] Velázquez E, Igual JM, Willems A, Fernádez MP, Muñoz E, Mateos PF, *et al. Mesorhizobium chacoense* sp. nov., a novel species that nodulates *Prosopis alba* in the Chaco Arido region (Argentina). Intern J Sys Evol Microbiol 2001; 51(3): 1011–21.

[45] Nour SM, Fernandez MP, Normand P, Cleyet-Marel J-C. *Rhizobium ciceri* sp. nov., consisting of strains that nodulate chickpeas (*Cicer arietinum* L.). Intern J Sys Bacteriol 1994; 44(3): 511–22.

[46] Chen WX, Li GS, Qi YL, Wang ET, Yuan HL, Li JL. *Rhizobium huakuii* sp. nov. isolated from the root nodules of *Astragalus sinicus*. Intern J Sys Bacteriol 1991; 41(2): 275–80.

[47] Jarvis BDW, Pankhurst CE, Patel JJ. *Rhizobium loti*, a new species of legume root nodule bacteria. Intern J Sys Bacteriol 1982; 32(3): 378–80.

[48] Nour SM, Cleyet-Marel J-C, Normand P, Fernandez MP. Genomic heterogeneity of strains nodulating chickpeas (*Cicer arietinum* L.) and description of *Rhizobium mediterraneum* sp. nov. Intern J Sys Bacteriol 1995; 45(4): 640–48.

[49] de Lajudie P, Willems A, Nick G, Moreira F, Molouba F, Hoste B, *et al.* Characterization of tropical tree rhizobia and description of *Mesorhizobium plurifarium* sp. nov. Intern J Sys Bacteriol 1998b; 48(2): 369–82.

[50] Gao J-L, Turner SL, Kan FL, Wang ET, Tan ZY, Qiu YH, *et al. Mesorhizobium septentrionale* sp. nov. and *Mesorhizobium temperatum* sp. nov., isolated from *Astragalus adsurgens* growing in the northern regions of China. Intern J Sys Evol Microbiol 2004; 54(6): 2003–12.

[51] Chen W, Wang E, Wang S, Li Y, Chen X. Characteristics of *Rhizobium tianshanense* sp. nov., a moderately and slowly growing root nodule bacterium isolated from an arid saline environment in Xinjiang, People's Republic of China." Intern J Sys Bacteriol 1995; 45(1): 153–59.

[52] Vinuesa P, Silva C, Martínez-Romero E, Werner D, León-Barrios M, Jarabo-Lorenzo A, *et al. Bradyrhizobium canariense* sp. nov., an acid-tolerant endosymbiont that nodulates endemic genistoid legumes (*Papilionoideae: Genisteae*) from the Canary Islands, along with *Bradyrhizobium japonicum* bv. genistearum, *Bradyrhizobium* genospecies alpha and *Bradyrhizobium* genospecies beta. Intern J Syst Evol Microbiol 2005; 55(2): 569–75.

[53] Kuykendall LD, Gaur YD, and Dutta SK. Genetic diversity among *Rhizobium* strains from *Cicer arietinum* L." Lett Appl Microbiol 1993; 17(6): 259–63.

[54] Kirchner O. Die Wurzelknöllchen der Sojabohne. *Beiräge zur Biologie der Pflanzen*, 1896; 7: 213–24.

[55]   Xu LM, Ge C, Cui Z, Li J, and Fan H. *Bradyrhizobium liaoningense* sp. nov., isolated from the root nodules of soybeans. Intern J Sys Bacteriol 1995; 45(4): 706–11.

[56]   Yao ZY, Kan FL, WangET, WeiGH, Chen WX. Characterization of rhizobia that nodulate legume species of the genus *Lespedeza* and description of *Bradyrhizobium yuanmingense* sp. nov. Intern J Sys Evol Microbiol 2002; 52(6): 2219–30.

[57]   Vandamme P, Goris J, Chen WM, de Vos P, Willems A. *Burkholderia tuberum* sp. nov. and *Burkholderia phymatum* sp. nov., nodulate the roots of tropical legumes. Sys Appl Microbiol 2002; 25(4): 507–12.

[58]   Dreyfus B, Garcia JL, Gillis M. Characterization of *Azorhizobium caulinodans* gen. nov., sp. nov., a stem-nodulating nitrogen fixing bacterium isolated from *Sesbania rostrata*. Intern J Sys Bacteriol 1988; 38(1): 89–98.

[59]   de Souza Moreira FM, Cruz L, de Faria SM, Marsh T, Martínez-Romero E, de Oliveira Pedrosa F, Pitard RM, *et al.* *Azorhizobium doebereinerae* sp. nov. microsymbiont of *Sesbania virgata* (Caz.) Pers. Sys Appl Microbiol 2006; 29(3): 197–206.

[60]   Chen W-M, Laevens S, Lee T-M, Coenye T, De Vos P, Mergeay M, Vandamme P. *Ralstonia taiwanensis* sp. nov., isolated from root nodules of *Mimosa* species and sputum of a cystic fibrosis patient. Intern Journ Sys Evol Microbiol 2001; 51(5): 1729–35.

[61]   Rivas R, Willems A, Subba-Rao NS, Mateos PF, Dazzo FB, Kroppenstedt RM, *et al.* Description of *Devosia neptuniae* sp. nov. that nodulates and fixes nitrogen in symbiosis with *Neptunia natans*, an aquatic legume from India. Sys Appl Microbiol 2003; 26(1): 47–53.

[62]   Valverde A, Velazquez E, Gutierrez C, Cervantes E, Ventosa A, Igual J-M. *Herbaspirillum lusitanum* sp. nov., a novel nitrogen-fixing bacterium associated with root nodules of *Phaseolus vulgaris*. Intern Journ Sys Evol Microbiol 2003; 53(6): 1979–83.

[63]   Jourand P, Giraud E, Béna G, Sy A, Dreyfus B, de Lajudie P, *et al.* *Methylobacterium nodulans* sp. nov., for a group of aerobic, facultatively methylotrophic, legume root-noduleforming and nitrogen-fixing bacteria. Intern Journ Sys Evol Microbiol 2004; 54(6): 2269–73.

[64]   Trujillo ME, Willems A, Abril A, Planchuelo A.-M, Rivas R, Ludena D, Mateos PF, *et al.* Nodulation of *Lupinus albus* by strains of *Ochrobactrum lupine* sp. nov." Appl Environ Microbiol 2005; 71(3): 1318–27.

[65]   Young JM, Kuykendall LD, Martínez-Romero E, Kerr A, Sawada H. (A revision of *Rhizobium* Frank 1889, with an emended description of the genus, and the inclusion of all species of *Agrobacterium* Conn 1942 and *Allorhizobium undicola* de Lajudie *et al.* 1998 as new combinations: *Rhizobium radiobacter, R. rhizogenes, R. rubi, R. undicola* and *R. vitis*. Intern J Sys Evol Microbiol 200151(1): 89–103.

[66]   Young, JM, Park D-C, Weir BS. Diversity of 16S rDNA sequences of *Rhizobium* spp. implications for species determinations. FEMS Microbiol Lett 2004; 238(1): 125–31.

[67]   Smit G, Kijne JW, Lugtenberg BJ. Roles of flagella, lipopolysaccharide, and a Ca_2-dependent cell surface protein in attachment of *Rhizobium leguminosarum* biovar. *vicae* to pea root hair tips. J Bacteriol 1989; 171: 569–72.

[68]   Smit G, Kijne JW, Lugtenberg BJ. Involvement of both cellulose fibrils and a Ca_2-dependent adhesin in the attchment of *Rhizobium leguminosarum* to pea root hair tips. J Bacteriol 1987; 169: 4294–301.

[69]   Smit G, Kijne JW, Lugtenberg BJ. Correlation between extracellular fibrils and attachment of *Rhizobium leguminosarum* to pea root hair tips. J Bacteriol 1986; 168: 821–7.

[70]   Diaz CL, Logman T, Stam HC, Kijne JW. Sugar-binding activity of pea lectin expressed in white clover hairy roots. Plant Physiol 1995; 109: 1167–77.

[71]   Diaz CL, Van Spronsen PC, Bakhuizen R, Logman GJJ, Lugtenberg EJJ, Kijne JW. Correlation between infection by *Rhizobium leguminosarum* and lectin on the surface of *Pisum sativum* roots. Planta 1986; 168: 350–9.

[72]   Hirsch AM. Role of lectins (and rhizobial exopolysaccharides) in legume nodulation. Curr Opin Plant Biol 1999; 2: 320–6.

[73]   de Ruijter, Rook NCAMB, Bisseling T, Emons AMC. Lipochito-oligosaccharides re-initiate root hair tip growth in *Vicia sativa* with high calcium and spectrin-like antigen at the tip. Plant J 1998; 13: 341–50.

[74]   Heidstra R, Geurts R, Franssen H, Spaink H, Van Kammen A, Bisseling T. Root hair deformation activity of nodulation factors and their fate on *Vicia sativa*. Plant Physiol 1994; 105: 787.

[75]   Sieberer B, Emons AMC. Cytoarchitecture and pattern of cytoplasmic streaming in root hairs of *Medicago truncatula* during development and deformation by nodulation factors. Protoplasma 2000; 214: 118–27.

[76]   Esseling JJ, Lhuissier FG, Emons AMC. Nod factor-induced root hair curling: continuous polar growth towards the point of Nod factor application. Plant Physiol 2003; 132: 1982–8.

[77]   Callaham DA, Torrey JG. The structural basis for infection of root hairs of *Trifolium repens* by *Rhizobium*. Can J Bot 1981; 59: 1647–64.

[78]   Oldroyd GED, Engstrom EM, Long SR. Ethylene inhibits the Nod factor signal transduction pathway of *Medicago truncatula*. Plant Cell 2001; 13: 1835–49.

[79]   Cardenas LL, Vidali, Dominguez J, Perez H, Sanchez F, Hepler PK, Quinto C. Rearrangement of actin microfilaments in plant root hairs responding to *Rhizobium etli* nodulation signals. Plant Physiol 1998; 116: 871–7.

[80]   Kohno T, Shimmen T. Ca$_2$+induced fragmentaion of actin filaments in pollen tubes. Protoplasma 1987; 141: 177.

[81]   Timmers AC, Auriac MC, Truchet G. Refined analysis of early symbiotic steps of the *Rhizobium-Medicago* interaction in relationship with microtubular cytoskeleton rearrangements. Development 1999; 126: 3617–28.

[82]   De Ruijter NCA, Rook MB, Bisseling T, Emons AMC. Lipochito-oligosaccharides re-initiate root hair tip growth in *Vicia sativa* with high calcium and spectrin-like antigen at the tip. Plant J 1998; 13: 341–50.

[83]   Somasegaran P, Hoben HJ. Gurgun V. Effects of inoculation rate, rhizobial strain competition and nitrogen fixation in chickpea. Ag Journ 1988; 80: 68-73.

[84]   Van Batenburg, Jonker FHDR, Kijne JW. *Rhizobium* induces marked root hair curling by redirection of tip growth: A computer simulation. Physiol Plant 1986; 66: 476–80.

[85]   Ridge RW. A model of legume root hair growth and *Rhizobium* infection. Sym 1993; 14: 359–73.

[86]   Emons AMC, Mulder B. Nodulation factors trigger an increase of fine bundles of subapical actin filaments in *Vicia* root hairs: Implication for root hair curling around bacteria. In Biology of Plant-Microbe Interactions, P.J.G.M. De Wit, T. Bisseling, and J.W. Stiekema, eds (St. Paul, Minnesota: The International Society of Molecular Plant-Microbe Interaction). 2000; pp. 272–276.

[87]   Callaham DA, Torrey JG. The structural basis for infection of root hairs of *Trifolium repens* by *Rhizobium*. Can J Bot 1981; 59: 1647–64.

[88]   Van den Bosch, Bradley KADG, Knox JP, Perotto S, Butcher GW, Brewin NJ. Common components of the infection thread matrix and intercellular space identified by immuno cytochemical analysis of pea nodules and uninfected roots. EMBO J 1989; 8: 335–42.

[89]   Van Spronsen, Bakhuizen PCR, Van Brussel AAN, Kijne JW. Cell-wall degradation during infection thread formation by the root nodule bacterium *Rhizobium leguminosarum* is a 2-step process. Eur J Cell Biol 1994; 64: 88–94.

[90]   Van Workum, Van Slageren WATS, Van Brussel AAN, Kijne JW. Role of exopolysaccharides of *Rhizobium leguminosarum bv. viciae* as host plant-specific molecules required for infection thread formation during nodulation of *Vicia sativa*. Mol Plant-Microbe Interact 1998; 11: 1233–41.

[91]   Carol RJ, Dolan L. Building a hair: tip growth in *Arabidopsis thaliana* root hairs. Philos Trans R Soc London Ser B 2002; 357: 815–21.

[92]   Hepler PK, Vidali L, Cheung AY. Polarized cell growth in higher plants. Annu Rev Cell Dev Biol 2001; 17: 159–87.

[93]   Smith LG. Cytoskeletal control of plant cell shape: getting the fine points. Curr Opin Plant Biol 2003; 6: 63–73.

[94]   Ketelaar T, Emons AMC. The cytoskeleton in plant cell growth: lessons from root hairs. New Phytol 2001; 152: 409–18.

[95]   Iwanami Y. Protoplasmic movement in pollen grains and pollen tubes. Phytomorphology 1956; 6: 288–95.

[96]   Gage DJ. Infection and invasion of roots by symbiotic, nitrogen-fixing rhizobia during nodulation of temperate legumes. Microbiol Mol Biol Rev 2004; 68(2): 280–300.

[97]   *Rhizobium*, Root Nodules and Nitrogen Fixation. Society for General Microbiology. 2002.

[98]   Parniske M, Downie JA. Plant biology: Locks, keys and symbioses. Nature 2003; 425: 569-70.

[99]   Cassab GI. Plant cell wall proteins. Annu Rev Plant Physiol Plant Mol Biol 1998; 49: 281–309.

[100] Cosgrove DJ, Li LC, Cho HT, Hoffmann-Benning S, Moore RC, Blecker D. The growing world of expansins. Plant Cell Physiol 2002; 43: 1436–44.

[101] Cosgrove DJ, Bedinger P, Durachko DM. Group I allergens of grass pollen as cell wall-loosening agents. Proceedings National Academy of Sciences, USA 1997; 94: 6559–64.

[102] Hirsch AM. Developmental biology of legume nodulation. New Phytol 1992; 122: 211–37.

[103] Quispel ANN. The Biology of Nitrogen Fixation. North-Holland Press, Amsterdam 1974.

[104] Stacey G, Evans H, Burris R. Biological Nitrogen Fixation. Chapman and Hall, New York 1992.

[105] Verma DPS. Molecular Signals in Plant Microbe Communication. CRC Press, Boca Raton, FL 1992.

[106] Spaink HP, Kondorosi A, Hooykaas PJJ. The *Rhizobiaceae*: Molecular Biology of Model Plant-Associated Bacteria. Kluwer Academic Publishers, Dordrecht, The Netherlands 1998.

[107] Downie JA. Signaling strategies for the nodulation of legumes by rhizobia. Trends Microbiol 1994; 2: 318–24.

[108] Oke V, Long SR. Bacteroid formation in the Rhizobium-legume symbiosis. Curr Opin Microbiol 1999; 2: 641–46.

[109] Phillips DA, Joseph CM, Yang GP, Martinez-Romero E, Sanborn JR, Volpin H. Proceedings National Academy of Sciences, USA. 1999; 96: 12275–80.

[110] MacLean AM, Finan TM, Sadowsky MJ. Genomes of the symbiotic nitrogen fixing bacteria of legumes. Plant Physiol 2007; 144: 615–22.

[111] Crossman LC, Castillo-Ramirez S, McAnnula C, Lozano L, Vernikos GS, Acosta JL, *et al.* A common genomic framework for a diverse assembly of plasmids in the symbiotic nitrogenfixing bacteria. PLoS one 2008; 3: e2567.

[112] Fischer H-M. Genetic regulation of nitrogen fixation in rhizobia. Microbiol Rev 1994; 58: 352–86.

[113] Hernandez JA, Curatti L, Aznar CP, Perova Z, Britt RD, Rubio LM. Metal trafficking for nitrogen fixation: NifQ donates molybdenum to NifEN/NifH for the biosynthesis of the nitrogenase FeMo-cofactor. Proceedings of the National Academy of Sciences USA. 2008; 105: 11679–684.

[114] Rubio LM, Ludden PW. Biosynthesis of the iron–molybdenum cofactor of nitrogenase. Annu Rev Microbiol 2008; 62: 93–111.

[115] Bobik C, Meilhoc E, Batut J. FixJ; a major regulator of the oxygen limitation response and late symbiotic functions of *Sinorhizobium meliloti*. J Bacteriol 2006; 188: 4890–902.

[116] Martinez M, Colombo M-V, Palacios J-M, Imperial J, Ruiz-Arg ̈ ueso T. Novel arrangement of enhancer sequences for NifA-dependent activation of hydrogenase gene promoter in *Rhizobium leguminosarum* bv. viciae. J Bacteriol 2008; 190: 3185–191.

[117] Lindemann A, Moser A, Pessi G, Hauser F, Friberg M, Hennecke H, Fischer H-M. New target genes controlled by the *Bradyrhizobium japonicum* two-component regulatory system RegSR. J Bacteriol 2007; 189: 8928–943.

[118] Phillips DA, Streit WR. *In* G Stacey, NT Keen, eds, Plant Microbe Interactions. Chapman and Hall, New York, 1996; pp 236–71.

[119] Juan E. Gonza ́lez, Melanie M. Marketon. Quorum Sensing in Nitrogen-Fixing Rhizobia. Microbiol Mol Biol Rev 2003; 574–92.

[120] De ́narie ́ J, Debelle ́ F, Prome ́ J. *Rhizobium* lipo-chitooligosaccharide nodulation factories: signaling molecules mediating recognition and morphogenesis. Annu Rev Biochem 1996; 65: 503–35.

[121] Long SR. Rhizobium Symbiosis: Nod Factors in Perspective. Plant Cell 1996; 8: 1885–98.

[122] Schultze M, Kondorosi A. Regulation of symbiotic root nodule development. Annu Rev Genet 1998; 32: 33–57.

[123] Spaink HP, Kondorosi A, Hooykaas PJJ (1998) The Rhizobiaceae: Molecular Biology of Model Plant- Associated Bacteria. Kluwer Academic Publishers, Dordrecht, The Netherlands

[124] Verma DPS (1992) Molecular Signals in Plant MicrobeCommunication. CRC Press, Boca Raton, FL

[125] Nap JP, Bisseling T. Developmental biology of a plant- prokaryotic symbiosis: the legume root nodule. Science 1990; 250: 948-54.

[126] Brewin NJ. Plant cell wall remodelling in the *Rhizobium*–Legume symbiosis. Crit Rev Plant Sci 2004; 23: 293–316.

[127] Scheres B, Wiel CVD, Zalensky A, Horvath B, Spaink H, van Eck H, *et al.* The ENOD12 gene product is involved in the infection process during the pea-*Rhizobium* interaction. Cell 1990; 60: 281-94.

[128] Pichon M, Journet E-P, Dedieu A, de Billy F, Truchet G, Barker DG. *Rhizobium meliloti* elicits transient expression of the early nodulin gene ENOD12 in the differentiating root epidermis of transgenic alfalfa. Plant Cell 1992; 4: 1199-211.

[129] Yang WC, Katinakis P, Hendriks P, Smolders A, de Vries F, Spree J, *et al.* Characterization of GmENOD40, a gene showing novel patterns of cell-speci®c expression during soybean nodule development. Plant Journ 1993; 3: 573-85.

[130] Journet EP, Pichon M, Dedieu A, de Billy F, Truchet G, Barker DG. *Rhizobium meliloti* Nod factors elicit cell-speci®c transcription of the ENOD12 gene in transgenic alfalfa. Plant Journ 1994; 6: 241-49.

[131] Cook D, Dreyer D, Bonnet D, Howell M, Nony E, Vandenbosch K. Transient induction of a peroxidase gene in *Medicago truncatula* precedes infection by *Rhizobium meliloti*. Plant Cell 1995; 7: 43-55.

[132] Minami E, Kouchi H, Carlson RW, Cohn JR, Kolli VK, Day RB, Ogawa T, Stacey G. Cooperative action of lipo-chitin nodulation signals on the induction of the early nodulin, ENOD2, in soybean roots. Mol Plant-Microbe Interaction 1996; b 9: 574-83.

[133] Lohar DP, Sharopova N, Endre G, Pe ˜ nuela S, Samac D, Town C, eta l. Transcript analysis of early nodulation events in *Medicago truncatula*. Plant Physiol 2006; 140: 221–34.

[134] Bauer P, Ratet P, Crespi MD, Schultze M, Kondorosi A. Nod factors and cytokinins induce similar cortical cell division, amyloplast deposition and MsENOD12A expression patterns in alfalfa roots. Plant Journ, 1996; 10: 91-105.

[135] Hirsch AM, Bhuvaneswari TV, Torrey JG, Bisseling T. Early nodulin genes are induced in alfalfa root outgrowths elicited by auxin transport inhibitors. Proceedings of the National Academy of Sciences, USA, 1989; 86: 1244-48.

[136] Cooper JB, Long SR. Morphogenetic rescus of *Rhizobium meliloti* nodulation mutants by trans-zeatin secretion. Plant Cell, 1994; 6: 215-25.

[137] Hirsch AM, Fang Y. Plant hormones and nodulation: what's the connection. Plant Mol Biol, 1994; 26: 5-9.

[138] Downie JA, Walker SA. Plant responses to nodulation factors. Curr Opin Plant Biol, 1999; 2: 483–89.

[139] Schauser L, Roussis A, Stiller J, Stougaard J. A plant regulator controlling development of symbiotic root nodules. Nature, 1999; 402: 191–95.

[140] Brewin NJ. Development of the legume root nodule. Annu Rev Cell Biol 1991; 7: 191–226.

[141] Fisher RF, Long SR. *Rhizobium*-plant signal exchange. Nature 1992; 357: 655–60.

[142] Long SR. *Rhizobium*-legume nodulation: life together in the underground. Cell 1989; 56: 203–14.

[143] Long SR, Staskawicz BJ. Prokaryotic plant parasites. Cell 1993; 73: 921–35.

[144] Rolfe B G, Gresshoff PM. Genetic analysis of legume nodule initiation. Annu Rev Plant Physiol Plant Mol Biol 1988; 39: 297–319.

[145] Verma DPS, Long SR. The molecular biology of *Rhizobium* legume symbiosis. Int Rev Cytol 1983; 14: 211–45.

[146] Hirsch AM, Long SR, Bang M, Haskins N, Ausubel FM. Structural studies of alfalfa roots infected with nodulation mutants of *Rhizobium meliloti*. J Bacteriol 1982; 151: 411–19.

[147] Carlson RW, Price NP, Stacey G. The biosynthesis of rhizobial lipo-oligosaccharide nodulation signal molecules. Mol Plant-Microbe Interact 1994; 7: 684–95.

[148] Long SR. *Rhizobium* symbiosis: nod factors in perspective. Plant Cell 1996; 8: 1885–98.

[149] Long SR, Atkinson EM. *Rhizobium* sweet-talking. Nature 1990; 344: 712–13.

[150] Long SR. *Rhizobium*-legume nodulation: life together in the underground. Cell 1989; 56: 203–14.

[151] Perret X, Staehelin C, Broughton WJ. Molecular basis of symbiotic promiscuity. Microbiol Mol Biol Rev 2000; 64: 180–201.

[152] Fisher RF, Brierley HL, Mulligan JT, Long SR. Transcription of *Rhizobium meliloti* nodulation genes. Identification of a *nodD* transcription initiation site *in vitro* and *in vivo*. J Biol Chem 1987; 262: 6849–55.

[153] Mulligan JT, Long SR. Induction of *Rhizobium meliloti nodC* expression by plant exudate requires *nodD*. Proc. Natl. Acad. Sci. USA. 1985; 82: 6609–13.

[154] Peters NK, Frost JW, Long SR. A plant flavone, luteolin, induces expression of *Rhizobium meliloti* nodulation genes. Science 1986; 233: 977–80.

[155] D'Haeze W, Holsters M. Nod factor structures, responses, and perception during initiation of nodule development. Glycobiology 2002; 12:79R–105R.

[156] Hirsch AM, Lum MR, Downie JA. What makes the rhizobia-legume symbiosis so special? Plant Physiol 2001; 127: 1484–92.

[157] Atkinson EM, Palcic MM, Hindsgaul O, Long SR. Biosynthesis of *Rhizobium meliloti* lipooligosaccharide Nod factors: NodA is required for an *N*-acyltransferase activity. Proc Natl Acad Sci USA. 1994; 91: 8418–22.

[158] Denarie J, Cullimore J. Lipo-oligosaccharide nodulation factors: a new class of signaling molecules mediating recognition and morphogenesis. Cell 1993; 74: 951–54.

[159] Denarie J, Debelle F, Rosenberg C. Signaling and host range variation in nodulation. Annu Rev Microbiol 1992; 46: 497–531.

[160] Ehrhardt DW, Atkinson EM, Faull KF, Freedberg DI, Sutherlin DP, Armstrong R, *et al. In vitro* sulfotransferase activity of NodH, a nodulation protein of *Rhizobium meliloti* required for host-specific nodulation. J Bacteriol 1995; 177: 6237–45.

[161] Fisher RF, Long SR. *Rhizobium*-plant signal exchange. Nature 1992; 357: 655–60.

[162] Schwedock JS, Liu C, Leyh TS, Long SR. *Rhizobium meliloti* NodP and NodQ form a multifunctional sulfate-activating complex requiring GTP for activity. J Bacteriol 1994; 176: 7055–64.

[163] Caetano-Anolles G, Gresshoff PM. Plant genetic control of nodulation. Annu Rev Microbiol 1991; 45: 345–82.

[164] Lerouge P, Roche P, Faucher C, Maillet F, Truchet G, Prome JC, *et al.* Symbiotic host-specificity of *Rhizobium meliloti* is determined by a sulphated and acylated glucosamine oligosaccharide signal. Nature 1990; 344: 781–84.

[165] Roche P, Debelle F', Maillet F, Lerouge P, Faucher C, Truchet G, *et al.* Molecular basis of symbiotic host specificity in *Rhizobium meliloti*: *nodH* and *nodPQ* genes encode the sulfation of lipo-oligosaccharide signals. Cell 1991; 67: 1131–43.

[166] Stougaard J. Genetics and genomics of root symbiosis. Curr Opin Plant Biol 2001; 4: 328–35.

[167] Bauer WD. Infection of legumes by rhizobia. Annu Rev Plant Physiol 1981; 32: 407–49.

[168] Carlson RW, Krishnaiah BS. Structures of the oligosaccharides obtained from the core regions of the lipopolysaccharides of *Bradyrhizobium japonicum* 61A101c and its symbiotically defective lipopolysaccharide mutant, JS314. Carbohydr Res 1992; 231: 205–19.

[169] Leigh, J. A., D. L. Coplin. Exopolysaccharides in plant-bacterial interactions. Annu Rev Microbiol 1992; 46: 307–346.

[170] Petrovics G, Putnoky P, Reuhs B, Kim J, Thorp TA, Noel KD, *et al.* The presence of a novel type of surface polysaccharide in *Rhizobium meliloti* requires a new fatty acid synthase-like gene cluster involved in symbiotic nodule development. Mol Microbiol 1993; 8: 1083–94.

[171] Gonza´lez JE, Reuhs BL, Walker GC. Low molecular weight EPS II of *Rhizobium meliloti* allows nodule invasion in *Medicago sativa.* Proc Natl Acad Sci USA 1996; 93: 8636–41.

[172] Gonza´lez JE, York GM, Walker GC. *Rhizobium meliloti* exopolysaccharides: synthesis and symbiotic function. Gene 1996; 179: 141–46.

[173] Finan TM, Hirsch AM, Leigh JA, Johansen E, Kuldau GA, Deegan S, *et al.* Symbiotic mutants of *Rhizobium meliloti* that uncouple plant from bacterial differentiation. Cell 1985; 40: 869–77.

[174] Leigh JA, Reed JW, Hanks JF, Hirsch AM, Walker GC. *Rhizobium meliloti* mutants that fail to succinylate their calcofluor-binding exopolysaccharide are defective in nodule invasion. Cell 1987; 51: 579–87.

[175] Leigh JA, Signer ER, Walker GC. Exopolysaccharide deficient mutants of *Rhizobium meliloti* that form ineffective nodules. Proc Natl Acad Sci USA 1985; 82: 6231–5.

[176] Yang C, Signer ER, Hirsch AM. Nodules initiated by *Rhizobium meliloti* exopolysaccharide mutants lack a discrete, persistent nodule meristem. Plant Physiol 1992; 98: 143–51.

[177] Dawson M. ”Nitragin“ and the nodules of Leguminous plants. Philos. Trans. R. Soc. London Ser. B. 1900; 192: 1–28.

[178] Ward HM. On the turbercular swellings on the roots of *Vicia faba.* Philos. Trans. R. Soc. London Ser. 1887; B 178: 539–62.

CHAPTER 4

Recent Advances in Sago Palm (*Metroxylon Sagu* Rottboell) Micropropagation

Annabelle U. Novero[*]

College of Science and Mathematics, University of the Philippines Mindanao; Mintal, Davao City 8022 Philippines

Abstract: *Metroxylon sagu* (sago palm) has long been considered as one of the oldest sources of food for humans because of the presence of huge amounts of starch in its trunk. Sago palm is an important food source in Papua New Guinea, Malaysia and Indonesia. Sago palm is increasingly gaining acceptance as an important food source in southern regions of Vietnam, Thailand and the Philippines. Hence, more research efforts toward sago palm breeding and conservation are being conducted. The increasing global pressure to explore non-traditional sources of food and fuel dictates an urgent focus on sago palm (*Metroxylon sagu* Rottboell) biotechnology research. This paper reports the results of a starch yield assessment test among various sago palm ecotypes: spiny and non-spiny in either mesic or hydric environments to determine the most suitable sources of explants for *in vitro* experiments. Although non-spiny sago palms from a mesic area produced the highest mean starch yield of 64.3 kg, statistical analysis showed no significant differences in mean starch yield between the four ecotypes. An account of published reports on sago palm micropropagation is also summarized to give an overview of the status of sago palm biotechnology research. Breakthroughs in micropropagation will continue to be invaluable until a rapid method of multiplication of planting materials has been achieved.

Keywords: Arecaceae, *in vitro* regeneration, Malasia, Indonesia.

INTRODUCTION

Sago palm (*Metroxylon sagu* Rottboell), a palm plant belonging to the family Arecaceae is an important crop in its center of diversity, New Guinea [1]. Its distribution extends westward into Indonesia and Malaysia where the majority of palm stands are cultivated. There are approximately 2.4 million hectares planted to sago palm. Indonesia, Papua New Guinea and Malaysia are the largest sago palm-growing areas (Table **1**). Sago palm is also found in other Southeast Asian countries such as the southern regions of Vietnam, Thailand and the Philippines where they exist in limited natural stands, being in the outer periphery of sago palm distribution [2]. Rauwerdink [1] and Flach [3] have extensively documented the history and characteristics of the genus *Metroxylon*, including the species *sagu*.

Unlike other important palms such as coconut and date, the sago palm grows in a clump and suckers profusely like banana. Thus, sago palm is mainly reproduced vegetatively. There exists a dominant mother palm surrounded by numerous younger palms. Fig. **1** shows a sago palm stand in Agusan del Sur, southern Philippines. The palm is rarely produced *via* seedlings because seeds are recalcitrant and germination rate is very low [4]. The palm has a life span of 10-20 years and flowers on the average at 10-13 years [2]. Flach and Schuiling [5] reported that just before flowering, a sago palm trunk which averaged 1000 kg accumulated an average maximum starch content of about 150 kg. This average starch content was arrived at after analyzing several hundreds of sago palm trunks in West Malaysia. A mother palm soon dies after flowering. Natural sago palms stands are often limited to growth in swampy areas, thus lessening their starch productivity. If palms are grown in a suitable environment using well-planned agricultural practices, starch yield of up to 25 tons per hectare per year could be obtained [6].

*Address correspondence to Annabelle U. Novero: College of Science and Mathematics, University of the Philippines Mindanao; Mintal, Davao City 8022 Philippines; E-mail: anovero@upmin.edu.ph

Aakash Goyal and Priti Maheshwari (Eds)

Table 1: Major sago palm-growing areas of the world as of 2008 [7]

Country	Sago Palm Planting Area, million ha.
Indonesia	1.128
Papua New Guinea	0.953
Malaysia	0.330
Total	2.411

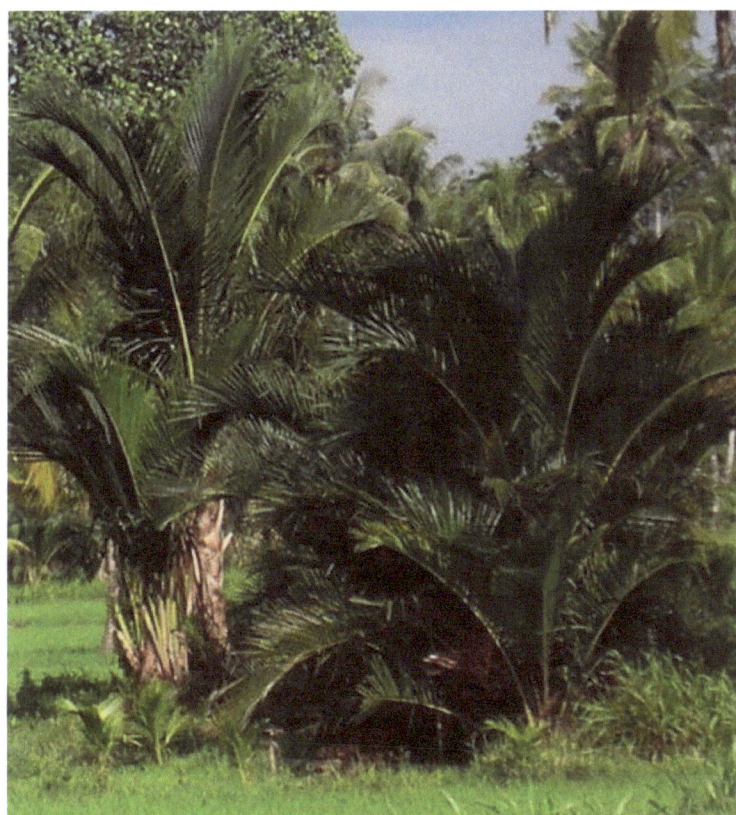

Figure 1: A sago palm stand in Agusan del Sur, Philippines.

Starch from the sago palm has many uses such as food and feed as well as in the production of glucose and dextrose. Sago palm starch and starch derivatives also have applications in the paper, textile and wood industries among many others [3, 8]

Secondary uses include the use of the sago palm's various parts such as the trunk and leaves for building houses. Table **2** summarizes some of the reported uses of sago palm in countries where it is found.

Table 2: Some of the many uses of sago palm

Source	Application or Use	References
Pith	Source of starch, fiber, food and feed	[2, 3, 5, 6, 8, 10, 12]
Fronds	For thatching roofs	[3, 8]
Rachis	House building	[3, 8]
Trunk	House building	[3, 8]
Starch derivatives	Production of ethanol, lactic acid, fermentable sugars, biodegradable plastics, *etc.*	[3, 6, 8]

INVESTIGATIONS ON SAGO PALM GENETIC VARIATION

Breeding programs for sago palm have not yet been reported probably because it takes an average of ten years for the palm to flower. In addition to this, seed germination is very low if they do germinate at all. There have been extensive research efforts in Malaysia and Indonesia in the last 20-30 years mainly focused on cultural practices toward improved starch yield. Studies on genetic variation were mainly to elucidate certain morphological characteristics in relation to starch yield. Some of these quantitative morphological variables are trunk girth, trunk length, leaf number, absence or presence of spines, number and distance of leaf scars. The presence or absence of spine has earlier been used to segregate sago palms into varieties. Fig. **2** shows spines present on the rachis of sago palms growing in a wild stand in Agusan del Sur, Philippines.

In southern Thailand, variation in fifteen populations of sago palm were compared using isozyme analysis [9]. The variations were investigated based on zymograms of peroxidase, esterase, acid phophatase and sorbitol dehydrogenase using polyacrylamide gel electrophoresis (PAGE). A low enzyme variation between populations, suggesting very close relationships between the palm plants sampled.

In north Sulawesi and northern Maluku, Indonesia, eleven local varieties were investigated [10]. A wide diversity of bands on the back of petiole and rachis were found on spineless varieties. In relation to starch yield, short spine may be good indicators of starch yield in Indonesia as farmers have reported more starch derived form spiny varieties. However, farmers prefer the ease of handling spineless varieties, thus, this claim remains unsubstantiated. The differences in pith dry matter yield were mainly due to differences in trunk diameter. There was no correlation of pith dry matter yield to other characteristics such as plant height, trunk length, leaf length and width, scar number and distance.

Kjaer *et al.,* [11] reported the results of their investigation of the morphological variation of sago palms in seven localities in Papua New Guinea. Ten morphological variables studied were number of leaves, number of leaflets, width of leaflets, length of leaflets, length of petiole, width of petiole, length of rachis, thickness of petiole, and presence or absence of spines. Amplified fragment length polymorphism (AFLP) was used to analyze genetic variation using selective primer combinations. The primer pairs EcoR1-ACT/MseI-CAT yielded 41 bands 16 of which were polymorphic. The primer pairs EcoR1-ACC/MseI-CTT produced 26 bands 16 of which were polymorphic [11]. Quantitative analysis of results indicated that vegetative morphological characteristics, most especially the presence or absence of spines have no relation to the underlying genetic variation.

In South Central Seram, Maluku, Indonesia, Ellen [12] reported eleven landraces of sago palm recognized by the Nuaulu tribe. Some of the main characteristics used to classify the palm plants are type and color of spine, color and texture of flesh and color of bark.

DETERMINATION OF STARCH YIELD OF SAGO PALM IN RELATION TO ECOTYPE

In Agusan del Sur, Philippines, limited wild stands of sago palm can be found growing in dry (mesic) and wet (hydric) areas. The sago palm, not yet considered an crop is often cut down and burned to give way to the expansion of rice fields. Spiny and spineless palms are observed in both ecological environments. The starch yields of spiny and non-spiny sago palms in mesic and hydric areas were assessed to determine which palm ecotype has the highest starch yield. This was to determine the ecotype most suitable for use in propagation experiments. A total of nine sago palm trunks were collected. The spiny-hydric ecotype was not sampled due security concerns in the area where this ecotype is found.

Sago palm trunks approximately nine years old based on the number of leaf scars, but not yet in the stage of inflorescence were the samples for sago starch extraction. The trunks were cut down, sectioned and debarked individually. Sago pith of each trunk was stripped separately. The sago strips were then sun-dried and milled. The crude flour was subsequently sieved through a 60-mesh filtering apparatus to obtain pure sago starch (without the fiber). The starch was transferred correspondingly to labeled sacks. The total starch content per mature plant of ecotypes was weighed.

The Kruskal-Wallis test at 95% confidence level was utilized in comparing the mean starch content of the sago ecotypes used as source of explants for micropropagation. Non-spiny sago palms from a mesic area had the highest mean starch content of 64.3 kg, followed by non-spiny sago palms in hydric area with 63 kg mean starch content, and spiny sago palms in mesic area with 41.3 kg mean starch content (Table **3**).

Figure 2: Spines present on the rachis of sago palm plants growing in Agusan del Sur, Philippines.

Between the collected non-spiny varieties, the total extracted starch was quite low in the hydric area compared to palms in the mesic environment. Previous studies noted that less starch content is present in sago trunks in wet surroundings, especially when flooding occurs. Inundation of sago palm stands could cause oxygen deficiency in the roots of the plants. Lack of oxygen reduces the production of ATP needed in the synthesis of starch [13]. The availability of carbon dioxide for photosynthesis will be limited for plants even partly submerged in water. This is due to slow diffusion of carbon dioxide and carbonic acid (CO_2 and HCO_3) when water becomes stagnant like in the swamps [14]. Thus, with photosynthesis being affected, the production of starch will be limited. However, no significant differences were found among the mean starch yields of the collected sago samples upon the application of Kruskal-Wallis Test at 95% confidence level using SPSS 17.0 (Tables **4a** and **4b**). Thus, it could be deduced that the source of explants for *in vitro* culture can be collected from any type of environment or any sago palm type in Agusan del Sur, Philippines.

Table 3: Total starch content (kg) of sago palms sampled. Palms were approximately nine years old based on number of leaf scars

Ecotype	Starch Content (kg)			Total	Mean
	1	2	3	(kg)	(kg)
Non-spiny- Hydric	78.5	78	32.5	189	63
Non-spiny- Mesic	82	57	54	193	64.3
Spiny-Mesic	51	41	32	124	41.3
			Grand Total	506	

Table 4a: Kruskal-Wallis test of the total starch content and sago ecotype using SPSS version 17.0

Sago Ecotype	N	Mean Rank
1.0	3	5.67
2.0	3	6.67
3.0	3	2.67
Total	9	

Table 4b: Test Statistics [a,b] of the mean starch yield

	Starch Yield
Chi-Square	3.467
Df	2
Asymp.Sig.	0.177

[a] Kruskal Wallis Test; [b] Grouping Variable: Sago ecotype.

STUDIES ON SAGO PALM PROPAGATION

Sago palm is propagated by separation of suckers from the parent palm. Owing to the long life cycle of sago palm which takes an average of 15 years, a need to find alternative means of propagation was deemed necessary by sago palm growers. The first published account of successful micropropagation was that of Alang and Krishnapillay [15] from Univeristi Pertanian Malaysia. Nodular callus formation from young leaf tissues was obtained after five months in culture on modified Murashige and Skoog medium supplemented with high levels of 2,4-dichlorophenoxyacetic acid (2,4-D; 125-175 mg L^{-1}). For about 20 years hence, no further research on sago palm micropropagation was reported in scientific journals although it was no secret that sago starch production in Malaysia has began in commercial scale in 1986, with three organizations at the helm. In Sarawak alone, sago palm plantations covered 28,000 hectares [7]. There is no account as to what percentage of sago palm planting is attributed to plantlets propagated *in vitro*.

Riyadi *et al.,* [16] published accounts of their research efforts on sago palm micropropagation in the Indonesian journal *Menara Perkebunan*. In this study, somatic embryos were produced from embryogenic callus initiated from leaf apical tissues of young suckers on modified Murashige and Skoog medium supplemented with 0.1 mg L^{-1} kinetin and 5 mg L^{-1} 2,4-D. Globular callus formation was reported in as early as four weeks. The number of plantlets regenerated ranged in number from 7.2 – 13 from the various media combinations tested [17]. There was also no report of the plant regeneration efficiency of the reported protocol.

Limited availability of uniform quality planting materials (suckers) hinders the the mass propagation and development of sago palm plantations in Indonesia. Tissue culture efforts led to the production of embryogenic callus ontained from shoot apical tissues grown on Murashige and Skoog's medium supplemented with 10 mg L$_{-1}$ 2,4-D and 1.0 mg L^{-1} kinetin. Shoots were formed at NAA levels of 1.0-2.0 mg L^{-1} [18].

In establishing a new micropropagation system, many factors affecting the growth and development of plant tissues must be considered. In addition to finding the optimum combination of plant growth regulators in culture media, factors such as light intensity, photoperiod, pH and osmotic concentration are also vital. Osmotic potential is considered to be a significant factor in inducing somatic embryogenesis in some crop scpecies. Much of the osmotic potential in tissue culture is attributable to sugars added to culture media [19].

We [20] investigated the role of osmotic potential of media on the early growth stages of sago palm *in vitro*. Significant effects on the growth rate of explants after four weeks of culture under different sucrose and D-sorbitol combinations were noted. A modified Murashige and Skoog (MS) medium supplied with 22.5 g sucrose and 7.5 g sorbitol which created a low osmotic potential environment (-0.457 MPa) caused increases in weights of the explants. This osmotic potential was found optimal for the initial growth of sago palm explants *in vitro*. In contrast, explants cultured in intermediate osmotic potential media (-0.380 MPa) obtained little increase in weight. Results suggested that owing to the sago palm's ecology where it is highly adapted to flooding even by saline water, explants likewise preferred a low osmotic potential *in vitro*.

Seeds of the sago palm are recalcitrant. Thus, there is a need for continuous and sustained effort toward sago palm genetic conservation. Jong [4] has conducted numerous studies on sago palm seed germination. The formation of parthenocarpic fruits was often noted from sterile palms. Further, seeds were found to germinate only when fully ripe as exhibited by their large size.

A study on the improvement of percentage of seed germination of sago palm was conducted *in vitro via* embryo rescue. A high embryo survival rate of 83% was obtained from embryos inoculated on MS medium with 5 mg L^{-1} BAP [21].

The increasing recognition of sago palm as a potential food source creates a demand for the availability of seedlings. Sago palm is commonly propagated through seedlings and suckers but both are only available when the mother palm plant is several years old. Sago palm plants can be generated by seeds. However, the plants flower only every 10 to15 years. The difficulty is compounded by the fact that the sago trunk needs to be cut down in order to harvest the starch. The percentage of seed germination is also very low. Therefore, unconventional methods such as propagation through *in vitro* techniques are highly desirable. In addition to a faster rate of producing new and uniform planting material, *in vitro* techniques will continue to be useful in the establishment of a heterogeneous plant population.

ACKNOWLEDGEMENTS

The author would like to thank the Department of Science and Technology- Philippine Council for Advanced Science and Technology Research and Development (DOST-PCASTRD) and the University of the Philippines for funding the starch yield experiment reported in this paper. The assistance of Dr. Dulce Flores, Mr. Efren Tutor, Ms. Joan Acaso and Aileen Grace Delima in the conduct of the starch yield test is highly appreciated.

ABBREVIATIONS USED

PAGE : Polyacrylamide gel electrophoresis

AFLP : Amplified fragment length polymorphism

2,4-D : 2,4-dichlorophenoxyacetic acid

NAA: 1-Naphthaleneacetic acid

MS : Murashige and Skoog

BAP: Benzyl amino purine

REFERENCES

[1]   Rauwerdink JB. An essay on Metroxylon, the sago palm. Principes 1986; 30: 165-180.

[2]   Hisajima, H. In: Metoxylon sagu Rottb. (Sago Palm). In: Bajaj YPS, ed. Biotechnology in forestry and agriculture v. 34. Springer-Berlag 1996; pp. 217-230.

[3]   Flach, M. Sago palm. Metroxylon sagu Rottb. In: Promoting the conservation and use of underutilized and neglected crops. 13. Institute of Plant Genetics and Crop Plant Research. Rome, Italy; 1997.

[4]   Jong FS. Studies on the seed germination of sago palm metroxylon sagu) towards greater advancement of the sago industry in the 90s. In: Ng TT, Tie YL, Kueh HS (Eds.) Proc 4th Int Sago Symp, Kuching, Sarawak, Malaysia 1991; pp 88-93.

[5]   Flach M, Schuiling DL. Revival of an ancient starch crop: a review of the agronomy of the sago palm. Agroforestry Systems 1989; 7: 259-281.

[6]   Karim, AA, Pei-Lang Tie, Manan DMA, Zaidul ISM. Starch from the sago (Metroxylon sagu) palm tree- properties, prospects, and challenges as a new industrial source for food and other uses. Comprehensive Reviews in Food Science and Food Safety 2009; 7 (3): 215-228.

[7]   Bin Sundin N, Rahman KAAA. Composite sago waste in sound absorber panel making.Proceedings of the 2nd ICACA, Universiti Malaysia Sarawak 2008; 2: 91-92.

[8]   Flach, M. The Sago Palm. FAO Plant Production and Protection Paper 47, AGPC/MISC/80. FAO, Rome; 1983.

[9]   Boonsermuk S, Hisajima S, Ishizuka K. Isozyme variation in sago palm (*Metroxylon* spp.) in Thailand. Japanese Journal of Tropical Agriculture 1995; 39 (4): 229-235.

[10]   Ehara H, Susanto S, Mizota C, Hirose S, Matsuno T. Sago palm (*Metroxylon sagu,* Arecaceae) production in the Eastern archipelago of Indonesia: variation in morphological characteristics and pith dry matter yield. Economic Botany 2000; 54(2): 197-206.

[11]   Kjaer A, Barford AS, Asmussen CB, Seberg O. Investigation of genetic and morphological variation in the sago palm (*Metroxylon sagu*; Arecaceae) in Papua New Guinea. Annals of Botany 2004; 94: 109-117.

[12]   Ellen R. Local knowledge and management of sago palm (*Metroxylon sagu* Rottboell) diversity in South Central Seram, Maluku, Eastern Indonesia. Journal of Ethnobiol 2006; 26 (2): 258-298.

[13]   Rolletschek, H., Weschke,W.,H. Weber, U.Wobus, and L.Borisjuk. .Energy state and its control on seed development: starch accumulation is associated with high ATP and steep oxygen gradients within barley grains. J. Exp. Bot. 2004; 55(401): 1351-9.

[14]   Smith, M.R. Oxygen Production During Photosynthesis in Aquatic Plant *Myriophyllum hippuroides*. Science One Research projects . The University of British Columbia 2009; Available from: https://circle.ubc.ca/handle/2429/8042.

[15]   Alang ZC, Khrisnapillay B. Somatic embryogenesis from young leaf tissues of the sago palm- *Metroxylon sagu*. Plant Tissue Culture and Letters 1987; 4(1): 32-34.

[16]   Murashige T, Skoog F. A revised medium for rapid growth and bio assays with tobacco tissue cultures. Physiol Plant 1962; 15: 473-497.

[17]   Riyadi I, Tahardi JS, Sumaryono. The development of somatic embryos of sago palm (*Metroxylon sagu* Rottb.) on solid media. Menara Perkebunan 2005; 73 (2): 35-43.

[18]   Tajuddin T, Karyanti K, Minaldi M, Haska N. the development of ex vitro and *in vitro* culture of sago palm (Metroxylon sagu Rottb.). In: Toyoda Y, Okazaki M, Quevedo M, Bacusmo J, Eds. Sago: its Potential in Food and Industry, Tokyo, TUAT Press, 2001\7; pp. 231-235.

[19]   George, EF. Plant propagation by tissue culture – Part 1: The Technology. Exergetics Limited Edington, Wiltshire, United Kingdom, 1993.

[20]   Novero A, Delima AG, Acaso J, Baltores LM. The influence of osmotic concentration of media on the growth of Sago Palm (*Metroxylon sagu* Rottb.) *in vitro*. Aust J Crop Sci 201 4(6): 453-456.

[21]   Ibisate MT, Abayon EI. Regeneration and conservation of the sago palm in Panay Island, Philippines through *in vitro* techniques. In: Toyoda Y, Okazaki M, Quevedo M, Bacusmo J (eds). Sago: its potential in food and industry. Proc. 9th Intl. Sago Symp. Visayas State Univ., Philippines 2007; pp. 195-200.

CHAPTER 5

Tools for Generating Male Sterile Plants

Sudhir P. Singh[1,2,*], Joy K. Roy[1,2], Dinesh Kumar[3] and Samir V. Sawant[1]

[1] *National Botanical Research Institute, Council of Scientific and Industrial Research, Lucknow, India-226 001;* [2]*Current Address:National Agri-Food Biotechnology Institute, Department of Biotechnology, Mohali, India and* [3]*University of Lucknow, Lucknow, India*

Abstract: Most higher plant species are hermaphroditic and development of efficient pollination control systems is highly desirable in maintaining economical and convenient outbreeding in plants for hybrid development. Over the last two decades, several pollination control systems have been developed by genetic-engineering of nuclear or cytoplasmic encoded genes. Some of the strategies are in practical use or ready to be used in agriculture and many of them have significant importance in future hybrid-breeding programs. For the successful application of these systems, future research should take into account efficient and economical production of male-sterile female parent lines and biosafety majors.

Keywords: Male sterility, gene expression system.

INTRODUCTION

Male sterility is the failure of plants to produce functional male gametes. The phenotype of male sterility may include one or more of the following aberrations: complete absence of male organs, undeveloped normal sporogenous tissues, failure of meiosis, abortion of pollen at any step of its development, failure of anther dehiscence, inability of mature pollen to germinate on compatible stigma, and barriers other than incompatibility preventing pollen from reaching ovule [1-3]. The conversion of stamens to different type of floral organs also represents a male sterility condition [4].

Male sterility in plants has received considerable attention because of its potential value in breeding and hybrid development programs. Moreover, it is of tremendous importance in molecular and developmental studies of stamen and pollen grains and evolutionary studies on the origin of dioecy [4].

The processes of development of microspores, pollen release followed by gametic union and seed formation are time sequenced–programmed gene-controlled functions. Breakdown of any one of these functions occurs due to mutation of the genes controlling them [5]. One such spontaneous event, mutation of either nuclear or cytoplasmic genes or both, causes male sex abolition and leads to the evolution of new sex types, the male sterile. Naturally occurring male sterile plants or male sterile female lines developed through plant breeding rely on two genetic means of pollination control: (i) genic male sterility (GMS), which is often governed by a single recessive gene and (ii) cytoplasmic male sterility (CMS), which is mainly caused by mitochondrial abnormalities and demonstrates a maternal inheritance pattern. The CMS systems, where a nuclear gene for fertility restoration is known, are termed as cytoplasmic genic male sterile (CGMS) systems. CGMS has been greatly exploited for the production of hybrids in crop plants [6]. However, development of male sterile lines through conventional breeding faces several challenges and there are limitations of their application, discussed in next sections of the chapter. Biotechnology interventions have added several new possibilities to obtain male-sterile plants. The genetically engineered genes for crops seems promising to induce agricultural growth. However, an efficient fertility restoration system is yet to be developed and validated. Many public and private sectors are engaged in agro-biotech research and development programs worldwide. Biotechnology research involves substantial investments hence the trend is towards joint ventures between public and private. Despite the immense possibilities of

*****Address correspondence to Sudhir P. Singh:** National Agri-Food Biotechnology Institute, Department of Biotechnology, Mohali, India; Phone: +91- 05222297944; Fax:+91- 05222297941; E-mail:pratapsudhir@yahoo.co.in

transgenic crops in increasing yield and improving quality, by using the molecular tools of fertility control, there are serious concerns regarding bio-safety. Environmentalists and the public, in general, have genuine concerns over the transgenic technology. Bio-safety and regulatory approvals especially in the case of genes from non-plant sources remain a major challenge. In the biotechnological strategies, use of genes of plant origin should be promoted as it will face less regulatory issues. In the present review our major focus is to discuss the efforts to achieve male sterility in plants.

DISCOVERY OF MALE STERILITY IN PLANTS

First documentation of male sterility came in 1763 when Kolreuter observed anther abortion within species and specific hybrids, after 69 years of the detection of sex in higher plants by Camerarius (1694). Male-sterility-inducing cytoplasms have been known for over 100 years [9]. In 1921, Bateson and Gairdner [10] reported that male sterility in flax was inherited from the female parent. Chittenden and Pellow observed in 1927 [11] that male sterility in flax was due to an interaction between the cytoplasm and nucleus. In 1943, Jones and Clarke [12] established that male sterility in onion is conditioned by the interaction of the male-sterile (S) cytoplasm with the homozygous recessive genotype at a single male-fertility restoration locus in the nucleus. The authors also described the technique used today to exploit cytoplasmic-genic male sterility (CMS) for the production of hybrid seeds.

CLASSIFICATION OF MALE STERILITY IN PLANTS

Kaul [13] examined a number of events leading male sterility in plants and suggested a comprehensive classification. We present here a modified classification of male sterility.

PHENOTYPIC BASIS OF MALE STERILITY

Irrespective of genetic constitution, certain phenotypic abnormalities in anther development, form and function lead to the following types of male sterility.

Structural Male Sterility

The male reproductive organs are either completely absent or deformed or malformed so that no microsporogenous tissue is developed.

Sporogenous Male Sterility

The male reproductive organs develop normally but pollen are either completely absent or non-functional or extremely scarce.

Functional Male Sterility

Viable pollen are formed but they are incapable of fertilization due to some barriers. The main barriers are anther indehiscence faulty or no exine formation, inability of pollen to migrate to stigma or affect fertilization.

GENOTYPIC BASIS OF MALE STERILITY

Depending on the nuclear and cytoplasmic genome constitution, male sterile lines rely on two genetic means of pollination control:

Genic Male Sterility (GMS)

GMS is ordinarily governed by a single recessive gene, ms [14], but dominant genes are also known [15].

Cytoplasmic Male Sterility (CMS)

CMS is caused by rearrangements of mitochondrial genome and shows a maternal inheritance pattern [16]. It is common in higher plants [17]. The CMS, where a nuclear gene for restoring the fertility is known are called as cytoplasmic genic male sterile (CGMS). This is the most common approach for the production of hybrids in crop plants [6].

Environment Sensitive Male Sterility

Rick [18], Martin and Crawford [19] and Shi [20, 21] were the first to report on the occurrence of temperature and light sensitive reversible male sterile mutants in tomato and pepper. This is a type of GMS where the recessive gene causes male sterility at temperature higher or lower to the critical point. Low temperature has been shown to induce male sterility in rye [22, 23]. In case of rice, both low [24] and high temperature [25] have been reported to induce male sterility. In sorghum also, temperature dependent male sterility has been reported [26]. Photoperiod sensitive male sterility also exists. In this case the expression of male sterility related gene (genic) is affected by photoperiod. This type of male sterility is reported in rice [21]. Several photoperiod sensitive, temperature sensitive, photoperiod-influenced temperature sensitive, and *vice versa* have been identified worldwide [27].

ARTIFICIALLY INDUCED MALE STERILITY

Male sterility can artificially be induced by following three modes:

Chemical Induced Male Sterility

Male sterility can be artificially induced into flowering plants by chemical treatment. Some chemicals used to produce male sterility in crops are Sodium 2,3-dichloroisobutyrate [28], gibberellic acid [29], naphthalene acetic acid, maleic hydrazide, Ethrel [30], FW 450, sodium 1-*p*-chlorophenyl-1,2-dihydro-4,6-di-methyl-2-oxonicotinate (RH-531) [31, 32], mitomycin and streptomycin [33], disodium methanearsonate (DSMA) [34], Bialaphos (glutamine synthetase inhibitor) [35], Sodium methyl arsenate, and Zinc methyl arsenate.

Or X-Rays Inactivation

Melchers *et al.,* [36] treated mesophyll protoplasts of *Lycopersicon esculentum* with iodoacetamide to inactivate mitochondria and irradiated protoplasts of *Solanum tuberosum* with Ƴ or X-rays to inactivate nuclei. The two protoplasts were fused to produce male sterile heterologous hybrids.

TRANSGENIC MALE STERILITY

Male sterility may be achieved by delivery of candidate gene/s into the genome of a plant by genetic engineering. This type of male sterility has been discussed in detail in later part.

The classification of male sterility is summarized in Table **1**.

Table 1: Classification of male sterility

Phenotypic Basis of Male Sterility		**Genotypic Basis of Male Sterility**	
Sl. No.	**Types**	**Sl. No.**	**Types**
1	Structural male sterility: Completely absent or deformed or malformed male organs	1	Genic male sterility (GMS): Governed by (recessive or dominant) nuclear gene
2	Sporogenous male sterility: Normal male organs, pollen is either completely absent or non-functional or extremely scarce	2	Cytoplasmic male sterility (CMS): Caused by rearrangements of mitochondrial or sometimes plastid genome
3	Functional male sterility: Normal male organs with viable pollen, physical or physiological barriers in fertilization barriers		

Table 1: cont….

Environment Sensitive Male Sterility		Artificially Induced Male Sterility	
1	Temperature sensitive: Low or high temperature induces male sterility	1	Chemical induced male sterility
2	Photoperiod sensitive: Expression of male sterility trait is affected by photoperiod	2	ϒ or X-rays inactivation: Use of ϒ or X-rays to inactivate nuclei of male gamete
3	Photoperiod-influenced temperature sensitive	3	Transgenic male sterility: Delivery of candidate gene/s into the genome of a plant by genetic engineering
4	Temperature -influenced photoperiod sensitive		

USE OF MALE STERILE PLANTS

The occurrence of male sterility was initially considered as an undesirable anomaly. Therefore, the selection value of male sterile plants was taken to be negligible and most of them were eliminated through natural selection because of either no or little seed set. However, as the appreciation of hybrid vigour grew where male sterility plays important role, methods were urgently required that would eliminate the tedium of hand emasculation. This requirement prompted male sterility to become an asset and a breeder's tool for hybrid development [13]. Hybrid plants display superior plant growth, higher yield, better quality, and a pronounced stress resistance. Crop uniformity is another advantage of hybrid plants when the parents are homozygous; this leads to improved crop management. Therefore, hybrid plants have become increasingly important in various commercial food crops around the world. In crops such as maize, sunflower, sorghum, sugar beet, cotton, and many vegetables, hybrids account for a large share of the market. Not only the USA and Europe, but also many developing countries rely on their food production to a large extent on hybrids. Hybrid markets account for nearly 40 per cent of the global commercial seed business of about US$ 15 billion [37]. In crops such as maize, sunflower, sorghum, sugar beet, cotton, and many vegetables, hybrids account for a large share of the seed market. Hybrid seed technology is considered as major factor responsible for the dramatic rise in agricultural output during the last half of the 20th century [38]. The production of hybrid varieties of maize (from the thirties in the US), cotton (since 1970 in India) and rice (since 1976 in China) represents the most significant and successful breeding efforts of the twentieth century. A 6-fold increase was observed between 1930 and 1990 for US corn yield after the introduction of hybrid breeding, compared to uniform performances for selected open pollinated populations during the previous 60 years [38]. Hybrid seed is produced by cross-pollination of genetically different parental lines. The key to the economically viable production of hybrid seeds requires sufficient control of the pollination process to avoid selfing of the female line. In conventional method, male parts are manually removed from bisexual flower to convert it into female parent. Use of male sterile plants is very demanding to avoid manual emasculation of female parents, as ~40% of the total labour is expended for this task, and to avoid self pollination [36]. The Figs. **1** and **2** explains the process of hybridization involving tricky emasculation. Thus, male sterile plants help in reducing the labour cost, manual damages to the ovary, and other human errors. Another importance of male sterile plants, apart from hybrid breeding, is their use as a biological safety method for increasing number of genetically modified organisms used in field trials and agricultural production [14]. Male sterile plants also may be useful to floriculture industry, since male sterile plants have more flowers than fertile plants [39]. In ornamental crops, the inability of a plant to set seed could reduce the need for dead heading to extend the flowering period, eliminate nuisance fruit, remove allergy producing pollen and increase flower longevity and prevent gene flow between genetically modified and related native plants [40, 41].

A further importance of male sterile plants, apart from hybrid breeding, is there use as a biological safety method for increasing number of genetically modified organisms used in field trials and agricultural production [14]. Male sterile plants also may be useful to floriculture industry, since male sterile plants have more flowers than fertile plants [39]. In ornamental crops, the inability of a plant to set seed could reduce the need for dead heading to extend the flowering period, eliminate nuisance fruit, remove allergy producing pollen and increase flower longevity and prevent gene flow between genetically modified and related native plants [40, 41].

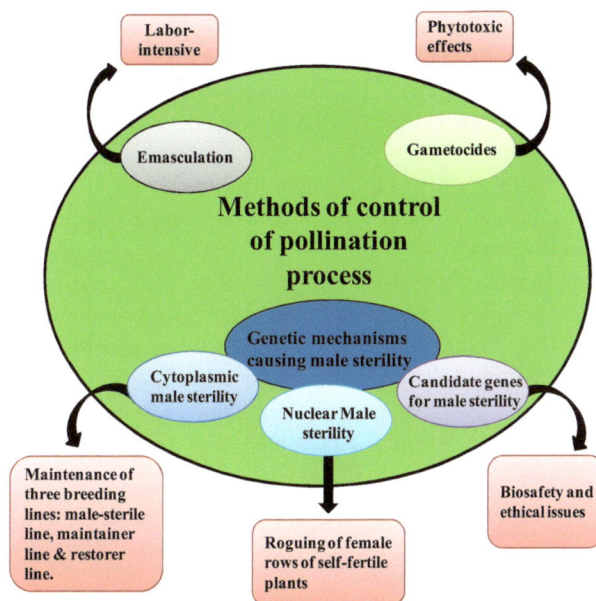

Figure 1: Methods of control of pollination processes and their drawbacks

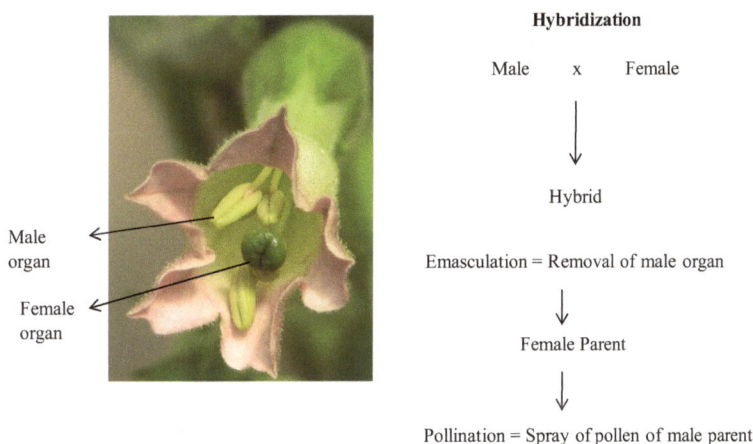

Figure 2: A bisexual flower (of tobacco) and schematic presentation of hybridization.

BIOTECHNOLOGICAL TOOLS IN DEVELOPING MALE STERILE PLANTS

Area of biotechnology has added new possibilities of several pollination control systems, broadening the potential to produce hybrid seeds for many crops [42]. In recent time, transgenic approach is getting popularity for developing male sterile plants [8]. Since the first transgenic male sterility system was described [7], many strategies to produce male-sterile plants have been reported. These male sterility biotechnological tools can be grouped into following classes:

SELECTIVE ABLATION OF MALE REPRODUCTIVE TISSUES

Systems that rely on the selective removal of tissues needed for the production of functional pollen by the tissue-specific expression of a gene encoding a protein able to disrupt cell function [7, 41, 43-61]. Inducible sterility was also obtained by expressing a gene encoding a protein that catalyzes the conversion of a pro-herbicide into a herbicide only in male reproductive tissues [62-64] or by engineering the male sterility gene in such a way that its expression can be induced by the application of a chemical [65]. Chemical regulation of the expression of a gene that can complement the male sterility (fertility restorer gene) or of a repressor of the male sterility gene can be used to induce fertility [51, 66-69]. Hird *et al.,* [70] demonstrated restoration of fertility by doing transgene silencing of ß-1,3-glucanase, mediated by T-DNA.

ALTERATION OF BIOCHEMICAL PATHWAYS

Another approach to engineer male sterility is to introduce a gene, or set of genes, able to alter the levels of metabolites needed for the production of viable pollen, such as amino acids [71] sugars [72], flavonols [73], jasmonic acid [74-76], biotin [77] or auxins [78]. Vander Meer *et al.,* [79] inhibited flavonoid biosynthesis by using antisense approach in petunia anthers and it resulted male sterility. Inducible fertility can be achieved for some of these systems by application of the missing metabolite [74-77, 80, 81]. Hofig *et al.,* [82] expressed grapevine stilbene synthase (STS) gene in tapetum and observed male sterility in *Pinus radiate.* They observed, in tapetum STS competes with the enzyme chalcone synthase for substrate malonyl-CoA and coumaroyl-CoA and this results cytotoxicity due to abolished flavonol biosynthesis. Ribarits *et al.,* [83] did metabolic engineering of glutamine to achieve reversible male sterility in plants.

TOXIC ACTION OF TWO DNA FRAGMENTS, IN COMBINATION

Male sterility can also be generated by the combined action of two genes brought together into the same plant by crossing two different parental lines each expressing one of the genes [84-88]. An example of this approach is the reconstitution of an active toxin from two inactive toxin fragments each encoded by a separate transgene. The combined expression of both transgenes in the male reproductive tissues can result in male sterility for example splitting of *Barnase* gene into two non-functional fragments, fused to trans-splicing inteins, jointly coding for male sterility in heterozygous plants.

NATURAL OR INDUCED MUTATION

The possibility of making use of natural or induced mutations that cause male sterility has also been described. In this case, the wild-type allele is used as a fertility restorer gene, and its expression is controlled by the application of chemicals [67].

DELAYED TAPETUM DEGENERATION

Takada *et al.,* [89, 100] adopted a strategy of delayed tapetum degeneration to induce pollen abortion. They found that constitutive or tapetum specific expression of melon ethylene receptor genes (ETR1/H69A and ERS1/H70A) delays tapetum generation followed by male sterility. Kawanabe *et al.,* [98] showed that expression of *Arabidopsis Bax Inhibitor-1 (AtBI-1)* gene, a homologue of *mammalian Bax Inhibitor-1* gene, in tapetum inhibited its degeneration and subsequently resulted pollen abortion. Recently our group observed male sterility by abrogation of cell death programme of tapetum, high level expression of *BECLIN 1* gene in tapetal cells of tobacco plant [103].

PREMATURE DISSOLUTION OF THE MICROSPOROCYTE CALLOSE WALL

Microsporocytes synthesize a cell wall consisting of callose, a β-1,3-linked glucan, between the cellulose cell wall and plasma membrane. After the completion of meiosis and the initiation of microspore exine wall formation, the callose wall is broken down by callase, a tapetally secreted P-1,3-glucanase activity [136], releasing free microspores into the locular space. Worrall *et al.,* [99] and Tsuchiya *et al.,* [104] adopted a strategy of premature (before meiosis) dissolution of the microsporocyte callose wall- by expressing a modified pathogenesis-related vacuolar β-1,3-glucanase in tapetum- followed by sterile pollen development.

ANTHER INDEHISCENCE

A different approach of male sterility induction is anther indehiscence. Beals and Goldberg [105] utilized a new cell ablation strategy to ablate specific anther cell types involved in the dehiscence process of anthers. Recently, Fernando *et al.,* [106] expressed gamma carbonic anhydrase 2 in *Arabidopsis* and observed male sterile phenotype due to anther indehiscence.

EXPRESSION OF MITOCHONDRIAL OPEN READING FRAMES, DETERMINING CMS

Discovery of mitochondrial open reading frames (ORFs) determining CMS led development of male sterility induction system with nuclear genome transgenic approach. Many ORFs have been uncovered in different plants *viz* Maize, Petunia, Sunflower, Common bean, Radish *etc.* [137]. Some of the CMS associated genes like *atp9* [107] *orf239* [109], *orf456* [110], *orf129* [111], *orfH522* [6] were introduced with mitochondrial targeting sequence, placed under the control of stamen specific promoter, into the plant's nuclear genome and transgenic expression of those genes resulted male sterility.

PLASTID ENCODED ENGINEERED MALE STERILITY

Ruiz and Daniel [9] for the first time reported engineering of plastid encoded cytoplasmic male sterility. They expressed acinetobacter b-ketothiolase (polyhydroxyl alkanoates) or *phaA* gene in tobacco plastid. The transgenic lines were normal except for the male-sterile phenotype. Reversibility of the male-sterile phenotype was observed under continuous illumination, resulting in viable pollen and copious amount of seeds.

SUPPRESSION OF EXPRESSION OF PROTEINS IMPORTANT FOR MALE GAMETOPHYTE DEVELOPMENT

Another approach is suppression of genes specific to or required for anther or pollen development through targeted sense or antisense or RNA interference down regulation. Expression of antisense SnRK1 protein kinase- regulator of carbon metabolism- caused abnormal pollen development and male sterility in transgenic barley [138]. Yui *et al.,* [113] performed antisense inhibition of mitochondrial pyruvate dehydrogenase E1 alpha subunit in anther tapetum and observed male sterility. Mou *et al.,* [112] and Preston *et al.,* [114] suppressed expression of PEMAT and MYB32 genes by using antisense silencing approach and observed male sterility. Luo *et al.,* [57] used antisense of RTS gene to induce male sterility in Rice. Sandhu *et al.,* [115] performed suppression of a nuclear gene, *Msh1* which is supposed to be involved in substoichiometric shifting (SSS) of mitochondrial genome, in tobacco and tomato by using RNAi strategy and observed pollen sterility through CMS. A transcription factor, AtMYB103, is essential for pollen development by regulating tapetum development [119]. Block of glutamine synthetase activity in the developing pollen can induce complete male sterility [118]. Down-regulation of the *AtMYB103* gene in tapetum commenced male sterility [117, 119] and fertility was restored by expressing *AtMYB103* [117]. Xing and Zachgo [122] silenced Agamous-like18 (AGL18), a MADS-box gene, and attained male sterility in *Arabidopsis*. Rice UDP-glucose pyrophosphorylase (UGPase) genes are essential for pollen development and suppression of UGPAse genes is shown to be a tool to induce male sterility in rice [123, 124]. Ma *et al.,* [128] reported *Zm401* as one of the key growth regulators in anther development and showed that knockdown of *Zm401* significantly affected the development of the microspore and tapetum, and finally male-sterility. Wang and Li [129] found that suppression of acyl-CoAsynthetase (ACS) gene, *GhACS1*, severely affected tapetum and consequently blocked normal microsporogenesis in early anther development. Characterization of *PS2* gene by Gorguet *et al.,* (2009) suggested that repression of *PS-2* homologues may be a potential way to introduce male sterility in plants. Wang *et al.,* [130] observed that decreased expression of programmed cell death gene, *OsPDCD5*, caused by antisense technology could induce pollen sterility in photoperiod-sensitive rice.

A similar approach is male sterility commencement by introducing mutation in certain genes. Wan *et al.,* [120] demonstrated that mutation in one lectin receptor-like kinase (At3g53810) resulted defects in pollen development which led to male sterility in *Arabidopsis*. Xing and Zachgo [122] showed that loss of function of two Glutaredoxins genes, *ROXY1* and *ROXY2*, resulted in defective sporogenous cell formation in anthers and defective tapetum differentiation. Some reports demonstrated role of an autophagy related gene, ATG6 in pollen germination and observed that disruption of ATG6 gene led to male sterility [125-127]. Yang *et al.,* [131] characterized a novel gene of *Arabiodopsis*, *THERMOSENSITIVE MALE STERILE1 (TMS1)* and observed that a knockout mutation in *TMS1* greatly retarded pollen tube growth.

EXPRESSION OF REGULATORY PROTEINS

Kurek *et al.*, [132] obtained male sterility by expressing prolyl isomerase domains of FKBP73 (cyclophilin and parvulin families genes) in transgenic rice and suggested association of the domains with a novel interaction with anther specific target proteins. Expression of a fission yeast cell-cycle regulator, Spcdc25, in male reproductive tissues of wheat results in reduction of pollen viability [133]. SET-domain proteins are presumed to be epigenetic regulators of gene expression and chromatin structure. Thorstensen *et al.*, [134] reported that inducible over-expression of a SET domain protein from Arabidopsis, ASHR3, resulted aborted anthers development and degenerated pollen, leading to complete male sterility. Tzeng *et al.*, [135] observed short stamens after expressing lily putative p70[s6k] *LS6K1* gene in *Arabidopsis*.

COMMERCIAL APPLICATIONS

Although a number of biotechnology based male sterility stystems have been developed (Table **2**), only the *Barnase/Barstar* system, developed by the Aventis Crop Science Company (formerly Plant Genetic Systems) in 1990s has been predominantly used in practice. Aventis Crop Science used this male sterility system to develop Canola (*Brassica napus*) hybrids (Table **3**). Bejo Zaden BV Company developed genetically engineered *Cichorium intybus* (Chicory) by using *Barnase* system (Table **3**). Bayer Crop Science developed transgenic male sterile lines of maize.

Table 2: Biotechnological strategies for generating male sterile lines

Sl No.	Strategy	Candidate Gene for Male Sterility	Source of Candidate Gene	Plant	References
1	Early degradation of tapetum layer which provides nutrition and structural components to pollen	*Barnase*	*Bacillus amyloliquefac-iens*	Tobacco,	[26, 89, 90]
				Brassica napus	[90, 91]
				Arabidopsis	[90, 92-94]
				Wheat	[91]
				Brassica juncea	[95]
				Agrostis stolonifera	[93]
				Oryza sativa	[93]
				Solanum lycopersicon	[90]
				Maize	[96]
				Brassica campestris	[59]
				Pepper	[97]
				Creeping bent grass, *Kalanchoe blossfeldiana*	[41]
		RNase T-I	*Aspergillus oryzae*	*Brassica napus*	[45]
		Diphtheria toxin a-chain	pLAT59-DTM	*Arabidopsis*,	[46, 55, 60]
				Cabbage (*B. oleracea*)	[54]
				Tobacco	[52]
		Ribosome Inactivating Protein	*Dianthus sinensis*	*Arabidopsis*	[98]
		BAX	Mouse	Tobacco	[99]
2	Premature Dissolution of the Microsporocyte Callose Wall	*β-1,3-glucanase*	Tobacco	Tobacco	[70]
			Brassica napus,	*Arabidopsis*	[98]

			Arabidopsis		
3	Disruption of tapetal PCD	*Arabidopsis* Bax inhibitor-1	*Arabidopsis*	Tobacco	[89, 100]
		Ethylene receptor gene *Cm-ETR1/H69A*	Melon	Tobacco	[101]
		mutated ethylene receptor gene (*Cm-ERS1/H70A*	Melon	Lettuce (*Lactuca sativa*)	[102]
				Arabidopsis	[61]
		Cysteine proteases gene BoCysP1	*Brassica* sp.	Tobacco	[103]
		AtBeclin 1	*Arabidopsis*	Tobacco	[62]
4	Male reproductive organ specific expression of an enzyme which converts a non toxic compound into toxic compound	*Cytochrome P450$_{sul}$*	*Sfreptomyces griseolus*	Tobacco	[63]
		argE	*Escherichia coli*	*Arabidopsis thaliana*	[64]
		Phosphonate monoester hydrolase	*Burkholderia caryophilli*	Maize	[65, 68]
5	Chemical induced expression of a candidate gene for male sterility	*Barnase*	*Bacillus amyloliquefaci-ens*	Maize	[51]
		DNA adenine methylase (DAM) gene	*Escherichia coli*	Tobacco	[69]
		glutathione-S-transferase II (GSTII	Maize	Tobacco	[99]
6	Premature dissolution of the microsporocyte callose wall	pathogenesis-related vacuolar β-1,3-glucanase	Tobacco	Tobacco	[104]
		pathogenesis-related endo- β-1,3-glucanase	Soyabene	*Petunia hybrida*	[79]
7	Altering the levels of metabolites needed for the production of viable pollen	*Antisense chalcone synthase (chs) gene*	*Petunia hybrida*	Tobacco	[78]
		Expression of *rolB gene*	*Agrobacterium rhizogenes*	*Arabidopsis*	[75]
		Mutation of fad3-2, fad7-2, fad8 genes	*Arabidopsis*	Tobacco	[77]
		Expression of Avidin	Chicken	*Arabidopsis*	[76]
		T-DNA insertion of 12-oxophytodienoate reductase gene	*Arabidopsis*	Tobacco	[72]
		Antisense Nin88	Tobacco	Tobacco, Potato	[71]
		Antisense acetolactate synthase (ALS)	Potato	*Pinus radiate*	[82]
		Expression of *stilbene synthase (STS) gene*	Grape vine	Tobacco	[83]
		Inactivation of *cytoplasmic glutamine synthetase*	Tobacco	Tobacco and *Brassica napus*	[84]
8	Two genes- brought together into the same plant by crossing two different parental lines- in combination produce cytotoxic product in male organ	*Indole acetamide hydrolase (IamH)* and *Indole acetamide synthase (IamS)*	*Agrobacterium tumefaciens* and *Pseudomonas savastanoi*	Tobacco	[85]
		amino-terminal barnase and *carboxy-terminal barnase*	*Bacillus amyloliquefaci-ens*	Tomato	[86]
				Triticum astivum	[87]
				Arabidopsis	[88]
				Tomato	[85]
		Avr9 and *Cf9*	*Cladosporium fulvum* and *Lycopersicon esculentum*	Tobacco	[105]
9	Anther indehiscence	*Barnase-Barstar*	*Bacillus amyloliquefaci-ens*	*Arabidopsis*	[106]
		Gamma carbonic anhydrase 2	*Arabidopsis*	Tobacco	[107]

Table 2: cont….

10	Engineered cytoplasmic male sterility	*atp9* (unedited)	Wheat	*Arabidopsis*	[108]
				Tobacco	[109]
		orf239	Common Bean	Tobacco	[9]
		acinetobacter b-ketothiolase (polyhydroxyl alkanoates) or phaA	*Acinetobacter*	*Arabidopsis*	[110]
		orf456	*Capsicum annuum*	Tobacco	[111]
		orf129	Wild Beet	Tobacco	[6]
		orfH522	Sunflower	*Arabidopsis*	[112]
11	Suppression of genes specific to anther or pollen development	Antisense *phosphoethanolamine N-methyltransferase*		Sugar Beet	[113]
		Antisense *Mitochondrial pyruvate dehydrogenase E1 alpha subunit*		*Arabidopsis*	[114]
		T-DNA insertion *AtMYB32* gene		*Oryza sativa*	[57]
		Antisense of *RTS* gene		Tomato, Tobacco	[115]
		RNAi of *MSH1* gene		*Arabidopsis*	[116]
		transposon insertion of *dysfunctional tapetum1 (dyt1)* gene		*Arabidopsis*	[117]
		AtMYB103EAR (AtMYB103 gene with 12-amino-acid EAR suppressor)		Tobacco	[118]
		Dominant negative mutant of *glutamine synthetase* gene		*Arabidopsis*	[119]
		Ethyl-methane sulfonate (EMS) mutation in *AtMYB103* gene		*Arabidopsis*	[120]
		T-DNA insertion of *Lectin receptor-like kinases*		*Arabidopsis*	[121]
		T-DNA insertion of *Drl1-1* and *Drl1-2*		*Arabidopsis*	[122]
		RNAi of *Agamous-like18 (AGL18)*		Rice	[123, 124]
		RNAi of *UDP-glucose pyrophosphorylase (UGPase)*		*Arabidopsis*	[125-127]
		T-DNA insertion of *ATG6/AtBECLIN 1*		*Arabidopsis*	[122]
		T-DNA insertion of *ROXY1* and *ROXY2* genes		Cotton	[128]
		RNAi of *zm401*		Cotton	[129]
		RNAi of *GhACS1* gene		Rice	[130]
		Antisense of *OsPDCD5*		*Arabidopsis*	[131]
		T-DNA insertion of *THERMOSENSITIVE MALE STERILE1*		Rice	[132]
12	Expression of regulatory proteins	prolyl isomerase domains of FKBP73	Wheat	Wheat	[133]
		Spcdc25	Yeast	*Arabidopsis*	[134]
		trxG protein ASHR3	*Arabidopsis*	*Arabidopsis*	[135]
		LS6K1	*Arabidopsis*	*Arabidopsis*	[135]

The US forest biotechnology company ArborGen has developed male sterile eucalyptus, by using *Barnase* gene, and has got the permission from USDA/APHIS (United States Department of Agriculture/Animal and

Plant Health Inspection Service) to undertake open field trial. In India, M/s Proagro PGS (India) Ltd. is developing superior hybrid cultivars of Mustard and Cauliflower by using the Barnase tool. Delhi University (India) has developed male-sterile and restorer lines, harbouring the *Barnase/Barstar* sytem, for hybrid seed production [139].

Table 3: Commercial use of biotechnological tool for male sterility in crop (taken from AGBIOSE database, http://www.agbios.com/dbase.php)

1.	*Brassica napus* (Argentine Canola)		
	Event	**Company**	**Description**
	MS1, RF1	Aventis CropScience (formerly Plant Genetic Systems)	MS lines contained the *barnase* gene, RF lines contained the *barstar* gene and both lines contained the phosphinothricin N-acetyltransferase (PAT) encoding gene from *Streptomyces hygroscopicus*.
	MS1, RF2	do	do
	MS8xRF3	Bayer CropScience (Aventis CropScience(AgrEvo)	do
	PHY14, PHY35	Aventis CropScience (formerly Plant Genetic Systems)	Male sterility was *via* insertion of the barnase ribonuclease gene, fertility restoration by insertion of the barstar RNase inhibitor, PPT resistance was *via* PPT-acetyltransferase (PAT) from *Streptomyces hygroscopicus*.
	PHY36	do	do
2.	*Cichorium intybus* (Chicory)		
	Event	**Company**	**Description**
	RM3-3, RM3-4, RM3-6	Bejo Zaden BV	Male sterility was *via* insertion of the barnase ribonuclease gene from Bacillus amyloliquefaciens; PPT resistance was *via* the bar gene from *Streptomyces hygroscopicus* which encodes the PAT enzyme.
3.	*Zea mays* L. (Maize)		
	MS3	Bayer CropScience (Aventis CropScience(AgrEvo)	Male sterility caused by expression of the barnase ribonuclease gene from Bacillus amyloliquefaciens; PPT resistance was *via* PPT-acetyltransferase (PAT).
	MS6	do	do

BOTTLENECKS OF MALE STERILITY SYSTEMS

Several bottlenecks are involved during the development and maintenance of male sterile lines through conventional breeding. The first requirement is identification of a genetic source of male sterility. Second is for the propagation of male-sterile phenotypes. In case of CMS, which is a maternally inherited male sterility trait, for commercial seed production involves maintenance of three breeding lines: a male-sterile line (female parent), a maintainer line that is isogenic to the male-sterile line but contains fully functional mitochondria and a restorer line which has nuclear genes (Rf genes) for fertility restoration. Often, the CMS lines, and their corresponding restorer lines are not available in genotypically diverse backgrounds. Sometimes, male sterility genes are linked with other agronomically undesirable genes. The successful use of cytoplasmic male-sterility for commercial hybrid seed production requires a stable male-sterile cytoplasm, an adequate pollen source, and an effective system of getting the pollen from the male parent to the male-sterile female [140]. Also the cytoplasmic-genetic system of male sterility requires three lines to produce a single crossed hybrid; the A line (male-sterile), B line (male-fertile maintainer), and R line (male-fertile with restorer genes). Three-way crosses produced with cytoplasmic-genetic male sterility involves maintenance and production of four lines, an A and B line of one inbred and male-fertile inbreds of the other two [140]. Different methods of control of pollination processes and their shortcomings have been diagrammatically represented in Fig. **3**.

In case of GMS, when governed by single recessive gene, male sterile plants (ms ms) are maintained by crossing it with heterozygous male fertile plants (Ms ms) which results in a segregation of male sterile and male fertile plants in a ratio of 1:1 in diploid species. The female line thus contains both male sterile and male fertile plants; latter must be identified and removed (rouging) before pollen shedding. Roguing of male fertile plants from the female line is costly as a result of which hybrid seeds become costlier [15]. This can only be done efficiently if the recessive male-sterility allele is segregated together with a selectable or screenable marker [45]. Moreover commercial utility of GMS systems is limited by the expense of clonal propagation of male sterile plants. In addition the use of natural male sterile and/or self-incompatible lines is limited to the particular crop varieties which contain this attribute.

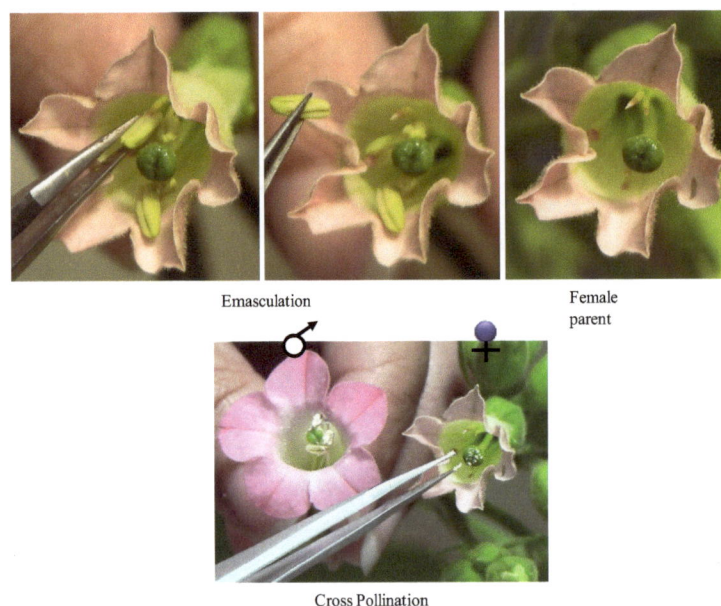

Emasculation   Female parent

Cross Pollination

Figure 3: Technical demonstration of emasculation and artificial cross pollination.

Chemically induced male sterility has several limitations: effect of the chemical is at certain specific stages of male gametophyte development therefore pollen abortion is incomplete; these chemicals may cause female sterility; phytotoxic effects of some chemicals are also reported.

Various candidate genes have been utilized for rendering plants male sterile by expressing them during anther development. The list of genes includes *BARNASE* [7], Fig. **4a-b**; *MAMMALIAN UNCOUPLING PROTEIN* [44], *PATHOGENESIS-RELATED GLUCANASE* [99], Fig. **4c-d**; *DEACETYLASE GENES* [63], *PHOSPHONATE MONOESTERASE* [64], *AVIDIN* [67], *DNA ADENINE METHYLASE* (Cigan and Albertson, 2000), *RIBOSOME INACTIVATING PROTEIN* [52], *DIPHTHERIA TOXIN A-CHAIN* [55], Fig. **4e-f**; *BAX* [98], *ARABIDOPSIS BAX INHIBITOR-1* [98], *CYTOKININ OXIDASE* [141], *STILBENE SYNTHASE* [82], *ARABIDOPSIS BECLIN 1* [103] *etc.* Biosafety and regulatory approvals, especially in the case of genes from non-plant sources remain a major challenge. For example even low level of leaky expression of cytotoxic genes, such as *BARNASES*, in non-targeted plant parts might affect plant growth negatively [142, 143].

CONCLUSION

Biotechnology has explored novel technologies for obtaining male sterile plants that has been proved useful in developing hybrid plants (Table **3**) and several male sterility systems (Table **2**), which in future, would be useful for hybrid seed production. However, there are some limitations in the potential application of commercial hybrid production: propagation of the male-sterile female parent line, impaired development of vegetative plant body, biosafety and regulatory matters. Future research should take into account to develop more efficient male sterile technologies to overcome these limiting factors in the large-scale production of hybrids.

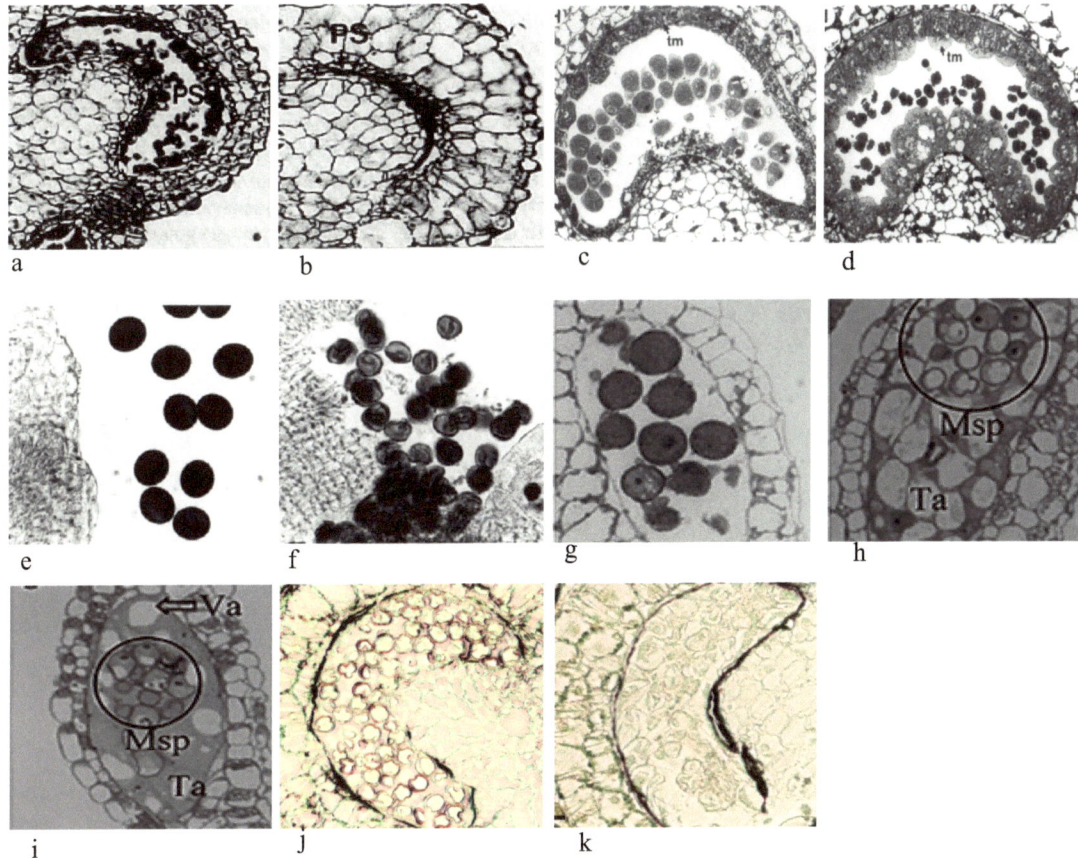

Figure 4: (a, c) Anther of wild tobacco. (b) Anther section of male sterile anther from transgenic tobacco transformed with TA_{29}-Rnase gene (*Nature, 347, 1990*). (d) Anther section of an A9- pathogenesis-related (PR) glucanase gene transformed male sterile tobacco plant (*The Plant Cell, 4, 1992*). (e) Mature pollen grain of Wild-type *Arabidopsis thaliana*. (f) Mature pollen of *Arabidopsis thaliana* transformed with A9- temperature-sensitive diphtheria toxin A chain (DTA) (*Plant Biotechnology Journal, 1, 2003*). (g) Anther section of wild *Arabidopsis*. (h) Anther section of transgenic *Arabidopsis* with construct Osg6b- *Arabidopsis Bax Inhibitor-1* (Kawanabe et al. 2006). (i) Anther section of transgenic *Arabidopsis* with construct Osg6b-*Bax* (Kawanabe et al. 2006). (j) Anther section of wild tobacco (k) Anther section of transgenic tobacco with two-component-*AtBeclin 1* (Singh *et al.*, 2009). PS: Pollen Sac, Tm: Tapetum, Msp: Microspore

ACKNOWLEDGEMENTS

Sudhir P. Singh is grateful to CSIR and NABI (DBT) for conferring fellowships. Authors acknowledge the guidance given by Dr Rakesh Tuli (Executive Director, NABI, Mohali, India).

ABBREVIATIONS

MS: Male sterile (dominant)

ms: male sterile (recessive)

Ms: Male sterile (heterozygous)

CMS: Cytoplasmic male sterility

GMS: Genic male sterility

CGMS: Cytoplasmic genic male sterility

REFERENCES

[1]   Gableman WH. Male sterility in vegetable breeding. Brook Haven symposia in Biology No. 1956; 9: 113-22.

[2]   Frankel R and Galun E. Pollination mechanisms, reproduction, and plant breeding. Heidelberg: Springer-Verlag 1977.

[3]   Mehdi M and Anwar A. Role of genetically engineered system of male sterility in hybrid production of vegetables. Journal of Phytology 2009; 1(6): 448–60.

[4]   Sawhney VK and Shukla A. Male Sterility in Flowering Plants: Are Plant Growth Substances Involved? American Journal of Botany 1994; 81(12): 1640-7.

[5]   Kaul M L H and Murthy T G K. Mutant genes affecting higher plant meiosis. Theor Appl Genet 1985; 70: 449–466.

[6]   Nizampatnam NR, Doodhi H, Narasimhan YK, Mulpuri S, Viswanathaswamy DK. Expression of sunflower cytoplasmic male sterility-associated open reading frame, orfH522 induces male sterility in transgenic tobacco plants. Planta 2009; 229: 97-1001.

[7]   Mariani C, Beuckeleer MD, Truettner J, Leemans J and Goldberg RB Induction of male sterility in plants by a chimeric ribonuclease gene. Nature 1990; 347: 737–41.

[8]   Reynaerts A, Wiele HD, Sutter GD and Janssens J. Engineered genes for fertility control and their application in hybrid seed production. Scientia Horticulturae 2003; 55: 125-39.

[9]   Ruiz O and Daniell H. Engineering cytoplasmic male sterility *via* the chloroplast genome by expressing β-Ketothiolase. Plant Physiol 2005; 138: 1232-46.

[10]   Bateson W, Gairdner AE. Male sterility in flax subject to two types of segregation. J Genet. 1921; 11: 269–75.

[11]   Chittenden RJ, Pellow CA. A suggested inter pretation of certain cases of anisogeny. Nature 1927;119: 10–12.

[12]   Jones HA and Clarke AE. Inheritance of male sterility in the onion and the production of hybrid seed, Proc. Am Soc Hort Sci 1943; 43:189–94.

[13]   Kaul M L H. Male Sterility in Higher Plants. (Springer, Berlin), pp. 1988; 758-75.

[14]   Hirshe J, Engelke T, Voller D, Gotz M, Roitsch T. Interspecies compatibility of the anther specific cell wall invertase promoters from Arabidopsis and tobacco for generating male sterile plants. Theor Appl Genet 2009; 118: 235-245.

[15]   Singh BD. Plant Breeding: Principles and Methods, 6th edn. Kalyani Publication, Ludhiana, India. 2000.

[16]   Hanson MR and Bentolila S. Interactions of mitochondrial and nuclear genes that affect male gametophyte development. Plant Cell 2004; 16: 154–69.

[17]   Chase CD. 2007 Cytoplasmic male sterility: a window to the world of plant mitochondrial nuclear interactions. Trends Genet. 2007;23: 81-90.

[18]   Rick CM. Genetics and development of nine male sterile tomato mutants. Hilgardia 1948, 18:599-633.

[19]   Martin JA and Crawford JH. Several types of sterility in Capsicum frutescens, Proc Amer Soc Hort Sci 1951; 57: 335-8.

[20]   Shi MS. Preliminary report of breeding and utilization of late japonica natural double-purpose line. J Hubei Agric Sci 1981;7: 1-3.

[21]   Shi MS. The discovery and study of the photosensitive recessive male sterile rice. Scientia Agricultura Sinica 1985; 2, 44-8.

[22]   Scoles GJ and Evans LE The genetics of fertility restoration in cytoplasmic male sterile rye. Can J Genet Cytol 1979a; 21: 417-22.

[23]   Scoles GJ and Evans LE. Pollen development in male fertile and cytoplasmic male sterile rye. Can J Genet 1979b; 57: 2782-90.

[24]   Hayase H, Satake T, Nishiyama I and Ito N. Male sterility caused by cooling treatment at the meiotic stage in rice plants II. The most sensetice stage to cooling and fertilizing ability of pistil. Proc Z Crop Soc Jpn 1969; 38: 706-11.

[25]   Satake T and Yoshida S. High temperature induced sterility in Indica rice at flowering. Jpn J Crop Sci 1978; 47: 6-7.

[26]   Zhang KT and Fu HY. Effect of high temperature treatment on male sterility in sorghum. Acta Genet. PR China. 1982; 9: 71-77.

[27]   Siddiq EA and Jauhar Ali A. Innovative male sterility systems for exploitation of hybrid vigour in crop plants. PINSA B. 1999; 65 (6): 331-50.

[28]   Wit F. Chemically induced male sterility, a new tool in plant breeding? Euphytica 1960; 9: 1-9.

[29]   Colombo N and Favret EA. The effect of gibberellic acid on male fertility in bread wheat. Euphytica 1996; 91: 297-03.

[30]   Parmar KS, Siddiq EA and Swaminathan MS Chemical induction of male sterility in rice. Indian J of Genet Plant Breed 1979; 39: 529-41.

[31] Urushizaki S, Katsuta M, Yamamura S, Ota Y and Saka H. Nippon Sakumotsugaku Kaishi (in Japanese). 1987; 56 (Suppl. I): 136-137.

[32] Yamamura S and Saka H Nippon Sakumotsugaku Kaishi (in Japanese). 1988; 57 (Suppl. I): 107-8.

[33] Jan CC, Rutger JN. Mitomycin C- and streptomycin-induced male sterility in cultivated sunflower. Crop Science Madison:Crop Science Society of America 1988; 28: 792–5.

[34] Sakaki M and Oshio H. Shokubutsu no Kagakuchousetsu (in Japanese). 1998; 23: 52-7.

[35] Kimura Y, Shimada A and Nagai T. Effects of glutamine synthetase inhibitors on rice sterility. Biosci Biotech Biochem 1994; 58(4): 669-73.

[36] Melchers G, Mohri, Y, Watanabe K, Wakabayashi S And Harada K. One step generation of cytoplasmic male sterility by fusion of mitochondrial-inactivated tomato protoplast with nuclear-inactivated Solanum protoplast. Proc Natl Acad Sci 1992; 89: 6632-836.

[37] Wijk, J van. , "Hybrids, Bred for Superior Yields or for Control?" Biotechnology and Development Monitor 1994; 19: 3-5.

[38] Stuber CW. Heterosis in plant breeding. Plant Breeding Rev. 1994; 12: 227–51.

[39] Zhao D, Wang G, Speal B and Ma H. The EXCESS MICROSPOROCYTES1 gene encodes a putative leucine-rich repeat receptor protein kinase that controls somatic and reproductive cell fates in the Arabidopsis anther. Genes Dev 2002; 16: 2021-31.

[40] Gardner N, Felsheim R, Smith AG. Production of male-and female-sterile plants through reproductive tissue ablation. J Plant Physiol 2009; 166(8): 871-81.

[41] Garcia-Sogo B, Pineda B, Castelblanque L, Anton T, Medina M, Roque E, *et al.* Efficient transformation of Kalanchoe blossfeldiana and production of male-sterile plants by engineered anther ablation. Plant Cell Rep 2010; 29: 61–77.

[42] Perez-Prat E and Campagne MML hybrid seed production and the challenges of propagating male sterile plants. Trends in Plant Science 2002; 7(5): 199-203.

[43] Koltunow AM, Truettner J, Cox KH, Wallroth M and Goldberg RB. Different temporal and spatial gene expression patterns occur during anther development. Plant Cell 1990; 2: 1201–24.

[44] George BI, Jonathan BSW, James GA and Walter SW. Hybrid seed production. Patent publication no. WO/1990/008830. 1990.

[45] Denis M, Delourme R, Gourret JP, Mariani C and Renard M. Expression of engineered nuclear male sterility in Brassica napus. Plant Physiol 1993; 101: 1295-1304.

[46] Thorsness MK, Kandasamy MK, Nasrallah ME and Nasrallah JB. Genetic ablation of Xoral cells in Arabidopsis. Plant Cell 1993; 5: 253–61.

[47] Roberts MR, Boyes E and Scott RJ. An investigation of the role of anther tapetum during microspore development using genetic cell ablation. Sex Plant Reprod 1995; 8: 299–307.

[48] Twell D. Diphtheria toxin-mediated cell ablation in developing pollen: vegetative cell ablatio   n blocks generative cell migration. Protoplasma 1995; 187: 144–54.

[49] Zhan XY, Wu HM, Cheung AY. Nuclear male sterility induced by pollen-specific expression of a ribonuclease. Sex Plant Reprod 1996; 9: 35–43.

[50] Block MD, Debrouwer D and Moens T. The development of a nuclear male sterility system in wheat. Expression of the barnase gene under the control of tapetum specific promoters. Theor Appl Genet 1997; *95*: 125-31.

[51] Cigan AM and Albertsen MC. Reversible nuclear genetic system for male sterility in transgenic plants. Patent no. US6072102. 2000.

[52] Cho HJ, Kim S, Kim M and Kim BD. Production of Transgenic Male Sterile Tobacco Plants with the cDNA Encoding a Ribosome Inactivating Protein in Dianthus sinensis L. Mol Cells 2001; 11(3): 326-33.

[53] Jagannath A, Bandyopadhyay P, Arumugam N, Gupta V, Burma PK and Pental D. The use of a Spacer DNA fragment insulates the tissue-specific expression of a cytotoxic gene (barnase) and allows high-frequency generation of transgenic male sterile lines in Brassica juncea L. Molecular Breeding 2001; 8: 11–23.

[54] Lee YH, Chung KH, Kim HU, Jin YM, Kim HI and Park BS. Induction of male sterile cabbage using a tapetum-specific promoter from Brassica campestris L. ssp. Pekinensis. Plant Cell Rep 2003; 22: 268–73.

[55] Guerineau F, Sorensen AM, Fenby N and Scott RJ. Temperature sensitive diphtheria toxin confers conditional male-sterility in Arabidopsis thaliana. Plant Biotechnology J 2003; 1: 33-42.

[56] Gomez MD, Beltrán JP and Canas LA. The pea END1 promoter drives anther-speciWc gene expression in diVerent plant species. Planta 2004; 219: 967–81.

[57]   Luo H, Lee JY, Hu Q, Nelson-Vasilchik, K Eitas, TK Lickwar, C Kausch, AP Chandlee, JM and Hodges TK RTS, a rice anther-specific gene is required for male fertility and its promoter sequence directs tissue-specific gene expression in different plant species. Plant Mol Biol 2006; 62: 397-408.

[58]   Roque E, Gómez MD, Ellul P, Wallbraun M, Madueño F, Beltrán JP *et al.* The PsEND1 promoter: a novel tool to produce genetically engineered male-sterile plants by early anther ablation. Plant Cell Rep 2007; 26: 313–25.

[59]   Cao B, Meng C, Lei J and Chen G. The pTA29-barnase chimeric gene transformation of Brassica campestris L. subsp. chinensis Makino var. parachinensis mediated by Agrobacterium. Sheng Wu Gong Cheng Xue Bao 2008; 24(5): 881-6.

[60]   Liu Z and Liu Z. The second intron of AGAMOUS drives carpel and stamen-specific expression sufficient to induce complete sterility in Arabidopsis. Plant Cell Rep 2008; 27: 855–63.

[61]   Konagaya K, Ando S, Kamachi S, Tsuda M and Tabei Y. Efficient production of genetically engineered male sterile Arabidopsis thaliana using anther-specific promoters and genes derived from Brassica oleracea and B. rapa. Plant Cell Rep 2009; 27(11): 1741-54.

[62]   O'Keefe DP, Tepperman JM, Dean C, Leto KJ, Erbes DL and Odell JT. Plant expression of a bacterial cytochrome P450 that catalyzes activation of a sulfonylurea pro-herbicide. Plant Physiol. 1994; 105: 473–82.

[63]   Kriete G, Niehaus K, Perlick AM and Broer I. Male sterility in transgenic tobacco plants induced by tapetum-specific deacetylation of the externally applied non-toxic compound N-acetyl-L-hosphinothricin. Plant J 1996; 9: 809–18.

[64]   Dotson SB, Lanahan MB, Smith AG, Kishore GM. A phosphonate monoester hydrolase from Burkholderia caryophilli PG2982 is useful as a conditional lethal gene in plants. Plant J 1996; 10 (2): 383–92.

[65]   Goff SA, Crossland LD and Privalle LSControl of gene expression in plants by receptor mediated transactivation in the presence of a chemical ligand. Patent no. US5880333. 1999.

[66]   Ward ER, Ryals A, Miflin BJ. Chemical regulation of transgene expression in plants. Plant Mol Biol 1993; 22: 361–6.

[67]   Albertsen MC, Beach LR, Howard J, Huffman GA and Taylor L. Control of male fertility using externally inducible promoter sequences. US Patent 5432068. 1995.

[68]   Knight ME, Jepson I, Daly A and Bayliss M. Hybrid seed production. Patent Publication no. WO1999/042598. 1999.

[69]   Bridges IG. *et al.* Plant gene construct encoding a protein capable of disrupting the biogenesis of viable pollen. Patent no. US6172279. 2001.

[70]   Hird DL, Worrall D, Hodge R, Smartt S, Paul W and Scott RJ. The anther-specific protein encoded by the Brassica napus and Arabidopsis thaliana A6 gene displays similarity to ß-1,3 glucanases. Plant J 1993; 4: 1023-33.

[71]   Dirks R, Trinks K, Uijtewaal B, Bartsch K, Peeters R, Hofgen R *et al.* Process for generating male sterile plants. US Patent 6262339. 2001.

[72]   Goetz M, Godt DE, Guivarc'h A, Kahmann U, Chriqui D and Roitsch T. Induction of male sterility in plants by metabolic engineering of the carbohydrate supply. Proc Natl Acad Sci U.S.A. 2001;98: 6522–7.

[73]   Derksen J, Wezel RV, Knuiman B, Ylstra B, Tunen AJV. Pollen tubes of flavonol deficient petunia show striking alterations in wall structure leading to tube disruption. Planta 1999; 207: 575–81.

[74]   Browse JA. Conditionally male-fertile plants and methods and compositions for restoring the fertility thereof. Patent Publication No. WO 97/10703. 1997.

[75]   McConn M and Browse J. The critical requirement for linoleic acid in pollen development, not photosynthesis, in an Arabidopsis mutant. Plant Cell 1996; 8: 403–16.

[76]   Sanders PM *et al.* The Arabidopsis DELAYED DEHISCENCE1 gene encodes an enzyme in the jasmonic acid synthesis pathway. Plant Cell 2000, 12: 1041–61.

[77]   Albertsen MC, Howard JA, and Maddock S. Induction of male sterility in plants by expression of high levels of avidin. US Patent 5962769. 1999.

[78]   Spena A, Estruch JJ, Prinsen E, Nacken W, Van Onckelen H, and Somme H. Anther-specific expression of the rolB gene of Agrobacterium rhizogenes increases IAA content in anthers and alters anther development and whole flower growth. Theor Appl Genet 1992; 84: 520–7.

[79]   Vander Meer IM, Stam ME, van Tunen AJ, Mol JN and Stuitje AR Antisense inhibition of flavonoid biosynthesis in petunia anthers results in male sterility. Plant Cell 1992; 4(3): 253–62.

[80]   Taylor LP and Mo Y. Methods for the regulation of plant fertility. Patent Application no. WO /18142. 1993.

[81]   Van Tunan AJ, Van Der MIM and Mol Josephus NM. Male-sterile plants, methods for obtaining male-sterile plants and recombinant DNA for use therein. Patent no. EP 0513884. 1992.

[82]   Höfig KP, Möller R, Donaldson L, Putterill J and Walter C. Towards male sterility in Pinus radiata – a stilbene synthase approach to genetically engineer nuclear male sterility. Plant Biotechnology Journal 2006; 4: 333-43.

[83]   Ribarits A, Mamun ANK, Li S, Resch T, Fiers M, Heberle-Bors E, *et al.* Combination of reversible male sterility and doubled haploid production by targeted inactivation of cytoplasmic glutamine synthetase in developing anthers and pollen. Plant Biotechnology J 2007; 5: 483-94.

[84]   Fabijanski SF and Arinson PG. Binary cryptocytotoxic method of hybrid seed production. Patent no. US5426041. 1995.

[85]   Gutterson N and Ralston E. Two component plant cell lethality methods and compositions. US Patent. 6392119. 2002.

[86]   Burgess G, Ralston EJ, Hanson WG, Heckert M, Ho M, Jenq T, *et al.* A novel two component system for cell lethality and use in engineering nuclear male sterility in plants. Plant J 2002; 31(1): 113-25.

[87]   Gils M, Marillonnet S, Werner S, Grützner R, Giritch A, Engler C, *et al.* A novel hybrid seed system for plants. Plant Biotechnology Journal 2008; 6: 226-35.

[88]   Kempe K, Rubtsova M and Gils M. Intein-mediated protein assembly in transgenic wheat:production of active barnase and acetolactate synthase from split genes. Plant Biotechnology J 2009; 7: 283-97.

[89]   Takada K, Ishimaru K, Minamisawa K, Kamada H and Ezura H. Expression of a mutated melon ethylene receptor gene Cm ETR1/H69A affects stamen development in Nicotiana tabacum. Plant Sci 2005a; 169: 935-42.

[90]   Bartsch K. Genes coding for amino acid deacetylases with specificity for N-acetyl-Lphosphinothricin, their isolation and their use. US Patent 6177616. 2001.

[91]   Camerarius RJ. De Sexu Plantarum epistola. In Mobius M (ed) Oswalds Klassiker der exakten Wiss Nr 105. Englmann, Leipzig DDR. 1694;1899.

[92]   Gorguet B, Schipper D, van Lammeren A, Visser RG and van Heusden AW. ps-2, the gene responsible for functional sterility in tomato, due to non-dehiscent anthers, is the result of a mutation in a novel polygalacturonase gene. Theor Appl Genet 2009; 118: 1199–1209.

[93]   Sawhney VK and Shukla A. Male Sterility in Flowering Plants: Are Plant Growth Substances Involved? American Journal of Botany 1994; 81(12): 1640-7.

[94]   Tang L, Chu H, Kin Yip W, Yeung EC and Clive L. An anther-specific dihydroflavonol 4-reductase-like gene (DRL1) is essential for male fertility in Arabidopsis. New Phytologist 2008; 181: 576–87.

[95]   Gomez-Casati D, Busi MV, Gonzalez-Schain N, Mouras A, Zabaleta EJ and Araya A. A mitochondrial dysfunction induces the expression of nuclear-encoded complex I genes in engineered male sterile Arabidopsis thaliana. FEBS Lett 2002; 532: 70–4.

[96]   Sun Q, Zhang Y, Rong T, Dong S, Ma D and Zhang C Establishment of transgenic accept or and transformation of barnase gene by particle gun in maize inbredline 18-599 (white). Front Agric China 2008; 2: 37–43.

[97]   Liu J, Yu Y, Lei J, Chen G and Cao B. Study on Agrobacterium-Mediated Transformation of Pepper with Barnase and Cre Gene. Agricultural Sciences in China 2009; 8: 947-55.

[98]   Kawanabe T, Ariizumi T, Kawai-Yamada M, Uchimiya H and Toriyama K. Abolition of the Tapetum Suicide Program Ruins Microsporogenesis. Plant Cell Physiol 2006; 47(6): 784–7.

[99]   Worrall D, Hird DL, Hodge R, Paul W, Draper J and Scott R. Premature dissolution of the microsporocyte callose wall causes male sterility in transgenic Tobacco. The Plant Cell 1992; 4: 759-71.

[100]   Takada K, Kamada H. and Ezura H Production of male sterile transgenic plants. Plant Biotechnology 2005b; 22: 469-76.

[101]   Takada K, Ishimaru K and Kamada H. Anther specific expression of mutated melon ethylene receptor gene Cm-ERS1/H70A affected tapetum degeneration and pollen grain production in transgenic tobacco plants. Plant Cell Rep 2006; 25: 936–41.

[102]   Takada K, Watanabe S, Sano T, Ma B, Kamada H and Ezura H. Heterologous expression of the mutated melon ethylene receptor gene Cm-ERS1/H70 A produces stable sterility in transgenic lettuce (Lactuca sativa). J. Plant Physiol 2007; 164: 514–20.

[103]   Singh SP, Pandey T, Srivastava R, Verma PC, Singh PK, Tuli R *et al.* BECLIN 1 from Arabidopsis thaliana, under the generic control of regulated expression systems, a strategy for developing male sterile plants. Plant Biotechnology Journal 2010.

[104]   Tsuchiya T, Toriyama K, Yoshikawa M, Ejiri S and Hinata K. Tapetum-specific expression of the gene for an endo-beta-1,3-glucanase causes male sterility in transgenic tobacco. Plant Cell Physiol 1995; 36(3): 487-94.

[105]   Beals TP and Goldberg RB. A novel cell ablation strategy blocks Tobacco anther dehiscence. The Plant Cell 1997; 9: 1527-45.

[106]   Fernando W, Victoria M, Alejandro C, Nahuel G, Mariano P, Mariana M, *et al.* Ectopic expression of mitochondrial gamma carbonic anhydrase2 causes male sterility by anther indehiscence. Plant Mol Biol 2009; 70: 471-85.

[107] Hernould M, Suharsono S, Litvak S, Araya A, Mouras A. Male-sterility induction in transgenic tobacco plants with an unedited atp9 mitochondrial gene from wheat. Proc Natl Acad Sci 1993; 90(6): 2370-4.

[108] Gomez-Casati D, Busi MV, Gonzalez-Schain N, Mouras A, Zabaleta EJ and Araya A. A mitochondrial dysfunction induces the expression of nuclear-encoded complex I genes in engineered male sterile Arabidopsis thaliana. FEBS Lett 2002; 532: 70–4.

[109] He S, Abad AR, Gelvin SB and Mackenzie SA. A cytoplasmic male sterility associated mitochondrial protein causes pollen disruption in transgenic tobacco. Proc Nat Acad Sci 1996; 93: 11763-8.

[110] Kim DH, Kang JG, Kim BD. Isolation and characterization of the cytoplasmic male sterility-associated orf456 gene of chili pepper (Capsicum annuum L.). Plant Mol Biol 2007; 63(4): 519-32.

[111] Yamamoto MP, Shinada H, Onodera Y, Komaki C, Mikami T and Kubo T. A male sterility associated mitochondrial protein in wild beets causes pollen disruption in transgenic plants. Plant J 2008; 54: 1027-38.

[112] Mou Z, Wang X, Fu Z, Dai Y, Han C, Ouyang J, et al. Silencing of phosphoethanolamine N-methyltransferase results in temperature-sensitive male sterility and salt hypersensitivity in Arabidopsis. Plant Cell 2002; 14(9): 2031-43.

[113] Yui R, Iketani S, Mikami T and Kubo T. Antisense inhibition of mitochondrial pyruvate dehydrogenase E1 alpha subunit in anther tapetum causes male sterility. Plant J 2003; 34(1): 57–66.

[114] Preston J, Wheeler J, Heazlewood J, Li SF and Parish RW. AtMYB32 is required for normal pollen development in Arabidopsis thaliana. Plant J 2004; 40(6): 979-95.

[115] Sandhu APS, Abdelnoor RV and Mackenzie SA. Transgenic induction of mitochondrial rearrangements for cytoplasmic male sterility in crop plants. Proc Natl Acad Sci 2007; 104(6): 1766-70.

[116] Zhang W, Sun Y, Timofejeva, L, Chen C, Grossniklaus U and Ma H. Regulation of Arabidopsis tapetum development and function by DYSFUNCTIONAL TAPETUM1 (DYT1) encoding a putative bHLH transcription factor. Development 2006; 133: 3085-95.

[117] Li SF, Iacuone S and Parish RW. Suppression and restoration of male fertility using a transcription factor. Plant Biotechnol J 2007; 5: 297-312.

[118] Mamun ANK Reversible male sterility in transgenic tobacco carrying a dominant-negative mutated glutamine synthetase gene under the control of microspore-specific promoter. Indian Journal of Experimental Biology 2007; 45: 1022-30.

[119] Zhang ZW, Zhu J, Gao JF, Wang C, Li H, Li H, et al. Transcription factor AtMYB103 is required for anther development by regulating tapetum development, callose dissolution and exine formation in Arabidopsis. Plant J 2007; 52: 528–38.

[120] Wan J, Patel A, Mathieu M, Kim S, Xu D and Stacey G. A lectin receptor-like kinase is required for pollen development. Plant Mol Biol 2008; 67: 469–82.

[121] Tang L, Chu H, Kin Yip W, Yeung EC and Clive L. An anther-specific dihydroflavonol 4-reductase-like gene (DRL1) is essential for male fertility in Arabidopsis. New Phytologist 2008; 181: 576–87.

[122] Xing S and Zachgo S. ROXY1 and ROXY2, two Arabidopsis glutaredoxin genes, are required for anther development. Plant J 2008; 53:790-801.

[123] Chen R, Zhao X, Shao Z, Wei Z, Wang Y, Zhu L, et al. Rice UDP-Glucose Pyrophosphorylase1 is essential for pollen callose deposition and its cosuppression results in a new type of thermosensitive genic male sterility. The Plant Cell 2007; 19: 847-61.

[124] Woo M et al. Inactivation of the UGPase1 gene causes genic male sterility and endosperm chalkiness in rice (Oryza sativa L.) Plant J 2008; 54(2): 190-204.

[125] Qin G, Ma Z, Zhang L, Xing S, Hou X, Deng J, et al. Arabidopsis AtBECLIN 1/AtAtg6/AtVps30 is essential for pollen germination and plant development. Cell Res 2007; 17(3): 249-263.

[126] Fujiki Y, Yoshimoto K and Ohsumi Y. An Arabidopsis homolog of yeast ATG6/VPS30 is essential for pollen germination. Plant Physiol 2007; 143(3): 1132-9.

[127] Harrison-Lowe NJ and Olsen LJ. Autophagy protein 6 (ATG6) is required for pollen germination in Arabidopsis thaliana. Autophagy 2008; 4: 339-48.

[128] Ma J, Yan B, Qu Y, Qin F, Yang Hao X, Yu J, et al. Zm401, a short-open reading-frame mRNA or noncoding RNA, is essential for tapetum and microspore development and can regulate the floret formation in Maize. J. of Cellular Biochem 2008; 105: 136–46.

[129] Wang X and Li X. The GhACS1 gene encodes an acyl-CoA synthetase which is essential for normal microsporogenes is nearly anther development of cotton. Plant J 2009; 57: 473-86.

[130] Wang Y, Zha X, Zhang S, Qian X, Dong X, Sun F and Yang J. Down-regulation of the OsPDCD5 gene induced photoperiod-sensitive male sterility in rice. Plant Science 2010; 178: 221-8.

[131] Yang KZ, Xia C, Liu XL, Dou XY, Wang W, Chen LQ, *et al.* A mutation in Thermosensitive Male Sterile 1, encoding a heat shock protein with DnaJ and PDI domains, leads to thermosensitive gametophytic male sterility in Arabidopsis. Plant J 2009; 57: 870-82.

[132] Kurek I, Dulberger R, Azem A, Tzvi BB, Sudhakar D, Christou P *et al.* Deletion of the C-terminal 138 amino acids of the wheat FKBP73 abrogates calmodulin binding, dimerization and male fertility in transgenic rice. Plant Molecular Biology 2002; 48: 369–81.

[133] Chrimes D, Rogers HJ, Francis D, Jones HD and Ainsworth C. Expression of fission yeast cdc25 driven by the wheat ADP-glucose pyrophosphorylase large subunit promoter reduces pollen viability and prevents transmission of the transgene in wheat. New Phytologist 2005;166: 185–92.

[134] Thorstensen T, Grini PE, Mercy IS, Alm V, Erdal S, Aasland R and Aalen RB. The Arabidopsis SET-domain protein ASHR3 is involved in stamen development and interacts with the bHLH transcription factor ABORTED MICROSPORES (AMS). Plant Mol Biol 2008; 66(1-2): 47-59.

[135] Tzeng T, Kong L, Chen C, Shaw C and Yang C. Overexpression of the Lily p70^{s6k} Gene in Arabidopsis Affects Elongation of Flower Organs and Indicates TOR-Dependent Regulation of AP3, PI and SUP Translation. Plant Cell Physiol 2009; 50(9): 1695-709.

[136] Steiglitz H. Role of β-1,Sglucanase in postmeiotic microspore release. Dev Biol 1977; 57: 87-97.

[137] Schnable PS and Wise RP. The molecular basis of cyto plasmic male sterility and fertility restoration. Trends Plant Sci 1998; 3: 175–80.

[138] Zhang Y, Shewry PR, Jones H, Barcelo P, Lazzeri PA, Halford NG. Expression of antisense SnRK1 protein kinase sequence causes abnormal pollen development and male sterility in transgenic barley. Plant J 2001; 28: 431–441

[139] Rangasamy N and Elumalai K Market opportunities and challenges for Agri-Biotech Products in India. Agricultural Economics Research Review 2009; 22: 471-81.

[140] Crossland LD, Tuttle A, Stein JI. Transgenic male sterile plants for the production of hybrid seeds. Patent No. US5659124. 1997.

[141] Hung S, Crossland LD, Cheikh N and Morris RO. Reversible male sterility in transgenic plants by expression of cytotoxinin oxidase. 2007; Patent No. US7230168.

[142] Skinner JS. *et al.* Options for genetic engineering of floral sterility in forest trees. In Molecular Biology of Woody Plants (Jain, S.M. and Minocha, S.C., eds), 2000; pp.135-153, Kluwer Academic Publishers.

[143] Ramos HJO, Souza EM, Soares-Ramos JRL and Pedrosa FO. A new system to control the barnase expression by a NifA-dependent promoter. J of Biotechnology 2005; 118: 9–16.

CHAPTER 6

Anti-Bacterial and Crystallographic Studies of Jatrophone, the Macrocyclic Diterpenoid from the Roots of *Jatropha Gossypifolia* l.

R. S. Satyan[1,*], Ajay Parida[1] and Babu Varghese[2]

[1]*Bioprospecting & Tissue Culture Lab, M. S. Swaminathan Research Foundation, III Cross Street, Taramani Institutional Area, Taramani, Chennai- 600 113, India and* [2] *X-Ray Crystallography Department, Sophisticated Analytical Instrumentation Facility (SAIF), Indian Institute of Technology (IIT), Chennai- 36, India*

Abstract: *Jatropha gossypifolia* L., popularly known as the belly-ache bush is a rich natural source of novel macrocyclic diterpenoids. Crude root and stem extracts of *J. gossypifolia* were tested against certain human bacterial pathogens. Among the treatments, lipophilic hexane extract of the dried roots inhibited the gram negative bacterial human pathogens *Corynebacterium diphtheriae*, *Bacillus cereus* and *Proteus mirabilis*. Effective inhibition (18 mm) was observed against *C. diphtheriae*, the activity being dose-dependent among the dilutions tested, with maximum efficacy at 500 ppm. *B. cereus* was resistant at lower dilutions (25-100 ppm), however, 250 and 500 ppm concentrations exhibited effective inhibition (8.5 and 9.1 mm). Marginal inhibition was observed against *P. mirabilis* (5.5 mm) at 500 ppm. Of the 7 column fractions tested, fraction-7 exhibited significant activity against *C. diphtheriae*, *B. cereus*, *P. mirabilis*, *Shigella flexneri*, *Klebsiella aerogenes* and *Staphylococcus aureus*. Gram-negative organisms tested, except *P. mirabilis* and *S. flexneri*, showed marginal inhibition for the crude extracts. Crystalline fraction-7 appreciably inhibited *K. aerogenes and S. aureus* at higher concentrations (1000, 500 and 250 ppm). X-Ray Diffraction Crystallography (XRD) analysis of the crystals confirmed that it is the known macrocyclic diterpenoid, Jatrophone, with a few other proven biological activities.

Keywords: *Jatropha gossypifolia, Corynebacterium diphtheriae, Bacillus cereus, Proteus mirabilis,* Jatrophone.

INTRODUCTION

From time immemorial, phytochemicals were tested for its efficacy against human pathogens. Antimicrobial phytochemicals can be divided into major classes of compounds such as simple phenols and phenolic acids, quinones, flavones, flavonoids and flavonols, coumarins, terpenoids and essential oils, alkaloids, lectins and mixtures [1]. Methanol fraction of *Peperomia galioides* and ethyl acetate fraction of *Anredera diffusa* and *Krameria triandra* exhibited similar activity as that of the crude extracts against a series of human pathogens [2]. Anthemic acid, a phenolic acid from *Matricaria chamomilla* was found to be active against *Mycobacterium tuberculosis*, *Salmonella typhimurium* and *Staphylococcus aureus* [3]. Compounds purified from plants act independently or synergistically in influencing the growth of microorganisms. There is a continuous and urgent need to discover new antimicrobial compounds with diverse chemical structures and novel mechanisms of action because there has been an alarming increase in the incidence of new and re-emerging infectious diseases [4].

RESURGENCE OF DIPHTHERIA

Following the introduction of routine immunization with diphtheria toxoid in the 1940's and 1950's, diphtheria incidence declined dramatically in countries of the industrialized world. At the beginning of the 1980's many of these countries were progressing toward elimination of the disease. However, since the mid 1980's there has been a striking resurgence of diphtheria in several countries of Eastern Europe. For 1993,

WHO received reports of 15, 211 diphtheria cases in Russia and 2,987 cases in Ukraine. Epidemiological patterns of diphtheria are changing in developing countries, and the disease seems to be following patterns seen in industrialized countries 30 to 40 years ago [5].

*Address correspondence to R. S. Satyan: # 30/2, Senthil Andavar Street, Dhanalakshmi Colony, Vadapalani, Chennai- 600 026, INDIA; Mob: +91-98406 83539, Home: +91-44-2362 0539, E-mail: satyanoid@gmail.com.

Diphtheria is still an important disease in children in the Philippines. Despite the fact that it is vaccine preventable, there were 2140 cases and 524 deaths of diphtheria reported in 1982 and more importantly, a total of 95.7% of the reported cases were in children aged less than 14 years [6] Apart from the diagnosis and high immunization with the DPT vaccine, phytochemicals either in crude form or as single molecules could be subjected to clinical trials in diphtheriae patients. This approach might give a better insight into the drug targets and its application strategy in the future.

ANTI-DIPHTHERIAL PHYTOCHEMICALS

Roots of *Nauclea latifolia*, a savannah medicinal plant, is rich in indole-quinolizidine alkaloids, glycoalkaloids and saponins that are most effective against *Corynebacterium diphtheriae, Pseudomonas aeruginosa, Streptobacillus* sp., *Streptococcus* sp., *Neisseria* and *Salmonella* sp. [7] Ethanolic extracts of *Narcissus tazetta* spp. *tazetta* and *Leucojum aestivum* at 2mg/ml concentrations exhibited antibacterial activity against *S. aureus, Pseudomonas pseudomalii, Vibrio cholerae, Enterobacter cloacae, Salmonella typhi,C. diphtheriae and C. hoffmanni* [8] Extracts of Jingjie (*Schizonepeta tenuifolia*) exhibited very strong anti-bacterial activity against *C. diphtheriae* [9]. Crude methanolic extracts of *Arceuthobium oxycedri* (DC) M. Bieb (Viscaceae) exhibited weak activity against *C. diphtheriae* [10].

Alpha-bisabolol, a component from *Matricaria recutita* essential oil was active against *C. diphtheriae* [11]. Out of six different crystal components extracted from the roots of Baikal Scullcap- *Scutellaria baicalensis* G. (Huang-chin), Baicalin ($C_{21}H_{18}O_{11}$), a glucoronide, inhibited the growth of *V. cholerae, S. typhi, E. coli, Shigella dysenteriae, C. diphtheriae, Proteus vulgaris, P. aeruginosa, M. tuberculosis, Streptococcus, Pneumococcus and Meningococcus* sp. Early work on the leaves of *Myrtus communis* yielded two new acylphloroglucinols, Myrtucommulone-A and Myrtucommulone-B. The former at 80 µg/ml was highly bactericidal against *S. aureus, S. albus, B. subtilis, B. pumilus, S. faecalis, C. diphtheriae* and *C. xerosis* [12].

EUPHORBIA AND JATROPHA SPECIES (FAMILY: EUPHORBIACEAE)

Jatropha, a euphorbiaceous plant, is a large genus of herbs, shrubs and trees; distributed in the tropical and sub-tropical areas of the world (Picture **1**). In India, 9 species have been recorded. Some of the species are grown in garden for their ornamental foliage and flowers. Jatropha species are known to produce a large variety of diterpenes of polycyclic and macrocyclic classes, that are responsible for their biological properties. Among the diterpenoids synthesized, compounds with jatrophone, jatrophane and lathyrane skeletons have been the subjects of intense research, due to their biogenetic relevance, structural complexity and biological activity [13].

The root wood and the yellow/red root bark of *J. podagrica* were proved to exhibit anti-microbial activity against human pathogens [14]. Fractionation and purification of crude rhizome extract of *J. elliptica* (Pohl.) Muell. Arg. yielded jatrophone and a mixture of jatropholones A and B, as the main compounds. They were tested against the snail *Biomphalaria glabrata*. Jatrophone showed an LC_{50} of 1.16 ppm as a molluscicide and an LC_{50} of 1.14 ppm for the assay on egg mass, while the mixture of jatropholones A and B presented an LC_{50} of 58.04 ppm as a molluscicide and was not active against the egg mass at a concentration up to 100 ppm [15].

The activity of jatrogrossidione, the main diterpene of *J. grossidentata*, and jatrophone from *J. isabellii* were detected against *Leishmania* and *Trypanosoma cruzi* strains *in vitro* as well as against *Leishmania amazonensis in vivo*. The former showed a strong *in vitro* leishmanicidal and trypanocidal activity with IC_{100} of 0.75 and 1.5-5.0 µg/ml. The IC_{50} of jatrogrossidione was <0.25 µg/ml against amastigote forms of *Leishmania* infecting macrophages, with toxicity at concentrations higher than 0.5 µg/ml. [16].

Jatropha Gossypifolia

Jatropha gossypifolia (belly-ache bush) a small shrub belonging to Euphorbiaceae, grows wild in different parts of India (Picture **1**). It exhibits various medicinal [17] and pesticidal properties [18]. Several diterpenoids like jatrophone and its analogues [19], Jatropholones A and B [20] and Jatrophatrione [21] and lignans [22], Isogadain and its analogues [23] were isolated from this species. A novel lignan, gossypidien was reported from the stem of *J. gossypifolia* [24].

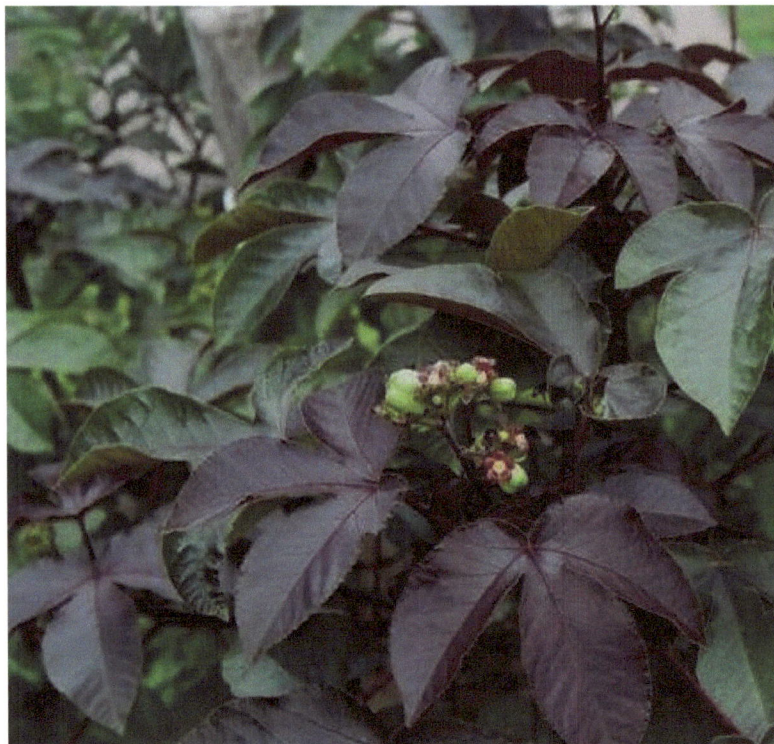

Picture 1: Jatropha gossypifolia.

JATROPHANE CLASS OF MACROCYCLIC DITERPENOIDS

Another class, the Jatrophane diterpenes, is of importance because of their unique structural character and their exclusive distribution in the family Euphorbiaceae [25]. The antimicrobial and antitumor assays of pubescenol, a new jatrophane diterpene, along with known *ent*-abietane lactones, helioscopinolide A and B, taraxerone, 24-methylenecycloartenol and vanillin, isolated from *Euphorbia pubescens* suggested that helioscopinolide A and B exhibited significant antibacterial activity against *S. aureus* whereas the diterpenes showed moderate growth inhibition of MCF-7, NCI-H460 and SF-268 human tumor cell lines [26].

The jatrophones and jatrophanes from *J. gossypifolia* has never been tested for their antimicrobial properties. Thus, the present study was focused on the elucidation of the antibacterial principles from *J. gossypifolia*, especially against highly infectious human pathogens, and the molecular characterization and confirmation through X-ray crystallography.

MATERIALS AND METHODS

Human Bacterial Pathogens

Escherichia coli ATCC 25922, *Staphylococcus aureus* ATCC 25923, *Shigella flexneri* ATCC 25875, *Shigella sonnei* ATCC 25931, *Shigella boydi* ATCC 35966, *Vibrio cholera* ATCC 11623, *Corynebacterium diphtheriae* ATCC 51696, *Salmonella typhi* ATCC 19943, *Salmonella paratyphi*-A & B, *Pseudomonas aeruginosa* ATCC 27853, *Proteus mirabilis* ATCC 29906, *Proteus vulgaris* ATCC 13315, *Klebsiella pneumonia* ATCC 15380, *Klebsiella aerogenes* ATCC 13882, *Enterococcus faecalis* ATCC 29212, *Bacillus cereus* 14579 and *Candida albicans* ATCC 10231 were obtained from the Department of Microbiology, Institute of Basic Medical Sciences (A.L.M.PGIBMS), Chennai. The organisms were inoculated in nutrient agar slant that were used as a source for further assays. Sterile glycerol stocks were maintained at –20 °C.

J. Gossypifolia Extraction and Bioassay with Crude Extracts

Roots, stem and leaf powder of *J. gossypifolia* were serially extracted with hexane and acetone. The materials were soaked in each of the solvents for 3 days. On the third day they were decanted and filtered using filter paper (Whatman No. 1). The process was repeated thrice to get maximum extractable. The pooled filtrate was condensed in a rotary evaporator (Buchi® R-200 rotavapor) and dried under air draft for three days to obtain the crude extracts (Table **1**). The extracts was stored at –5°C until use.

Table 1: Yield details of *J. gossypifolia* crude extracts

Plant Part	Dry Weight (in grams)	Solvent	Yield of Crude Extract (in grams)	Yield %
Root	500	Hexane	4.12	0.81
		Acetone	2.53	0.50
Stem	250	Hexane	1.03	0.41
		Acetone	1.55	0.62

Antibacterial Disc Diffusion Assay (Kirby and Bauer) Against Selected Human Pathogens

Preparation of Extract Discs

Sterile whatman filter paper discs (HiMedia®) are soaked in 25, 50, 100, 250, 500 and 1000 ppm concentrations of the crude extracts solution for 5 hrs. These were aseptically removed using sterile forceps and dried for 18 hrs under air draft to remove all traces of the solvent. Positive controls- Antibiotic discs (HiMedia®) and Negative controls- plain solvents were used for comparative performance.

Preparation of Starter Culture and Kirby and Bauer Bioassay

A loopful of culture was retrieved from the slants and inoculated in 2 ml of nutrient broth. The broth was incubated at 37°C/18 hrs. The inoculum turbidity was adjusted with McFarland standards. Sterile mueller-hinton agar plates were used for the bioassay. 100 µl of the starter culture was spread on the freshly made plates, using a glass spreader, to gain an even lawn. The discs fortified with various concentrations of the extracts were placed equidistant, such that, 7 discs were maintained in a single plate. Positive control (ready-made antibiotic discs, Himedia®) and negative controls (solvent) were employed in each plate for comparison. Triplicates were made and the plates were incubated at 37°C/ 24 hrs. The zone of inhibition around the discs was measured using vernier calipers and the average was tabulated.

Column Fractionation of Hexane extract of J. Gossypifolia Roots and Bioassay

Crude hexane extract of *J. gossypifolia* roots effectively inhibited some of the organisms and hence was chosen for further study. 2.7 g of the crude hexane was dissolved in 3 ml of chloroform and the admixture was made with 12.5 gm of SiO_2 (60-120 mesh). The mixture is thoroughly dried over night under air draft at 34°C. 200 g of SiO_2 (60-120 mesh) was packed with 100% hexane (bed length- 48 cm, diameter- 4.5 cm), equilibrated with nitrogen and allowed to stay for overnight. The admixture was added (2.5 cm height) and eluted with hexane and increasing concentrations of ethyl acetate (0.5 %, 1 %, 2 %, 5 %, 10 %, 15 %, 20 % up to 40 %). 7 fractions were eluted (Scheme **1**) based on the UV fluorescence (254 and 366 nm) that was condensed in rotary evaporator to yield oils, pigments and amorphous residues.

Assay with Fractions

25, 50 and 100 ppm concentrations of each fraction were assayed against the bacterial pathogens that were inhibited by the crude hexane extract. The assay procedure performed using the crude extracts was followed.

Purification of Fraction-7

Column fraction-7 exhibited marked inhibition of the assayed pathogens, hence was chosen for further study. 600 mg of the fraction was dissolved in 2.8 gm of SiO_2 (60-120 mesh) purified by column chromatography, (bed length- 22 cm, diameter- 1.8 cm), using petroleum ether: ethyl acetate (7:3) to yield an amorphous precipitate in yellowish oil. The precipitate was crystallized in a glass tube (5 cm- length and 3 mm- diameter) using 500 µl of chloroform to which, a drop of methanol was added. Single crystals were selected and subjected to bioassay with the strains in which inhibition was observed. One part of the crop was analyzed by single crystal XRD.

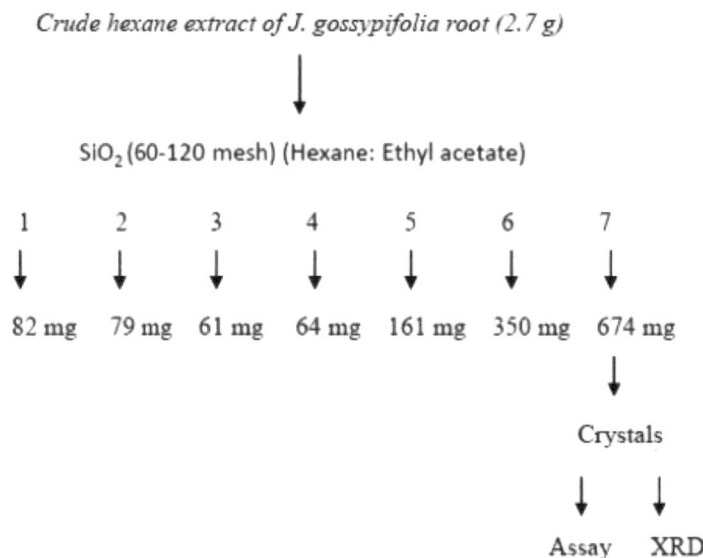

Crude hexane extract of J. gossypifolia root (2.7 g)

↓

SiO$_2$ (60-120 mesh) (Hexane: Ethyl acetate)

1	2	3	4	5	6	7
↓	↓	↓	↓	↓	↓	↓
82 mg	79 mg	61 mg	64 mg	161 mg	350 mg	674 mg

↓

Crystals

↓   ↓

Assay   XRD

Scheme 1: Column fractionation of hexane extract of roots of *J. gossypifolia*.

X RAY CRYSTALLOGRAPHY

XRD analysis of the crystals was done at Sophisticated Analytical Instrumentation Facility (SAIF), Indian Institute of Technology (IIT), Chennai. Intensity data for X ray structure analysis were collected using ENRAF NONIUS® CAD4-MV31 and Bruker® Kappa APEX II Single Crystal Diffractometer. Both copper and molybdenum probes were used according to the nature of the crystal. Unit cell parameters and orientation matrix were obtained by method of short vectors using 25 reflections collected through random search routine from different zones of the crystal. ω - 2 θ scan technique was used to measure individual intensities. The structure was solved by direct methods techniques using the program SIR 92. The structure was expanded for locating hydrogen atoms by difference Fourier techniques.

The structure was refined using "Full-Matrix Least Squares techniques". All non-hydrogen atoms were refined with anisotropic thermal parameters. Even though hydrogen atoms were located with difference Fourier map, they were finally fixed geometrically at chemically meaningful positions and were given riding model refinement. The refinement was continued until max.shift/esd converged to zero. X ray data collection parameters, geometric & thermal details of the structure *etc.* are also taken and the ORTEP picture and the packing diagram of the molecule in unit cell were derived based on the collected parameters.

RESULTS

Assay with Crude Extracts of *J. Gossypifolia*

Of the crude root and stem extracts tested against the given set of pathogens, hexane extract of the roots inhibited *C. diphtheriae*, *B. cereus* (Fig. **1**) and *P. mirabilis* (Table **2**). Effective inhibition was observed against *C. diphtheriae*, with maximum efficacy at 500 ppm (18 mm). *B. cereus* was resistant at lower dilutions (25-100 ppm). Nevertheless, 250 and 500 ppm of the crude hexane extract exhibited significant inhibition (8.5 and 9.1 mm). Mild inhibition was observed at 500 ppm against *P. mirabilis* (5.5 mm).

a. Effect of crude extract of the root of *J. gossypifolia* on *C. diphtheriae*; b. Effect of the crude extract compared with the standard antibiotic Methicillin (M[10]µg) on *B. cereus*

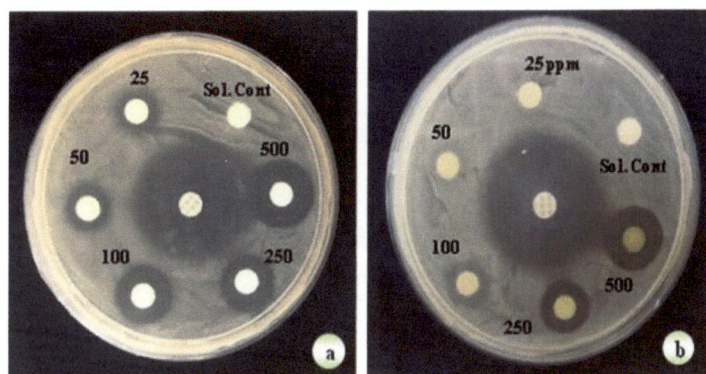

Figure 1: Antibacterial activity of hexane fraction-7 of the root extract of *J. gossypifolia.*

Table 2: Effect of crude extracts of *J. gossypifolia* against human pathogens

Plant Part	Conc. (ppm)	C. diphtheriae	P. mirabilis	S. flexneri	B. cereus
Root Hexane	**25**	6.0±0.2	3.5±0.2[gh]	-	-
	50	7.0±0.1	3.5±0.2[gh]	-	-
	100	7.8±0.1	4.0±0.1[ef]	-	-
	250	8.3±0.2	4.8±0.1[d]	-	8.5±0.2[f]
	500	9.0±0.2	5.5±0.2[c]	-	9.1±0.2[e]
Methicillin	**10µg/disc**	20±0.1	10±0.1[a]	10±0.1	20±0.1[d]
Solvent	-	-	-	-	-
Root Acetone	**25**	4.8±0.2	-	-	-
	50	6.0±0.2	-	-	-
	100	6.8±0.1	-	-	-
	250	6.3±0.2	3.25±0.1[h]	-	-
	500	7.5±0.2	3.8±0.2[fg]	-	-
Methicillin	**10µg/disc**	20±0.1	10±0.2[a]	10±0.2	25±0.2[a]
Solvent	-	-	-	-	-
Stem Hexane	**25**	-	-	-	-
	50	-	-	-	-
	100	-	-	-	-
	250	-	-	-	-
	500	-	-	-	-
Tetracycline	**30µg/disc**	24±0.2	9±0.2[b]	10±0.1	25±0.2[c]
Solvent	-	-	-	-	-
Stem Acetone	**25**	-	-	-	-
	50	-	-	-	-
	100	5.3±0.2	-	-	-
	250	6.1±0.2	Trace[e]	Trace	-
	500	7.3±0.1	4.2±0.2[b]	4.8±0.2	-
Tetracycline	**30µg/disc**	24±0.2	9±0.2[i]	10±0.1	25±0.2[b]
Solvent	-	-	-	-	-
CD (P=0.05)		N. S	0.3664	N. S	0.3112

- The average of triplicates is mentioned.
- The zone of inhibition is measured in mm.
- Different letters in each column differ significantly (5%) by LSD.

Column Fractionation and Assay of Crude Hexane Extract of *J. Gossypifolia*

Column chromatography of crude hexane extract of *J. gossypifolia* yielded 7 fractions (Fig. **2**). All the fractions were subjected to assay. Out of the 7 column fractions tested, fractions 1-6 did not exhibit any antimicrobial activity against the tested pathogens. Fraction-7 exhibited varying activities against *C. diphtheriae, B. cereus, P. mirabilis, S. flexneri, K. aerogenes* and *S. aureus* (Tables **3** and **4**). None of the Gram-negative organisms tested, except *P. mirabilis* and *S. flexneri* showed marginal inhibition. However, crystalline fraction-7 appreciably inhibited *K. aerogenes* and *S. aureus* at 1000, 500 and 250 ppm. *S. aureus* was sensitive to fraction-7 at MIC of 250 ppm. Crude acetone extract (25-500 ppm) of the roots showed marginal inhibition against *C. diphtheriae* and *P. mirabilis* (250 & 500 ppm).

Table 3: Effect of purified column fraction-7 of *J. gossypifolia* against human pathogens

Plant Part	Conc. tested	C. diphtheriae	P. mirabilis	S. flexneri	B. cereus
Root hexane Fraction-7	25	7.1±0.2^d	-	-	-
	50	8.5±0.2^c	-	-	-
	100	9.0±0.1^b	5.5±0.2^b	-	8.5±0.1^b
Gentamicin	10µg/disc	15±0.2^a	10±0.1^a	12±0.2	20±0.2^a
Solvent	-	-	-	-	-
CD (P=0.05)		0.5997	0.4697	N. S	0.0

- The average of triplicates is mentioned.
- Zone of inhibition is measured in mm.
- Different letters in each column differ significantly (5%) by LSD.

Table: 4: Effect of purified column fraction-7 of *J. gossypifolia* against human pathogens

Plant Part	Conc. tested	K. aerogenes	S. aureus
Root hexane Fraction-7	25	-	-
	50	-	-
	100	-	-
	250	4.9±0.2^e	4.9±0.1^e
	500	6.4±0.1^d	6.1±0.2^d
	1000	7.1±0.2^c	7.0±0.2^c
Cephadroxil	10µg/disc	10.6±0.2^b	15±0.1^a
Solvent Control	-	-	-
CD (P=0.05)		0.2815	

- The average of triplicates is mentioned.
- The zone of inhibition is measured in mm.
- Different letters in each column differ significantly (5%) by LSD.

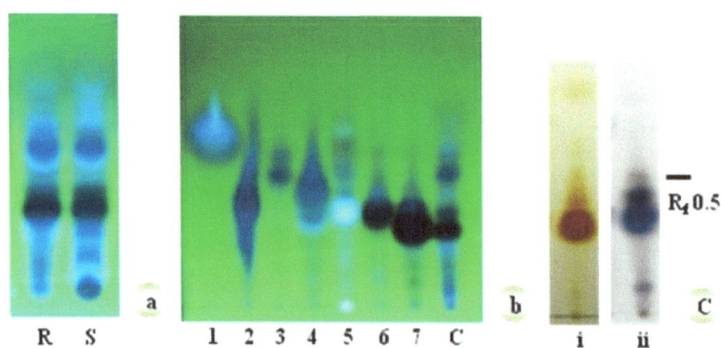

Figure 2: TLC Profile of the crude extract & column fractions of hexane extract of *J. Gossypifolia* roots.

a. Crude extract of the root (R) & stem (S); b. Partially purified column fractions (1-7) along with crude extract (c); c. (i) Fraction-7 developed in Iodine chamber & (ii) sprayed with anisaldehyde-H_2SO_4 (a terpenoid specific spray)

CHARACTERIZATION OF FRACTION-7 USING XRD

Bioactive fraction-7 was observed to be the major compound fraction of the crude extract. Fluorescence was observed under UV 254 nm as dark grey spot on the TLC plate (SiO_2 F_{254} precoated plates). The precipitate in the oil was supersaturated with hexane to which, a drop of MeOH was added and kept undisturbed in a glass vial (1 cm diameter) for 4 days at -4°C. Colorless crystals were formed that were separated and washed thrice with methanol. The final crop was dried on a reaction glass and subjected to XRD analysis. The crystal was characterized as Jatrophone (Fig. **3**), a known antileukemic macrocyclic diterpenoid through a comparative study with (Cambridge Crystallographic Datebase Center) CCDC. Summary of the collection data is mentioned in Table **5**.

Table 5: Summary of crystal data for Jatrophone

Chemical Formula	C_{20} H_{24} O_3
Formula Weight	312.39
Crystal System	Orthorhombic
Space Groups	P212121
Temperature	293(2) K
Wavelength	0.71073 A
Unit Cell Dimensions:	
a (Å)	9.6774(18) A alpha = 90°
b (Å)	10.8057(18) A beta = 90°
c (Å)	17.508(3) A gamma = 90°
V (Å)^3	1830.8 (6) A^3
Z	4, 1.133 Mg/m^3
Absorption Coefficient	0.075 mm^-1
F (000)	672
Index ranges	-9<=h<=8, -7<=k<=10, -16<=l<=16
Crystal size (mm)	0.3 X 0.3 X 0.2
Reflections collected	6231
Unique data	1669 [R (int) = 0.0294]
Completeness to theta	19.85 (99.6%)
Absorption correction	Semi-empirical from equivalents
Max. and min. transmission	0.987 and 0.945
Refinement method	Full-matrix least squares on F^2
Data/ Restraints/Parameters	1669/0/214
Goodness-of-fit	1.077
Final R indices [I>2sigma(I)]	R1 = 0.0357, wR2 = 0.0830
R indices (all data)	R1 = 0.0445, wR2 = 0.0884
Absolute Structure Parameter	0.7(18)
Extinction Coefficient	0.014(2)
Largest diff. Peak and hole	0.132 and –0.110 e. A^-3

DISCUSSION

Previous reports of the antimicrobial activity of *J. gossypifolia* have been vaguely demonstrated. Out of the solvent extracts from the dried leaves and roots of *J. gossypifolia* that were assayed against a set of bacteria and fungi, crude hexane extracts of fresh fruits showed bioactivity against all the tested fungi (*Aspergillus terreus, Fusarium oxysporum* and *Pestalotia palmarum*) and against 2 species of bacteria *i.e. S. aureus* and *B. subtilis* [27]. The present study proved that crude hexane extract and the crystallized jatrophone from the roots of *J. gossypifolia* were inhibitory to human pathogenic bacteria. *C. diphtheriae*, the causative agent of diphtheria in children is of importance and jatrophone exhibited a dose-dependent inhibition (25-100 ppm) against the species.

Since 1990, diphtheria has made a spectacular comeback in several European countries, with a high proportion of cases in adults. In developing countries, immunization of infants with diphtheria toxoid was introduced within the Expanded Programme on Immunization in the late 1970s. Coverage rose slowly to 46% in 1985 and 79% in 1992 [28]. Strengthening the phytochemical arsenal against *C. diphtheriae* and its related species would help us to opt for safer natural alternatives, as an add-on to the immunization regimens. Among the food poisoning bacterial pathogens, *B. cereus* is becoming one of the more important causes of lethality in the industrialized world. It produces one emetic toxin and three different enterotoxins that are highly toxigenic [29]. Fatal ourbreaks of *B. cereus* food poisoning in cooked/ fried rice and pasta are quite common [30]. Jatrophone at 100 ppm concentration inhibited *B. cereus*, the activity that was comparable to 50% efficacy of gentamicin at 10 μg.

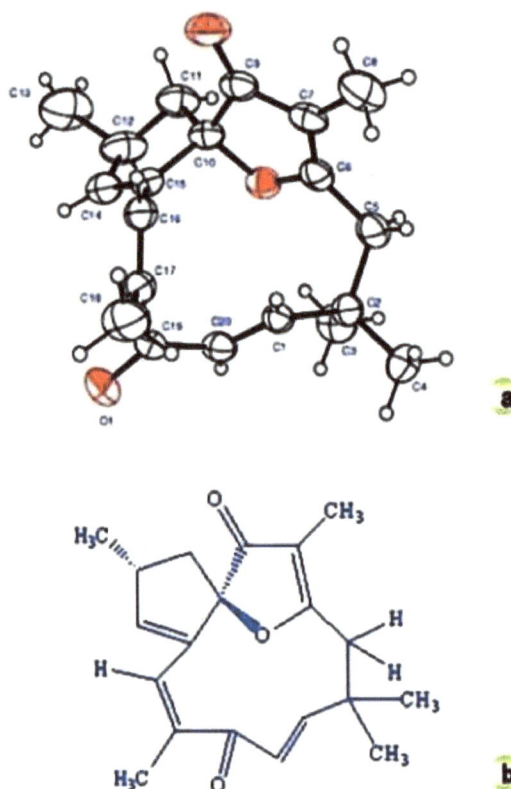

Figure 3: Crystallography of Bioactive fraction-7. **a.** ORTEP image of crystal (Molecular Formula: $C_{20}H_{24}O_3$); **b.** ChemDraw® image of the structure (IUPAC Nomenclature: (E)-(1R,3R)-3,7,11,11,14-Pentamethyl-16-oxa-tricyclo[11.2.1.0*1,5*] hexadeca-4,6,9,13-tetraene-8,15-dione)

JATROPHONES

In an earlier report, the yield of jatrophone was 6.15 g/Kg dry weight of *J. gossypifolia* roots [31], which is similar to the present study. Jatrophone is a compound with a 16 membered ring, which shares homology with the core ring in 'Macrolide' antibiotics that are widely used in the medical industry. However, the latter are rich in specific functionalities and are repeatedly modified to yield semi-synthetic products with greater efficacy.

Chemical modification of jatrophone for specific functionalities might yield derivatives that are more potent and site-specific. Macrolides are used to treat infections such as respiratory tract infections and soft tissue infections and it corroborates to the inhibition caused by jatrophone against *C. diphtheriae* and *K. aerogenes* in the present study. The total synthesis of (±)- jatrophone and its epimer (±)- epijatrophone has been successfully done to yield 0.83% and 0.28% respectively [32]. Hence, large-scale production of the compound could be done for mass production and application.

MODE OF ACTION

Macrocyclic natural products often display remarkable biological activities, and many of these compounds (or their derivatives) are used as drugs. The chemical diversity of these compounds is immense and may provide inspiration for innovative drug design [33]. The mechanism of action of the macrolides is inhibition of bacterial protein synthesis by binding reversibly to the subunit 50S of the bacterial ribosome, thereby inhibiting translocation of peptidyl-tRNA [34]. This action is mainly bacteriostatic, but can also be bactericidal in high concentrations. Macrolides tend to accumulate within leucocytes, and are therefore actually transported into the site of infection [35].

Crystal structures of macrolides complexed with the 50S ribosomal subunit have been emerging since 2001 [36-39] illustrating the stunning molecular mechanism of action of these compounds. Macrolides act as 'plugs' that block the progression of the nascent peptide through the ribosomal exit tunnel, which eventually leads to dissociation of the peptidyl-tRNA from the ribosome [40]. Three-dimensional structures are now available of 50S complexes with members of all important macrolide classes, including 14- and 16-membered lactones, such as erythromycin and tylosin, as well as the semi-synthetic 14-membered ketolides and 15-membered azalides [41].

Jatrophone, the 16 membered lactone-ringed macrocyclic antibiotics, yet to be commercialized, should have similar mode of action against the susceptible organisms. It has 2 chiral centers in its structure that might augment or nullify its bioactivities. The antimicrobial activity might be attributed to the furan 3-one and α, β-unsaturated ketone functionalities. Antibiotics have a long history in ribosome research and it is now, with the availability of crystal structures, that we can fully appreciate the astounding level of detail to which the intricacies of the ribosomal machinery have been elucidated in the past using small-molecule ligands as instruments of investigation. Even now, with three dimensional structures available of the ribosome and its components, small molecules remain irreplaceable tools for the study of ribosome function [42, 43].

However, in the present case docking experiments will give a better understanding of the elusive action of jatrophone against human bacterial infections. Further research on the class of molecules and their long-term toxicity studies would lead to a clear picture of its complexity and possibility for mass-utility drugs against infections caused by Gram positive and Gram negative organisms.

ACKNOWLEDGEMENTS

The authors are willing to thank Mr. M. M. Saravanan, laboratory technician (MSSRF) and Ms. Kalaivani, project student, for their unerring assistance.

ABBREVIATIONS USED

WHO: World health organization

DPT: Diphtheria pertussis toxoid

LC_{50}: Lethal concentration that effectively kills 50% of the population

Ppm: Parts per million

IC$_{50}$: The half maximal inhibitory concentration

MCF7: Michigan cancer foundation

NCI: H460- Non-small cell lung carcinoma

SF- 268: Glioblastoma cell line

ATCC: American type culture collection

A.L.M.PGIBMS: A. L. Mudaliar Post Graduate Institute of Basic Medical Sciences

Cm: Centimeter

Gm: Grams

Mg: Milligram

XRD: X-Ray diffraction crystallography

SiO$_2$: Silica gel

SAIF: Sophisticated analytical instrumentation facility

IIT: Indian institute of technology

ORTEP: Oak ridge thermal ellipsoidal plot

CD: Cumulative difference

LSD: Least standard deviation

N.S: Non significant

MIC: Minimum inhibitory concentration

UV: Ultra violet

TLC: Thin layer chromatography

MeOH: Methanol

CCDC: Cambridge crystallographic database center

μg: Micrograms

Kg: Kilogram

tRNA: Transfer ribonucleic acid

REFERENCES

[1]   Cowan M M. Plant Products as Antimicrobial agents. Clinical Microbiology Reviews 1999; 12: 564 - 582.
[2]   Neto CC, Owens CW, Langfield R, Comeau, AB, Onge J, Vaisberg AJ, *et al.* Antibacterial activity of some Peruvian medicinal plants from the Callejon de Huaylas. Journal of Ethnopharmacology 2002; 79: 133-8.

[3]   Bose PK. On some biochemical properties of natural coumarins. Journal of Indian Chemical Society 1958; 58: 367-375.

[4]   Rojas R, Bustamante B and Bauer J. Antimicrobial activity of selected Peruvian medicinal plants. Journal Of Ethnopharmacology 2003; 88: 199- 204.

[5]   Galazka AM, Robertson SE and Oblapenko GP. Resurgence of diphtheria. *European* Journal of Epidemiology 2005, 11(1): 95 – 105.

[6]   Trabajo EN, Tupasi TE and Kaneko Y. Immunity to Diphtheria in Children in a Rural Community of Baclayon Municipality, Bohol. *Philippine Journal of Microbiology and Infectious* Diseases 1989, 18(1): 22-24.

[7]   Deeni Y, Hussain H. Screening for antimicrobial activity and for alkaloids of *Nauclea latifolia*. Journal of Ethnopharmacology 1991; 35: 91 - 96.

[8]   Sener B, Bingo F, Erdogan I, Bowers WS, Evans PH. Biological activities of some Turkish medicinal plants. Pure and Applied Chemistry 1998; 70(2): 403 - 406.

[9]   Fung D, Lau CB. Schizonepeta tenuifolia: chemistry, pharmacology, and clinical applications. Journal Of Clinical Pharmacology 2002; 42: 30-36.

[10]   Zaidi MA, Huda A, Crow Jr., SA. Pharmacological Screening of Arceuthobium oxycedri (Dwarf Mistletoe) of Juniper Forest of Pakistan. Online Journal of Biological Sciences 2006; 6(2): 56-59.

[11]   Pauli A, Schilcher H. Specific Selection of Essential Oil Compounds for Treatment of Children's Infection Diseases. Pharmaceuticals 2004; I: 1 – 30.

[12]   Rotstein A, Lifshitz A, Kashman Y. Isolation and Antibacterial Activity of Acylphloroglucinols from Myrtus communis. Antimicrobial Agents and Chemotherapy 1974; 6(5): 539 - 542.

[13]   Appendino G, Jakupovic S, Tron GC, Jakupovic J, Milon V, Ballero M. Macrocyclic diterpenoids from *Euphorbia semiperfoliata*. Journal of Natural Products 1998; 61: 749 – 756.

[14]   Aiyelaagbe OO, Adesogan EK, Ekundayo O, Adeniyi BA. Antimicrobial activity of the roots of *Jatropha podagrica* (Hook.). Phytotherapy Research 2000; 14(1): 60 - 62.

[15]   Santos AFD, Euzébio A, Sant´Ana G. Molluscicidal activity of the diterpenoids jatrophone and jatropholones A and B isolated from *Jatropha elliptica* (Pohl.) Muell. Arg., Phytotherapy Research 1999; 13(8): 660 – 664.

[16]   Hirschmann SG, Razmilic I, Sanrain M. Antiprotozoal activity of jatrogrossidione from *Jatropha grossidentata* and *Jatropha isabellii*. Phytotherapy Research 1996; 10(5): 375 - 378.

[17]   Das B, Das R. Medicinal properties and chemical c   onstituents of *Jatropha. gossypifolia* Linn. Indian Drugs 1994; 31: 562 –567.

[18]   Manjunatha Reddy GV, Girish R, Uma MS, Srinivas N. Acaricidal activity of aqueous extracts from leaves and bark of *Cinnamomum* and *Jatropha* against two spotted spider mite, Tetranychus urticae Koch. Karnataka J. Agric. Sci. 2009; 22(3): 693-695.

[19]   Kupchan SM, Sigel CW, Matz MJ. Jatrophone, a novel macrocyclic diterpenoid tumor inhibitor from *J. gossypiifolia*, Journal of Chemical Society 1970; 92(14): 4476.

[20]   Purushothaman KK, Chandrasekharan S, Cameron A, Connolly JD, Labbé C, Maltz A, *et al.* Jatropholones A and B, new diterpenoids from the roots of *Jatropha gossipiifolia* (Euphorbiaceae). Crystal structure analysis of Jatropholone B.Tetrahedron Letters 1979; 979-980.

[21]   Mawardi R, Mohd IH and Toi YL. Jatropholone-A and Jatrophatrione, two diterpenes in *J. gosspifolia*. Pertanika 1990; 13(3): 405 - 408.

[22]   Chatterjee A, Das B, Pascard C, Prange T Crystal structure of a lignan from *Jatropha gossypifolia*. Phytochemistry 1981; 20(8): 2047-2048.

[23]   Das B, Rao SP, Srinivas KVNS. Isolation of Isogadain from *Jatropha gossypifolia*. Planta Medica 1996; 62: 1- 90.

[24]   Das B, Anjani G. Studies on phytochemicals, Part- 29: Gossypidien, a lignin from *Jatropha gossypifolia*. Phytochemistry 1999; 51(1): 115 - 117.

[25]   Hohmann J, Vasas A, Gunther G, Dombi G, Blazso G, Falkay G, *et al.* Jatrophane diterpenoids from *Euphorbia peplus*. Phytochemistry 1997; 51(5): 673-7.

[26]   Valente C, Pedro M, Duarte A, Nascimento MS, Abreu PM, Ferreira MJ. Bioactive diterpenoids, a new jatrophane and two ent-abietanes, and other constituents from *Euphorbia pubescens*. Journal of Natural Products 2004; 67(5): 902-4.

[27]   Madhumathi S, Mohan MSS, Radha R. Anti-microbial properties of crude fruit extracts of *Jatropha gossypifolia*. Journal of Medicinal and Aromatic Plant Sciences 2000; 22: 717-720.

[28]   Galazka AM, Robertson SE. Diphtheria: Changing patterns in the developing world and the industrialized world. European Journal of Epidemiology 1995; 11(1): 107-117.

[29] Granum PE, Lund T Bacillus cereus and its food poisoning toxins. FEMS Microbiol Lett. 1997; 157(2):223-8.

[30] Dierick K, Coillie EV, Swiecicka I, Meyfroidt G, Devlieger H, Meulemans A, *et al.* Fatal Family Outbreak of *Bacillus cereus*-Associated Food Poisoning. Journal of Clinical Microbiology 2005; 43(8): 4277-4279.

[31] Kupchan SM, Sigel CW, Matz MJ, Gilmore CJ, Bryan RF. Tumor Inhibitors: Structure and stereochemistry of Jatrophone, a novel macrocyclic diterpenoid tumor inhibitor. Journal of American Chemical Society 1976; 98(8): 2295 – 2300.

[32] Gyorkos AC, Stille JK, Hegedus LS. The total synthesis of (+)-epi-jatrophone and (+)-jatrophone using palladium-catalyzed carbonylative coupling of vinyl triflates with vinyl stannanes as the macrocycle-forming step. *J. Am. Chem. Soc.* 1990; *112* (23): 8465–8472.

[33] Wessjohann LA, Ruijter E, Garcia-Rivera D and Brandt W. What can a chemist learn from nature's macrocycles? – A brief, conceptual view. Molecular Diversity. 2005; 9: 171-186.

[34] Giambattista MDi, Vannuffel P, Sunazuka T, Jacob T, Omura S, Cocito C. Antagonistic interactions of macrolides and synergimycins on bacterial ribosomes. Journal of Antimicrobial Chemotherapy 1986; 18: 307-315.

[35] Honeybourne D, Kees F, Andrews JM, Baldwin D, Wise R. The levels of clarithromycin and its 14-hydroxy metabolite in the lung. Eur Respir J 1994; 7: 1275–1280.

[36] Schlunzen F, Zarivach R, Harms J, Bashan A, Tocilj A, Albrecht R, *et al.* Structural basis for the interaction of antibiotics with the peptidyl transferase centre in eubacteria. Nature 2001; 413:814-821.

[37] Hansen JL, Ippolito JA, Ban N, Nissen P, Moore PB, Steitz TA: The structures of four macrolide antibiotics bound to the large ribosomal subunit. Mol Cell 2002; 10:117-128.

[38] Schluenzen F, Harms JM, Franceschi F, Hansen HA, Bartels H, Zarivach R, Yonath A: Structural basis for the antibiotic activity of ketolides and azalides. Structure 2003; 11:329-338.

[39] Berisio R, Harms J, Schluenzen F, Zarivach R, Hansen HA, Fucini P, Yonath A: Structural insight into the antibiotic action of telithromycin against resistant mutants. J Bacteriol 2003; 185:4276-4279.

[40] Tenson T, Lovmar M, Ehrenberg M: The mechanism of action of macrolides, lincosamides and streptogramin B reveals the nascent peptide exit path in the ribosome. J. Mol. Biol. 2003; 330:1005-1014.

[41] Hermann T. Drugs targeting the Ribosome. Current Opinion in Structural Biology. 2005; 15:355–366.

[42] Starck SR, Qi X, Olsen BN, Roberts RW: The puromycin route to assess stereo- and regiochemical constraints on peptide bond formation in eukaryotic ribosomes. J Am Chem Soc 2003; 125:8090-8091.

[43] Fredrick K, Noller HF: Catalysis of ribosomal translocation by sparsomycin. Science 2003; 300:1159-1162.

CHAPTER 7

Biological Control of Plant Diseases by *Serratia* Species: A Review or a Case Study

Dipanwita Saha[1,*], Gargi Dhar Purkayastha[1] and Aniruddha Saha[2]

[1]Department of Biotechnology and [2]Department of Botany, University of North Bengal, Siliguri- 734013, India

Abstract: Plant pathogens pose a substantial threat to the production stability and the maintenance of quality of food, feed, fibre and now fuel. Certain bacterial strains have the capacity to prevent plant diseases in natural environments and may be used to replace chemical control measures now prevalent but potentially hazardous. Members of the genus *Serratia*, a gram negative bacterium which has been found to be frequently associated with rhizosphere of several plants has been studied for biocontrol mechanisms and application procedures. Some selected strains of *S. plymuthica*, *S. marcescens* and *S. liquefaciens* have been found to reduce disease severity to a desirable extent using specific application strategies. How *Serratia* achieves this ability for protection against pathogenic fungi has been discussed in detail. These strains produce antibiotics such as the red pigment prodigiosin and pyrrolnitrin. Besides they produce chitinases and siderophores which help to limit fungal growth. Regulatory processes at transcriptional and post transcriptional levels control the production of autoinducer signal molecules and the antibiotic pyrrolnitrin. Induced systemic resistance is another important mechanism involved in biological control of root pathogens by *Serratia* species.

Keywords: Biological control, *Serratia* spp., plant disease, antibiotics, chitinase, ISR.

INTRODUCTION

Almost all the cultivated crop plants on earth are prone to attack by a wide range of pathogens which cause severe diseases leading to substantial yield loss. Apart from our food, we depend on plants for feed, fibre and now for fuel also. Thus, a better control of pests and diseases are necessary to prevent socio-economic disasters especially in the developing countries. Soil-borne pathogens which include fungus, bacteria and nematodes often have overwhelming effects on field and are difficult to control by application of traditional methods like crop rotation and breeding of resistant varieties. These pathogens utilize the root exudates of host plant for survival and overcome the competition for nutrients and minerals with other microbial populations in the soil by penetrating the roots of host plant and residing there [1]. Over more than 100 years, use of chemical pesticides to combat the plant pathogens proved to be very effective for the farmers worldwide. But chemical solutions to plant disease problems have become unpopular due to several reasons such as non target environmental impacts [2], growing cost of pesticides, modified safety regulations [3, 4], development of pathogen resistance [5] and ineffectivity of chemicals in fastidious cases [4].

With increased pressure to reduce the dominance of chemicals accompanied by stricter regulations decreasing the available number of active molecules for use and a growing consumer demand for organic food, there has been considerable interest in the development of alternative ecofriendly methods in plant disease control. The term biological control has been used in plant protection studies to describe the use of live organisms in order to restrict the growth and proliferation of pests and pathogens. In plant pathology, biological control applies to the use of microbial antagonists for the suppression of diseases. These antagonists follow a variety of mechanisms that may directly affect the pathogen or may work indirectly by enhancing host resistance. Direct methods include competition for nutrients, space or infection sites; antibiosis; parasitism; production of cell wall degrading enzymes; and degradation of pathogenicity factors. Indirect methods include induced resistance and plant growth promotion [6, 7]. The microbial world is

*Address correspondence to Dipanwita Saha: Department of Biotechnology, University of North Bengal, Siliguri- 734013, India; Phone: +91-3532776385, Fax: +91- 353-2699001, E-mail: dsahanbu@yahoo.com

Aakash Goyal and Priti Maheshwari (Eds)

extremely rich in its diversity and is an endless reservoir of organisms that may be used to combat plant pathogens [8]. Literature studies reveal the utilization of a wide spectrum of bacteria as inoculums in plant health management practices to control or inhibit plant pathogens and stimulate plant growth [9-15].

These bacteria follow a variety of mechanisms which includes antagonism by producing antimicrobial substances such as antibiotics, cell wall degrading enzymes and hydrogen cyanide. Inhibition may also be achieved by indirect antagonism by secreting iron chelating siderophores and induction of resistance in host plants [6].

Although early interests on biological control was focused mostly on antagonistic mechanisms that were tested *in vitro*, inconsistency in field performance of biocontrol agents led to the understanding of the importance of *in vivo* studies. There are still only a few formulated products and those available do not always produce predictable results [16]. Development of advanced techniques in the recent years has provided new insights in the complex mechanisms that are involved in antagonism. Presently, biocontrol agents are often being tested directly in the field rather than relying on experiments done in plantlets under controlled conditions in the greenhouse. The focus is now on obtaining information on traits that contribute to disease suppression [17, 18], ecological fitness of antagonistic microorganisms [19-21] and genes that regulate disease suppression mechanism [22-24]. Unfortunately, in comparison to the wide variety of reported biocontrol strains, very limited number of bacteria has actually been studied in sufficient detail and successfully formulated and a majority have not been utilized to their capacity.

Most studies on the biocontrol of plant pathogens have been carried out on two genera, *Pseudomonas* and *Bacillus* which functions simultaneously as biocontrol and plant growth promoting agents. Members of these genera have been the subject of intense research at the genetic and biochemical level because of several reasons. *Pseudomonas* spp. are isolated easily from the rhizosphere, easy to culture as they utilize a variety of substrates and easy to manipulate genetically [7]. *Bacillus* spp. on the other hand are spore forming, hence they are considered better candidates to overcome the formulation problem that has long been a major hurdle in successful implementation of biological control [8]. However, there are numerous other genera which have shown excellent potential but their studies have received less attention in literature on biocontrol agents. In this short review we have highlighted the importance of the genus *Serratia* which has shown striking competence in controlling plant diseases. We focus on the recent progress in understanding the underlying mechanisms of beneficial traits of *Serratia* spp. and provide examples that illustrate the efficiency of the members of this genus.

THE GENUS *SERRATIA*

Bacterial species belonging to the genus *Serratia* occur worldwide and can be easily isolated from soil, water, digestive tract of animals, insects and plant material. In addition they occur as opportunistic pathogens in fish and are found in clinical specimens such as blood, wound exudates, peritoneal fluid and burn injury sites [25-27]. They are gram-negative, motile by peritrichous flagella, non spore forming, rod-shaped bacterium belonging to the Enterobacteriaceae family. They are facultative anaerobes possessing both respiratory and fermentative metabolism and are catalase positive. *Serratia* can be distinguished from other members of the same family by the production of three different enzymes DNase, lipase and gelatinase and they are also usually resistant to the antibiotics colistin and cephalothin [28-30]. Another special character is the production of the cell associated red pigment prodigiosin mainly by environmental isolates [31, 32] (Fig. **1**). *S. marcescens* strains are known nosocomial pathogens [27]; however, non pigmented strains are clinically more significant [33]. Other members such as *S. plymuthica* and *S. liquefaciens* are rarely reported as pathogens [34]. The biocontrol agent *S. plymuthica* did not display any adverse effect when tested on the model organism *Caenorhabditis elegans* [35].

Serratia spp. is frequently found to be associated with plants. Investigations have shown that rhizosphere inhabiting strains of *S. plymuthica*, *S. marcescens* and *S. liquefaciens* possessed antifungal activity against a number of fungal pathogens [36-43] (Fig **1**). Endophytic strains of *S. plymuthica* have been reported to induce resistance in plants [44]. Some strains have beneficial effect on plant growth and development [38,

45-47]. There are also reports on the antagonistic activity of *S. marcescens* and *S. entomophila* against root nematodes and pests [48-50].

APPLICATION OF *SERRATIA* FOR DISEASE CONTROL

Serratia strains have been used as effective biocontrol agents since the late 1980's and the frequency of such reports has increased steadily in the last two decades [36-45]. Several studies worldwide has shown that members of these genera are promising antagonists and markedly inhibited soil borne pathogens as well as foliar fungal diseases in a wide variety of crops (Table **1**). Some workers have used this to control post harvest diseases also. *Serratia* sp. is isolated most often from the rhizosphere and has been found to control plant diseases associated with the roots. Ordenlitch *et al.,* [37] isolated a potential biocontrol *S. marcescens* strain from soil that inhibited *Sclerotium rolfsii* to a level of 75% and *Rhizoctonia solani* to a level of 50% in bean under greenhouse conditions. Different methods were applied for inoculation of the antagonist to the soil and it was found that drench and drip application of bacterial suspension were more effective for the control of *S. rolfsii* than seed coating, spraying and soil mixing. Root colonization studies using a natural rifampicin mutant of the isolated strain of *S. marcescens* which was equally effective as the wild-type isolate revealed that the proximal portion of the root contained highest density of bacteria which decreased considerably until the tips where the population was found to increase again.

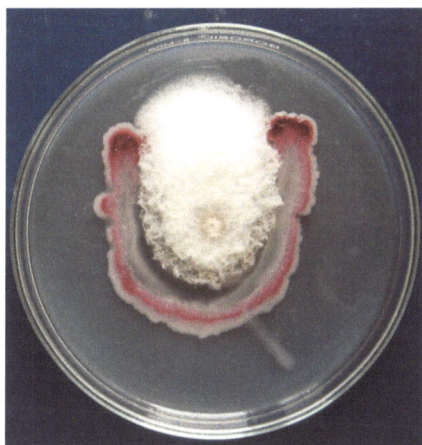

Figure 1: *Serratia* strain showing *in vitro* inhibition of growth of the fungus *Colletotrichum* sp. in dual culture test.

Table 1: *Serratia* strains reported as biocontrol agents

Biocontrol Strain	Tested Plant (disease)	Target Pathogen	References
S. marcescens B2	Carnation (wilt)	*Fusarium oxysporum*	[36]
	Broad bean (chocolate spot disease)	*Botrytis* sp.	[51]
	Cyclamen	*Rhizoctonia solani,* *Fusarium oxysporum*	[52]
	Cyclamen	*Botrytis cinerea*	[83]
	Rice (Sheath blight)	*Rhizoctonia solani*	[41]
S. marcescens	Bean, peanut and chick pea	*Sclerotium rolfsi*	[37]
	Bean (damping off)	*Rhizoctonia solani*	[37]
S. marcescens 90-166	Cucumber (Fusarium wilt)	*Fusarium oxysporum*	[56]
	Cucumber (Bacterial angular leaf spot)	*Pseudomonas syringae*	[55]
	Cucumber (anthracnose)	*Colletotrichum orbicularae*	[53, 54]
S. marcescens 9M5	Kentucky blue grass (summer patch syndrome)	*Magnaporthe poae*	[58]

Table 1: cont....

S. marcescens F-1-1	Cucumber (damping off)	*Phytophthora capsici*	[80]
S. marcescens R-35	Citrus (*C. limonia*) (gummosis)	*Phytophthora parasitica*	[59]
S. marcescens NBRI1213	Betelvine (foot & root rot)	*Phytophthora nicotianae*	[45]
S. marcescens N45	Cucumber, muskmelon, cantaloupe, Pumpkin (Damping off)	*Pythium ultimum*	[42]
S. marcescens	Rice (blast)	*Pyricularia oryzae*	[60]
S. plymuthica R1GC4	Cucumber	*Pythium ultimum*	[44]
S. plymuthica HRO-C48	Strawberry (wilt) (root rot)	*Verticillium dahliae* *Phytophthora cactorum*	[39]
	Oil seed rape (Verticillium wilt)	*Verticillium dahliae*	[63]
	Cucumber (damping off)	*Pythium aphanidermatum*	[115]
	Tomato, beans	*Botrytis cinerea*	[115]
S. plymuthica A21-4	Pepper (Phytophthora blight)	*Phytophthora capsici*	[40]
S. plymuthica IC14	Cucumber	*Botrytis cinerea* *Sclerotinia sclerotiorum*	[107]
	Orange (green mould)	*Penicillium digitatum*	[90]
	(blue mould)	*Penicillium italicum*	[90]
S. plymuthica IC1270 (Previously *Enterobacter agglomerans*)	Beans	*Rhizoctonia solani*	[72]
	Cucumber	*Pythium aphanidermatum*	[72]
	Orange (green mould)	*Penicillium digitatum*	[90]
	(blue mould)	*Penicillium italicum*	[90]
S. plymuthica 3Re4-18 (also designated as B4)	Potato	*Rhizoctonia solani*	[67]
	Lettuce		[66]
	Sugarbeet		[65]
S. plymuthica 5-6	Potato	*Fusarium sambucinum*	[43]
S. plymuthica C1	Pepper (Phytophthora blight)	*Phytophthora capsici*	[94]

Akutsu *et al.,* [51] isolated a highly chitinolytic bacterium *S. marcescens* B2 from tomato phylloplane which markedly inhibited the growth of *Botrytis allii*, *B. hyssoidea*, *B. cinerea*, *B. fabae* and *B. tulipae*. The culture filtrate of the bacterium suppressed conidial germination and hyphal growth of *B. cinerea* and *B. fabae*. The chitinolytic activity was linked to the inhibitory effect of the bacterium. The bacterium reduced chocolate spot disease caused by *B. faba* to 40% when tested on broad bean leaf disks. In a further study, Someya *et al.,* [52] found that *S. marcescens* B2 inhibited fungal diseases caused by *Rhizoctonia solani* and *Fusarium oxysporum* f. sp. *cyclaminis* in cyclamen plants in the greenhouse. After application to soil, the strain B2 survived at approximately 10^6 to 10^7 CFU/g for 4 months under greenhouse conditions. The bacterium also suppressed the germination of *R. solani* sclerotia *in vitro*. The authors further observed the presence of chitinolytic enzymes and antifungal low-molecular-weight compounds in filtrates of *S. marcescens* B2. Later the same strain was found to inhibit the mycelial growth of the rice sheath blight pathogen *Rhizoctonia solani* AG-1 IA [41]. Suspensions of *S. marcescens* cells were sprayed on rice plants which were subsequently challenge inoculated with the pathogen. Incidence of sheath blight was effectively reduced on application of *S. marcescens* which remained in rhizosphere soil at a concentration of 10^8 CFUs g^{-1} soil for more than 4 weeks under glass house conditions. The authors suggested *S. marcescens* strain B2 has potential as an effective and persistent biological control agent for rice sheath blight [41].

S. marcescens strain 90-166 was observed to control several diseases by inducing systemic resistance against *Colletotrichum orbiculare* [53-54], *Pseudomonas syringae* pv. *Lachrymans* [55], *F. oxysporum* f. sp. *cucumerinum* [56] and cucumber mosaic virus [57]. Another isolate *S. marcescens* 9M5 was found to reduce incidence of summer patch on Kentucky bluegrass caused by *Magnaporthe poae* in growth chamber studies [58]. Queiroz and Melo [59] isolated *S. Marcescens* strain R-35 from washed root surface of healthy citrus plants. In greenhouse trials, the strain suppressed more than 50% of root rot disease in citrus (*C. limonia*) caused by *Phytophthora parasitica*. Jaiganesh *et al.,* [60] studied management of rice blast caused by *Pyricularia oryzae* using talc based formulations of *S. marcescens*. In field experiments, foliar spraying of 2.5kg/ha produced maximum disease reduction. Bacterial population in the phyllosphere was also found to be highest at the same dose. In pot experiments, maximum efficacy for controlling blast was achieved when talc-based inoculum was applied on seeds at 10g/kg.

Alström and Gerhardson [61] studied a bacterium, frequently isolated from roots of various plant species and identified it as *Serratia plymuthica* strain G15. The isolated strain exhibited strong antagonism against *Botrytis cinerea* and *Gerlachia nivalis* and moderate antagonism against *Rhizoctonia solani, Fusarium culmorum* and *Pythium* sp. In greenhouse experiments, the bacterium significantly increased growth of lettuce plants when applied to the roots under non-sterile conditions.

The use of *S. plymuthica* strain HRO-C48 as plant growth promoter and biocontrol agent to control *Verticillium* wilt and *Phytophthora* root rot in strawberry was reported by Kurze *et al.,* [39]. The authors selected the bacterial strain isolated from the rhizosphere of oilseed rape by Kalbe *et al.,* [38] since the bacterium possessed high chitinolytic activity *in vitro*, had the ability to produce plant growth hormone indole-3-acetic acid, had low level of resistance to antibiotics and was comparatively safer to human health and environment. In greenhouse experiments, bacterial treatment reduced the percentage of Verticillium wilt by 18.5% and Phytophthora root rot by 33.4%. Under field conditions, root dipping method of application of *S. plymuthica* HRO-C48 was very effective in controlling both the pathogens and increase in strawberry yield to a considerable amount. A commercial formulation of *S. plymuthica* HRO-C48 was developed and named as 'RhizoStar', produced by E-nema GmbH Raisdorf, Germany [62]. Müller and Berg [63] developed a biological protection strategy by testing several seed treatment techniques based on this plant-beneficial bacterium *S. plymuthica* HRO-C48 to control *Verticillium dahliae* in oilseed rape. They evaluated pelleting, film coating and bio-priming techniques through green house experiments considering the influence on the control activity, cell stability during storage and practical feasibility. Bio-priming and pelleting treatments showed statistically significant results with respect to biocontrol. Additionally, survival of HRO-C48 was the highest using bio-priming at 20°C, and pelleting at 4°C. The authors concluded that the application of biopriming was the most suitable method for inoculating oilseed rape seeds with HRO-C48.

S. plymuthica strains have been found to be frequently associated with plants (reviewed by De Vleesschauwer and Höfte [64]). *S. plymuthica* A21-4 isolated from rhizosphere of onion was evaluated both *in vitro* and *in vivo* for the control of Phytophthora blight of pepper [40]. The bacterium inhibited mycelial growth, germination of zoosporangia and cytospores, and formation of zoospores and zoosporangia of *Phytophthora capsici in vitro*. In pot and green house experiments, the bacterium successfully controlled pepper infection showing 83% control efficacy. *S. plymuthica* 3Re4-18 isolated from the endorhiza was found to be the most effective isolate among 2648 strains associated with potato rhizosphere in controlling soilborne pathogens *Verticillium dahliae* and *Rhizoctonia solani* [65]. The same strain (indicated as B4) was found to control *Rhizoctonia solani* in sugarbeet, lettuce [66] and potato [67]. In another instance, Gould *et al.,* [43] employed *Serratia grimessi* and *S. plymuthica* for suppressing dry rot of potato caused by *Fusarium sambucinum*. Results of *in vivo* studies indicated significant reduction (upto 77%) in dry rot formation by both the strains. However, due to potential risk to human health associated with the use of *S. grimessi*, the authors recommended the use of *S. plymuthica* as a biofungicide. The results indicate that *Serratia* strains hold great potential as broad spectrum biocontrol agents that may be applicable to a wide variety of plants.

Apart from *S. marcescens* and *S. plymuthica*, other biocontrol species reported were *S. liquefaciens* [36, 38] *S. rubidaea* [38] and *S. grimesii* [68, 69]. *S. liquefaciens* isolated from rhizosphere of carnation provided protection against *Fusarium*-wilt in carnation caused by *Fusarium oxysporum* f. sp. dianthi. *S. liquefaciens* was recovered from carnation stem segments along the stem up to the top after 60 days and up to 2.5cm after 120 days [36].

BIOCONTROL MECHANISMS USED BY *SERRATIA*

Serratia uses several strategies to antagonize the phytopathogens like the production of antibiotics, cell-wall degrading enzymes, competition for essential micronutrients and induction of systemic resistance in host plants. These strategies are often used simultaneously by a single strain. Table **2** presents an overview of the different mechanisms or metabolites involved in disease suppression by *Serratia* strains.

Table 2: Overview of the mechanisms or metabolites involved in the biological control of phytopathogens by *Serratia* strains

Serratia strains	Suggested Mechanism/Metabolites Involved in Biocontrol Action	References
S. marcescens B2	Chitinase, prodigiosin, ISR	[51, 83, 113]
S. marcescens 90-166	ISR	[112]
S. marcescens F 1-1	Prodigiosin	[80]
S. marcescens GPS 5	Chitinase, ISR	[106]
S. marcescens 9M5	Chitinase	[58]
S. marcescens NBRI1213	ISR	[45]
S. marcescens N4-5	Chitinase, protease, prodigiosin	[42]
S. plymuthica RIGC4	ISR	[44]
S. plymuthica A153	Antibiotics (pyrrolnitrin, haterumalides)	[73, 91]
S. plymuthica HRO-C48	Antibiotics (pyrrolnitrin, VOC), chitinase, siderophore, IAA, ISR	[93, 106, 115]
S. plymuthica IC 14	Antibiotic (pyrrolnitrin), chitinase, siderophore, IAA	[90, 107]
S. plymuthica IC 1270	Antibiotic (pyrrolnitrin), chitinase, protease, siderophore, IAA, ISR	[72, 89, 116]
S. plymuthica Re4-18	Antibiotic (VOC) chitinase, glucanase, siderophore	[64, 93]
S. plymuthica A21-4	Antibiotic (chlorinated macrolide)	[92]
S. plymuthica C1	Chitinase	[94]

ANTIBIOTICS

Antibiotics are chemically heterogeneous group of organic, low molecular weight compounds secreted by microorganisms which are deleterious to the growth and metabolism of other microorganisms at very low concentrations [70]. In the past few decades, several antibiotics have been isolated and identified from many biocontrol bacterial species. *Serratia* sp. has also been reported to produce a number of antibiotics including prodigiosin, pyrrolnitrin, and chlorinated macrolide haterumalides [71-74]. Prodigiosin (methyl-3-pentyl-6 methoxyprodigiosin), a tripyrrole red pigment antibiotic, is produced by three species of *Serratia* namely *S. marcescens*, *S. rubidaea* and *S. plymuthica* [75]. Apart from *Serratia*, prodigiosin and its derivatives are also produced by members of other genera which include *Pseudomonas*, *Vibrio*, *Alteromonas*, *Rugamonas* and Gram-positive actinomycetes like *Nocardia* sp., *Streptomyces longisporus* ruber and *Streptoverticillium* sp. [34, 76-78]. Prodigiosin is a secondary metabolite [78] with no defined physiological function; however, it has been proposed that prodigiosin functions in an energy-spilling process as a tightly regulated uncoupler of proton transport and ATP synthesis in oxidative phosphorylation [32]. Besides, prodigiosin contributes to survival under competition and possesses antifungal, antibacterial, antiprotozoal, immunosuppressive and anticancer properties [71, 79, 80]. The prodigiosin biosynthesis gene cluster from *S. marcescens* has been cloned and functionally expressed in heterologous hosts. The production of prodigiosin and several other phenotypes was shown to be regulated by N-acyl homoserine lactone mediated quorum sensing system [75, 81, 82].

Being antifungal in nature, prodigiosin plays important role in biological control by suppressing fungal phytopathogens. For instance, the red pigment purified from *S. marcescens* strain B2 inhibited spore germination of the grey mould pathogen *Botrytis cinerea*. Further a synergistic antifungal action of chitinolytic enzymes and prodigiosin was observed when both factors were applied in concert [83]. However, how far prodigiosin is effective in the environment is not clear. Studies have shown that certain plant associated bacteria inhibit biosynthesis of prodigiosin [84] and may also degrade it [18], thus causing hindrance in the biocontrol process. Therefore, in the rhizosphere environment where such interfering strains are present, biocontrol efficacy of prodigiosin producing bacteria is likely to get reduced.

Pyrrolnitrin (3-chloro-4-(2-nitro-3-chlorophenyl)-pyrrole), is an antibiotic with broad spectrum antifungal activity produced by several bacteria some of which have been competently utilized as biocontrol agents against phytopathogens. Several pyrrolnitrin producing strains of genera *Pseudomonas* [85] and *Burkholderia* [86] are reported as biocontrol agents. Biosynthetic gene cluster for pyrrolnitrin, prnABCD has been cloned and sequenced and functionally expressed in *Escherichia coli* [85, 87, 88] and was found to be responsible for disease suppression in plants [87]. *Serratia* spp. have been reported to produce pyrrolnitrin and the antibiotic has been linked to biocontrol capacity in several strains [38, 73, 89]. But under *in vivo* conditions, the involvement of pyrrolnitrin in biological control is not clearly defined. Mezaine *et al.,* [90] observed that a particular mutant strain of *S. plymuthica* IC1270 deficient in pyrrolnitrin production showed similar efficacy against *P. digitatum* as the parent strain on orange fruits. However, under *in vitro* conditions, the mutant strain IC1270-P1 lost its antifungal activity and no inhibition zone was observed when it was tested against *P. digitatum* or *P. italicum*. Ovadis *et al.,* [72] suggested that pyrrolnitrin production by the same strain was more important for the antifungal action than the chitinolytic activity. These results suggest that there is a clear variation in conditions that prevail *in vitro* and *in vivo*. Nevertheless, biocontrol strains are fortified with several antifungal metabolites that allow these bacteria to exercise biocontrol activity under differential environmental conditions.

The chlorinated macrolides, haterumalide NA, B, NE, and X, are polyketide substances with antifungal properties that were first purified from the cell free supernatant of the cultures of the soil bacterium *Serratia plymuthica* strain A153 [73, 91]. Haterumalides NA, B and NE inhibited apothecial formation in sclerotia and ascospore germination of *Sclerotinia sclerotiorum*, and spore germination of several other filamentous fungi as well as Oomycetes. The strain A153 also produced the antifungal metabolites pyrrolnitrin and 1-acetyl-7-chloro-1-H-indole, which in contrast to the haterumalides, did not inhibit the apothecial formation on sclerotia. Pyrrolnitrin, and haterumalide NA, B and NE effectively inhibited spore germination of tested filamentous fungi at concentrations ranging from 0.06 to 50 μg ml^{-1}, whereas 1-acetyl-7-chloro-1-H-indole inhibited spore germination only at concentrations above 50 μg ml^{-1} [73]. Similarly, another chlorinated macrolide named as macrocyclic lactone A21-4 ($C_{23}H_{31}O_8Cl$) was isolated from *S. plumuthica* strain A21-4 which significantly inhibited the formation of zoosporangia and zoospore and germination of cyst of *P. capsici* at concentrations lower than 0.0625 μg/ml. The compound was also effective against *Pythium ultimum*, *Sclerotinia sclerotiorum* and *Botrytis cinerea* [92]. Volatile organic compounds (VOC) are another important group of antifungal molecules produced by *Serratia* strains [93]. Among an array of wide variety of VOC produced by *S. plymuthica* strains HRO-C48 and 3Re4-18 and *S. odorifera* 4Rx13, the antimicrobial compound beta phenyl ethanol was found to be common and was suggested to be responsible for inhibitory effect of these strains on the mycelial growth of *Rhizoctonia solani* [93].

The production of an autoinducer signal N-acyl homoserine lactone (AHL) by the bacterial cell regulates the synthesis of secondary metabolites like antibiotics and exoenzymes and other cellular phenomenon like motility, biofilm formation and production of biosurfactant in *Serratia* [72, 74]. *S. plymuthica* strain HRO-C48 was found to produce several AHLs including N-butanoyl-HSL, N-hexanoyl-HSL and N-3-oxo-hexanoyl-HSL (OHHL). Requirement of the presence of AHLs for the production of pyrrolnitrin was confirmed by mutation studies conducted by Liu *et al.,* [74]. The authors found that a mutant AHL-4 *Serratia plymuthica* strain was incapable of producing pyrrolnitrin and thus lacked the ability to suppress several fungal pathogens. However, pyrrolnitrin production was restored in the same strain by incorporating splIR genes cloned in pUCP26 plasmid. The genes splI and splR are analogues of luxI and luxR genes present in several Gram-negative bacteria. Absence of the global regulator genes grrA, grrS and rpoS in a biocontrol *S. plymuthica* strain IC1270 also restricted the production of pyrrolnitrin and N-acylhomoserine lactone quorum sensing autoinducer molecules. As a result, the ability to antagonize two fungal pathogens *Rhizoctonia solani* and *Pythium aphanidermatum* reduced under greenhouse conditions [72].

CHITINASES

Chitinases, the most studied lytic enzyme produced by biocontrol bacteria, have immense potential for controlling many phytopathogenic fungi [7, 94]. Chitin, the major structural component of most fungal cell walls is 1, 4-beta linked polymers of N-acetyl D-glucoseamine. It is hydrolysed by chitinases that may be produced by several chitinolytic bacteria such as members of the genera *Serratia* [39, 83], *Bacillus* [95, 96], *Enterobacter* [97] and *Pseudomonas* [98, 99]. *Serratia* is one of the most competent chitin-degrading bacteria and several authors have reported the detection and isolation of chitinases from several strains of this genus [58, 83, 100, 101]. Altogether four different chitinases are found to be present in *Serratia* and the chitinase encoding genes has been cloned and sequenced [102-104]. Earlier studies have shown that isolated chitinases are involved in degradation of fungal hyphae and a positive correlation existed between chitinolytic activity and biological control of phytopathgens [37, 97, 102]. Chitinolytic enzymes endochitinase (58-kDa) and chitobiase (98-kDa) purified from strain B2 showed inhibitory effects on the spore germination of pathogenic fungus *Botrytis cinerea* [83]. Biological control disappeared equally in two mutants of *S. plymuthica* IC1270 (previously known as *Enterobacter agglomerans*) one of which lost only chitinolytic activity but not antibiotic or proteolytic activity and the other, which lost all the three activities indicating that chitinolytic enzymes contribute significantly to the antagonistic activity of the strain [97]. Frankowskii *et al.,* [105] purified two chitinolytic enzymes CHIT60 and CHIT100 from the biocontrol strain *S. plymuthica* HRO-C48. Both enzymes inhibited spore germination and germ tube elongation of the phytopathogenic fungus *Botrytis cinerea* at a concentration of 100 μg ml^{-1}, the endochitinase CHIT60 being relatively more effective among the two. Chitin supplementation in foliar sprays increased the efficiency (64%) of chitinolytic *Serratia marcescens* GPS5 in controlling late leaf spot of groundnut caused by *Phaeoisariopsis personata* [106]. Recently, Mehmood *et al.,* [101] characterized a 60 kDa chitinase from *Serratia proteamaculans* 18A1 and demonstrated that the purified enzyme has antifungal activity against the pathogenic fungi *Fusarium oxysporum* and *Aspergillus niger*.

Although earlier studies have confirmed the role of chitinase in biological control, later studies have indicated that several other factors having prominent roles are also involved. In a study, Kamensky *et al.,* [107] found that chitinolytic activity of *S. plymuthica* strain IC14 was not essential for biological control. A 58 KDa endochitinase was found to be the major chitinolytic enzyme. Two mutant derivative of IC14, one with increased chitinolytic activity and the other deficient in chitinolytic activity did not show appreciable difference in production of other antifungal compounds or in suppression of *B. cinerea* and *S. sclerotiorum* *in vitro* or in the greenhouse when applied on cucumber plants. Müller *et al.,* [108] observed that chitinolytic activity of *S. plymuthica* HRO-C48 was regulated by N-acyl homoserine lactone (AHL)-dependent quorum sensing systems. Exo and endo-chitinases have also been detected in *S. plymuthica* strain IC1270 and the production of ChiA endo-chitinase was regulated by the expression of global regulator genes [72]. Altogether the studies show that chitinases are a part of a multi-component mechanism that underlies biological control in the environment; hence a deficiency only in chitinase production has negligible impact on the biocontrol capabilities of an antagonistic strain.

SIDEROPHORES

All microorganisms including soil borne plant pathogenic fungi require iron for growth which they must obtain from their environment. However, the concentration of iron in most soils is too low to support the healthy growth of microorganisms [109]. To survive in such conditions, some soil bacteria secrete low molecular weight high iron affinity molecules called siderophores which chelate and sequester ferric ion [110]. Thus iron becomes less available to the pathogens in the soil which adversely affects their growth and proliferation. Siderophore producers have received a lot of attention because of potential utilization of these chelators in agriculture as biocontrol agents against plant diseases [110]. *Serratia* strains have been shown to secrete siderophores under *in vitro* conditions [38, 107, 111] but there is no direct evidence of siderophore being the major contributor in biological control. Müller *et al.,* [108] observed that siderophore production in *S. plymuthica* strain HRO-C48 is not influenced by quorum sensing mechanism mediated by acyl homoserine lactones. However, siderophores seemed to be involved in induction of resistance in cucumber by *Serratia marcescens* 90-166 since it was affected negatively by high iron concentration [112]. The authors observed that a mutant 90-166 strain that was defective in siderophore production failed to induce systemic resistance in cucumber against *Colletotrichum orbiculariae.* Ovadis *et al.,* [72] studied the regulatory mechanism of siderophore production by *S. plymuthica* strain IC1270. However, neither GrrA/GrrS nor RpoS affected siderophore production by strain IC1270 indicating that these global regulators probably do not have a role in siderophore production.

INDUCTION OF SYSTEMIC RESISTANCE IN HOST PLANTS

Plants naturally use several physical and chemical barriers to restrict the invasion of pathogens. The defence mechanisms involved in such cases are not limited to the tissues attacked by the pathogens but also the distant tissues gets involved in the same mechanism in order to increase the resistance for any further attack by the same pathogen. This mechanism is termed as systemic acquired resistance (SAR) which is characterized by the accumulation of salicylic acid (SA) and pathogenesis related (PR) proteins. A number of potential root-colonizing and fungal antagonistic rhizobacteria like *Serratia*, *Bacillus* and *Pseudomonas* have also been observed to elicit of host defence. These non pathogenic bacteria can suppress plant disease by the mechanism that was termed as 'induced systemic resistance' (ISR) which differed from the pathogen induced SAR [113]. Liu *et al.,* [56] tested *S. marcescens* strain 90-166 for its ability to induce systemic resistance against *Fusarium* wilt of cucumber caused by *Fusarium oxysporum* f. sp. cucumerinum using split-root assay. Results revealed delayed disease symptom development and reduced number of dead plants in comparison to non bacterized controls. In another experiment with the strain 90-166, the same authors showed that ISR activity in cucumber against *Colletotricum orbiculariae* increased over time while the bacterial population decreased suggesting that ISR activity and colonization capacity are not related [55]. Press *et al.,* [112] showed that SA is not involved in induction of resistance in cucumber by *S. marcescens* 90-166 against *C. orbiculariae.* They found that 90-166 mutants defective in SA production retained ISR activity against the pathogen at levels not significantly different from those of the SA positive parental strain. They further observed that systemic induction of resistance was affected by iron concentration in the environment. Disease control decreased in cucumber as the iron concentration supplied in the fertilizer solution increased. But other than the probable involvement of siderophores, exactly how low iron concentration triggers ISR in these plants still remains unclear.

Someya *et al.,* [114] studied induction of resistance by *S. marcescens* strain B2 which controlled rice blast caused by *Pyricularia oryzae* after leaf and root inoculations by bacterial suspension. They observed a decrease in lesion size on rice leaves in plants treated with strain B2 with the rhizosphere. Lipoxygenase (LOX) activity increased rapidly in the leaves, which was closely related to rice blast resistance. Kishore *et al.,* [106] tested chitinolytic *S. marcescens* GPS5 for the ability to activate defence related enzymes in groundnut leaves. There was an enhanced activity of chitinase, beta 1, 3-glucanase, peroxidase and phenylalanine ammonia lyase when groundnut plants were treated with *S. marcescens* GPS5 supplemented with chitin. *S. marcescens* strain NBRI1213 was found to induce activities of plant defence enzymes such as phenylalanine ammonia-lyase, peroxidase, and polyphenoloxidase both locally and systemically in betelvine plants against foot and root rot disease caused by *Phytophthora nicotianae* [45]. Additionally there was considerable increase in the level of phenolics in plants treated with NBRI1213 and inoculated with *Phytophthora nicotianae.* Major phenolics detected by HPLC were gallic, protocatechuic, chlorogenic, caffeic, ferulic, and ellagic acids. Induction of systemic resistance is generally not associated with accumulation of SA or PR proteins [113]. While SA was found clearly to remain uninvolved in induced resistance in cucumber by *S. marcescens* 90-166, in contrast, accumulation of higher levels of PR proteins was associated with the resistance induced by *S. marcescens* GPS5 in groundnut and by *S. marcescens* strain NBRI 1213 in betelvine. Such variations has also been recorded in induction of resistance by the pseudomonads, *Pseudomonas fluorescens* CHA0 and *P. fluorescens* WCS374 and WCS417 which indicates the involvement of SAR [113].

S. plymuthica R1GC4 was observed to induce systemic resistance in cucumber against *Pythium ultimum* [44]. The authors found significant ultrastructural and biochemical changes such as deposition of enlarged callose-enriched wall appositions at sites of potential pathogen penetration and the accumulation of an osmiophilic material in the colonized areas in cucumber root cells that correlated with restriction of fungal colony to the outermost root tissues of bacterized seedlings. Cytochemical investigations revealed that appositions were composed of callose, pectin, and cellulose. The authors provided evidence that susceptible cucumber plants treated with *S. plymuthica* reacted more rapidly and more efficiently to *Pythium* attack through the formation of physical and chemical barriers at sites of potential fungal entry. *S. plymuthica* HRO-C48 was also capable of inducing resistance in bean and tomato against grey mould disease caused by *Botrytis cinerea* [115]. The authors observed that plants treated with HRO-C48 had significantly reduced lesion areas on leaves in comparison to untreated controls. They further showed that AHL signalling was required for the ISR activity. In a recent study, De Vleesschauwer *et al.,* [116] observed enhanced resistance in foliar tissues in rice against the blast pathogen *Magnaporthe oryzae* following root treatment with *Serratia plymuthica* IC1270. This ISR activity was attributed to accumulation of reactive oxygen species which were identified as modulators of antagonistic defence mechanism.

PLANT GROWTH PROMOTION

Plant growth is increased by plant growth promoting rhizobacteria (PGPR) either indirectly by suppressing disease causing pathogens or directly by several mechanisms [7] of which solubilization of nutrients such as phosphates and release of the phytohormone IAA is most commonly observed in biocontrol bacteria. Many soil organisms can suppress disease as well as promote plant growth [117]. Characterization of biocontrol strains most often includes evaluation of plant growth promoting traits whose presence increases the overall quality of a strain. *Serratia* strains used in biological control have been found to secrete IAA [38, 107] and solubilize phosphates [118]. One of the reason behind the selection of the antifungal and chitinolytic rhizobacteria *S. plymuthica* HRO-C48 as a biocontrol agent in several studies [39, 63] is the ability of this strain to secrete IAA. Plant growth promoting ability of HRO-C48 was observed in phytochambers and was confirmed in greenhouse and field trials. An increased number of buds, blossoms and fruit in addition to intense root branching of plants treated with HRO-C48 were observed in greenhouse and field trials. Apart from reducing Phytophthora root rot and Verticillium wilt, dipping of plants in HRO-C48 prior to planting increased the yield by 60% and 296% in strawberry in field experiments on biological control against the pathogens *Phytophthora cactorum* and *Verticillium dahliae* respectively [39]. *S. marcescens* strain 90-166 which protected plants through induction of resistance, promoted growth of hormone mutants of Arabidopsis *in vitro* but failed to affect plant growth in green house conditions [119]. *S. marcescens* strain NBRI1213 which controlled mortality of betelvine (*Piper betel* L.) due to foot and root rot disease caused by *Phytophthora nicotianae* by 27%, attained an average 81%, 68%, 152% and 290% increase in shoot length, shoot dry weight, root length and root dry weight in comparison to untreated controls in greenhouse trials [45].

Phosphatase producing ability is considered as a beneficial trait for biocontrol organisms as they release inorganic phosphate which is an essential nutrient for the plants. Phosphatase production by *S. plymuthica* IC1270 was shown to be controlled by the participation of global regulators GrrS (sensor kinase GacA/GacS-like regulatory system and sigma S subunit of RNA polymerase [118].

CONCLUSION

Biological control has achieved prime importance in the recent years as it directly addresses the issue of environmental safety by helping to reduce the use of chemicals in agriculture. Microbial agents are preferred to chemicals because they are safer, more targeted, requirement is less, risk of resistance development is lower and may be integrated to conventional procedures [120]. In the past twenty five years, scientists throughout the globe have found numerous strains of *Serratia* to possess immense potential as biocontrol agents. However, very few of these strains have been explored commercially due to barriers such as lack of effective formulation and probably more because of their appearance as pathogens. But with the development of better techniques in risk assessment [35], the evaluation process now requires less time and may be employed without much difficulty before using the biocontrol strains in formulations. A successful viable formulation with gram negative bacteria is a major obstacle for their large scale use [8]. However, recent studies on this aspect [63] open up new possibilities for obtaining a stable inexpensive product that may allow the research marvels to reach the farmers plots.

Another safety issue is the impact of introduced strains on indigenous microbiota. Although rhizosphere itself is the origin of most biocontrol strains, their mass application to the plant roots could probably have negative effect on the natural microflora including plant beneficial bacteria. However, several studies have indicated that introduction of antagonistic strains in the rhizosphere causes only minor and transient effects on microbiota composition and hence pose no definable non target risk [62, 68]. Another important concern is inconsistent and variable performance by biocontrol agents caused by rhizosphere incompetence that reduces the beneficial effects and fails to produce desired results [18, 121]. Future research can therefore aim to analyze the complex interactions between the introduced strains, the pathogen, the plant and the biotic and abiotic components of the environment so that failures may be avoided and the true efficiency under field conditions may be practicable for developing an overall environment-friendly agriculture. Work can be more focused on self perpetuating biocontrol strains with high colonizing ability as they are long lasting and therefore cost-effective. Continued work on the favourable parameters required for production of antimicrobials or other vital molecules holds potential for better disease suppression by biocontrol strains targeted for specific crops. For example, optimum light requirements for prodigiosin production [122], amendments with chitin [106], or using improved formulation procedures [63] can be very effective. Further strain improvement can be targeted with those having otherwise valuable properties (such as formulation ease and exceptional plant colonizing abilities) by creating transgenic strains with multiple biocontrol characters [4] as has been done effectively with *Pseudomonas fluorescens* CHAO [123] Similarly, combined application of multiple strains [124-126] or combining inoculum application with environmental modulations like soil solarisation, appropriate crop rotation and soil aeration as tried with other bacteria [127-129] can also increase the spectrum of target diseases and open up new avenues for integrated pest management strategies using *Serratia* strains.

ACKNOWLEDGEMENTS

Research in bacteria mediated biological control in our laboratory is funded by University Grants Commission, Government of India and Department of Biotechnology, Government of West Bengal, in the form of major research projects.

ABBREVIATIONS USED

AHL: Acyl homoserine lactones

CFU: Colony forming units

DNase: Deoxyribonuclease

IAA: Indole acetic acid

ISR: Induced systemic resistance

PR: Pathogenesis related

SA: Salicyclic acid

SAR: Systemic acquired resistance

VOC: Volatile organic compounds

REFERENCES

[1]   Haas D, Défago G. Biological control of soil-borne pathogens by fluorescent pseudomonads. Nat. Rev Microbiol 2005; 3: 307-19.

[2]   Pimentel D. Amounts of pesticides reaching target pests: Environmental impacts and ethics. J Agr   environ Ethic 1995; 8: 17-29.

[3]   Gerhardson B. Biological substitutes for pesticides. Trends Biotechnol 2002; 20: 338-43.

[4]   Compant S, Duffy B, Nowak J, Clément C, *et al.* Use of plant growth-promoting bacteria for biocontrol of plant diseases: principles, mechanisms of action, and future prospects. Appl Environ Microbiol 2005; 71: 4951-59.

[5]   Van den Bosch F, Gilligan A. Models of fungicide resistance dynamics. Ann Rev Phytopathol 2008; 46: 123-47.

[6]   Pal KK, Gardener BMS. Biological control of plant pathogens. Plant Health Instructor 2006; available from: http://www.oardc.ohio-state.edu/mcspaddengardenerlab/Pal PHI-biologicalControl.pdf

[7]   Whipps J. Microbial interactions and biocontrol in the rhizosphere. Journal of Experimental Botany 2001; 52: 487-11.

[8]   Emmert EAB, Handelsman J. Biocontrol of plant disease: a (Gram-) positive perspective. FEMS Microbiology Letters 1999; 171: 1-9.

[9]   Huang Y, Wong PTW. Effect of *Burkholderia (Pseudomonas) cepacia* and soil type on the control of crown rot in wheat. Plant Soil, 1998; 203: 103-08.

[10]   Ross IL, Alami Y, Harvey PR, Achouak W, *et al.* Genetic diversity and biological control activity of novel species of closely related pseudomonads isolated from wheat field soils in south Australia. Appl Environ Microbiol 2000; 66: 1609-16.

[11]   Berg G, Fritze A, Roskot N, Smalla K. Evaluation of potential biocontrol rhizobacteria from different host plants of *Verticillium dahliae* Kleb. J Appl Microbiol 2001; 91: 963–71.

[12]   Zhang S, Moyne AL, Reddy MS, Kloepper JW. The role of salicylic acid in induced systemic resistance elicited by plant growth-promoting rhizobacteria against blue mold of tobacco. Biol Control 2002; 25: 288-96.

[13]   Sabaratnam S, Traquair JA. Formulation of a *Streptomyces* biocontrol agent for the suppression of *Rhizoctonia* damping-off in tomato transplants. Biol Control 2002; 23: 245-53.

[14]   Collins DP, Jacobsen BJ. Optimizing a *Bacillus subtilis* isolate for biological control of sugar Beet Cercospora leaf spot. Biol Control 2003; 26:153-61.

[15]   Xue QY, Chen Y, Li SM, Chen LF, *et al.* Evaluation of the strains of *Acinetobacter* and e*nterobacter* as potential biocontrol agents against Ralstonia wilt of tomato. Biol Control 2009; 48: 252-58.

[16]   Fravel DR. Commercialization and implementation of biocontrol. Ann Rev Phytopathol 2005; 43: 337-59.

[17]   Viebahn M, Smit E, Glandorf DCM, Wernars K, *et al.* Effect of genetically modified bacteria on ecosystems and their potential benefits for bioremediation and biocontrol of plant diseases – A review. In: Lichtfouse E, Ed. Sustainable agricultural reviews Vol. 2: climate change, intercropping, pest control and beneficial microorganisms. Springer, Netherlands. 2009; pp. 45-69.

[18]   Someya N, Akutsu K. Indigenous bacteria may interfere with the biocontrol of plant diseases. Naturwissenschaften 2009; 96:743-47.

[19]   Pujol M, Badosa E, Manceau C, Montesinos E. Assessment of the environmental fate of the biological control agent of fire blight, *Pseudomonas fluorescens* EPS62e, on apple by culture and Real-Time PCR methods. Appl Environ Microbiol 2006; 72: 2421-27.

[20]   Ryan PR, Dessaux Y, Thomashow LS, Weller DM. Rhizosphere engineering and management for sustainable agriculture Plant Soil 2009; 321: 363-83.

[21]   Edel-Hermann V, Brenot S, Gautheron N, Aimé S, *et al.* Ecological fitness of the biocontrol agent *Fusarium oxysporum* Fo47 in soil and its impact on the soil microbial communities. FEMS Microbiol Ecol 2009; 68: 37-45.

[22]   Heeb S, Haas D. Regulatory roles of the GacS/GacA two-component system in plant-associated and other gram negative bacteria. Mol Plant Microbe Int 2001; 14: 1351-63.

[23]   Wang Y, Huang X, Hu H, Zhang X, *et al.* QscR Acts as an Intermediate in *gacA*-Dependent regulation of PCA Biosynthesis in *Pseudomonas* sp. M-18. Curr Microbiol, 2008; 56: 339-45.

[24]   Barret M, Frey-Klett P, Boutin M, Guillerm-Erckelboudt AY, *et al.* The plant pathogenic fungus *Gaeumannomyces graminis* var. *tritici* improves bacterial growth and triggers early gene regulations in the biocontrol strain *Pseudomonas fluorescens* Pf29Arp. New Phytol 2009; 181: 435-47.

[25]   Clark RB, Janda JM. Isolation of *Serratia plymuthica* from a human burn site. J Clin Microbiol 1985; 21: 656-57.

[26] Reina J, Borrell N, Llompart I. Community-acquired bacteremia caused by *Serratia plymuthica*. Case report and review of the literature. Diagn Microbiol Infect Dis 1992; 15: 449-52.

[27] Hejazi A, Falkiner FR. *Serratia marcescens*. J Med Microbiol 1997; 46: 903-12.

[28] Grimont PAD, Grimont F. The genus *Serratia*. Ann Rev Microbiol 1978; 32: 221-48.

[29] Farmer III JJ, Davis BR, Hickman-Brenner FW, McWhorter A, *et al.* Biochemical identification of new species and biogroups of Enterobacteriaceae isolated from clinical specimens. J Clin Microbiol 1985; 21: 46-76.

[30] Barrow GI, Feltham RKA, Ed. Cowan and Steel's manual for the identification of medical bacteria. 3rd edition. Cambridge University Press1993.

[31] Giri AV, Anandkumar N, Muthukumaran G, Pennathur G. A novel medium for the enhanced cel growth and production of prodigiosin from *Serratia marcescens* isolated from soil. BMC Microbiology 2004; 4: 11. Available from: http://www.biomedcentral.com/1471-2180/4/11

[32] Haddix PL, Jones S, Patel P, Burnham S, *et al.* Kinetic analysis of growth rate, ATP, and pigmentation suggests an energy-spilling function for the pigment prodigiosin of *Serratia marcescens*. J Bacteriol 2008; 190: 7453-63.

[33] Carbonell GV, Della Colleta HHM, Yano T, Darini ALC, *et al.* Clinical relevance and virulence factors of pigmented *Serratia marcescens*. FEMS Immunol Med Microbiol 2000; 28: 143-49.

[34] Khanafari A, Assadi MM, Fakhr FA. Review of Prodigiosin, Pigmentation in *Serratia marcescens*. Online J Biol Sci 2006; 6: 1-13.

[35] Zachow C, Pirker H, Westendorf C, Tilcher R, *et al.* The *Caenorhabditis elegans* assay: a tool to evaluate the pathogenic potential of bacterial biocontrol agents. Eur J Plant Pathol 2009; 125: 367-76.

[36] Sneh B, Agami O, Baker R. Biological control of Fusarium-wilt in carnation with *Serratia liquefaciens* and *Hafnia alvei* isolated from rhizosphere of carnation. J Phytopathol 1985; 113: 271-76.

[37] Ordentlich A, Elad Y, Chet I. Rhizosphere colonization by *Serratia marcescens* for the control of *Sclerotium rolfsii*. Soil Biol Biochem 1987; 19: 747-51.

[38] Kalbe C, Marten P, Berg G. Members of the genus *Serratia* as beneficial rhizobacteria of oilseed rape with antifungal properties. Microbiol Res 1996; 151: 433-39.

[39] Kurze S, Bahl H, Dahl R, Berg G. Biological control of fungal strawberry diseases by *Serratia plumuthica* HRO-C48. Plant Dis 2001; 85: 529-34.

[40] Shen SS, Choi OH, Lee SM, Park CS. *In vitro* and *In vivo* activities of a biocontrol agent, *Serratia plymuthica* A21-4, against *Phytophthora capsici*. Plant Pathology J 2002; 18: 221-24.

[41] Someya N, Nakajima M, Watanabe K, Hibi T, *et al.* Potential of *Serratia marcescens* strain B2 or biological control of rice sheath blight. Biocontrol Sci Technol 2005; 15: 105-09.

[42] Roberts DP, McKenna LF, Lakshman DK, Meyer SLF, *et al.* Suppression of damping-off of cucumber caused by *Pythium ultimum* with live cells and extracts of *Serratia marcescens* N4-5. Soil Biol Biochem 2007; 39: 2275-88.

[43] Gould M, Nelson MN, Waterer D, Hynes RK. Biocontrol of *Fusarium sambucinum*, dry rot of potato, by *Serratia plymuthica* 5-6. Biocontrol Sci Technol 2008; 18: 1005-16.

[44] Benhamou N, Gagné S, Le Quéré D, Dehbi L. Bacterial-mediated induced resistance in cucumber: Beneficial effect of the endophytic bacterium *Serratia plymuthica* on the protection against infection by *Pythium ultimum*. Phytopathology 2000; 90: 45-56.

[45] Lavania M, Chauhan PS, Chauhan SVS, Singh HB, *et al.* Induction of plant defence enzymes and phenolics by treatment with plant growth–promoting rhizobacteria *Serratia marcescens* NBRI1213. Curr Microbiol 2006, 52: 363-68.

[46] Selvakumar G, Mohan M, Kundu S, Gupta AD, *et al.* Cold tolerance and plant growth promotion potential of *Serratia marcescens* strain SRM (MTCC 8708) isolated from flowers of summer squash (*Cucurbita pepo*). Lett Appl Microbiol, 2008; 46: 171-75.

[47] Koo So-Yeon, Cho Kyung-Suk. Isolation and characterization of a plant growth-promoting rhizabacterium. J Microbial Biotechnol 2009; 19: 1431-38.

[48] Johnson VW, Pearson JF, Jackson JA. Formulation of *Serratia entomophila* for biologica control of grass grub. New Zeal Plant Prot 2001; 54: 125-27.

[49] Prischmann DA, Lehman RM, Christie AA, Dashiell KE. Characterization of bacteria isolated from maize roots: Emphasis on *Serratia* and infestation with corn rootworms (Chrysomelidae: Diabrotica). Appl Soil Ecol 2008; 40: 417-31.

[50] Mohamed ZK, El-Sayed SA, Radwan TEE, Ghada SAEW. Potency evaluation of *Serratia marcescens* and *Pseudomonas fluorescens* as biocontrol agents for root-knot nematodes in Egypt J Appl Sci Res 2009; 4: 93-02.

[51] Akutsu K, Hirata A, Yamamoto M, Hirayae K, *et al.* Growth inhibition of *Botrytis* spp. By *Serratia marcescens* B2 isolated from tomato phylloplane. Ann Phytopathol Soc Japan 1993; 59: 18-25.

[52] Someya N, Kataoka N, Komagata T, Hirayae K, *et al.* Biological control of cyclamen soilborne diseases by *Serratia marcescens* strain B2. Plant Dis 2000; 84: 334-40.

[53] Wei G, Kloepper JW, Tuzun S. Induction of systemic resistance of cucumber to *Colletotrichum orbicularae* by select strains of plant growth promoting rhizabacteria. Phytopathology 1991; 81: 1508-12.

[54] Liu L, Kloepper JW, Tuzun S. Induction of systemic resistance in cucumber by plant growth promoting rhizobacteria: Duration of protection and effect of host resistance on protection and root colonization. Phytopathology 1995; 85: 1064-68.

[55] Liu L, Kloepper JW, Tuzun S. Induction of systemic resistance in cucumber against bacterial angular leaf spot by plant growth-promoting rhizobacteria. Phytopathology 1995; 85: 843-47.

[56] Liu L, Kloepper JW, Tuzun S Induction of systemic resistance in cucumber against *Fusarium* wilt by plant growth-promoting rhizobacteria. Phytopathology 1995; 85: 695-98.

[57] Raupach GS, Liu L, Murphy JF, Tuzun S, *et al.* Induced systemic resistance in cucumber an tomato against cucumber mosaic cucumovirus using plant growth promoting rhizobacteri (PGPR). Plant Dis 1996; 80: 891-94.

[58] Kobayashi DY, Guglielmoni M, Clarke BB. Isolation of the chitinolytic bacteria *Xanthomonas maltophila* and *Serratia marcescens* as biological control agents for summer patch disease of turfgrass. Soil Biol Biochem 1995; 27: 1479-87.

[59] Queiroz BPV, Melo IS. Antagonism of *Serratia marcescens* towards *Phytophthora parasitica* and its effects in promoting the growth of citrus. Braz J Microbiol 2006; 37: 448-50.

[60] Jaiganesh V, Eswaran A, Balabaskar P, Kannan C. Antagonistic activity of *Serratia marcescens* against *Pyricularia oryzae*. Not. Bot. Hort. Agrobot. Cluj 2007; 35: 48-54.

[61] Alström S, Gerhardson B. Characteristics of a *Serratia plymuthica* isolate from plant rhizospheres. Plant Soil 1987; 103: 185-89.

[62] Scherwinski K, Wolf A, Berg G. Assessing the risk of biological control agents on the indigenous microbial communities: *Serratia plymuthica* HRO-C48 and *Streptomyces* sp. HRO-71 as model bacteria. BioControl, 2007; 52: 87-112.

[63] Müller H, Berg G. Impact of formulation procedures on the effect of the biocontrol agent *Serratia plymuthica* HRO-C48 on Verticillium wilt in oilseed rape. BioControl 2008; 53: 905-16.

[64] De Vleesschauwer D, Höfte M. Using *Serratia plymuthica* to control fungal pathogens in plants. CAB Reviews: Perspectives in Agriculture, Veterinary Science, Nutrition and Natural Resources 2007; 2: No. 046. Available from: http://www.cababstractsplus.org/cabreviews.

[65] Berg G, Eberl L, Hartmann A. The rhizosphere as a reservoir for opportunistic human pathogenic bacteria. Environ Microbiol 2005; 7: 1673–85.

[66] Faltin F, Lottmann J, Grosch R, Berg G. Strategy to select and assess antagonistic bacteria for biological control of *Rhizoctonia solani* Kühn. Can J Microbiol 2004; 50: 811-20.

[67] Grosch R, Faltin F, Lottmann J, Kofoet A, *et al.* Effectiveness of three antagonistic bacterial isolates to suppress *Rhizoctonia solani* Kühn on lettuce and potato. Can J Microbiol 2005; 51: 345-53.

[68] Lottmann J, Heuer H, de Vries J, Mahn A, *et al.* Establishment of introduced antagonistic bacteria in the rhizosphere of transgenic potatoes and their effect on the bacterial community FEMS Microbiol Ecol 2000; 33: 41-49.

[69] Lottmann J, Heuer H, Smalla K, Berg G. Influence of transgenic T4 lysozyme-producing potato plants on potentially beneficial plant-associated bacteria. FEMS Microbiol Ecol 1999; 29: 365-7.

[70] Raaijmakers JM, Vlami M, de Souza JT. Antibiotic production by bacterial biocontrol agents. Antonie Van Leeuwenhoek 2002; 81: 537-47.

[71] Berg G. Diversity of antifungal and plant-associated *Serratia plymuthica* strains. J Appl Microbiol 2000; 88: 952- 60.

[72] Ovadis M, Liu X, Gavriel S, Ismailov Z, *et al.* The global regulator genes from biocontrol strain *Serratia plymuthica* IC1270: Cloning, sequencing, and functional studies. J Bacteriol 2004; 186: 4986-93.

[73] Levenfors JJ, Hedman R, Thaning C, Gerhardson B, *et al.* Broad-spectrum antifungal metabolites produced by the soil bacterium *Serratia plymuthica* A 153. Soil Biol Biochem 2004; 36: 677-85.

[74] Liu X, Bimerew M, Ma Y, Muller H, *et al.* Quorum-sensing signaling is required for production of the antibiotic pyrrolnitrin in a rhizospheric biocontrol strain of *Serratia plymuthica*. FEMS Microbiol Lett 2007; 270: 299-05.

[75] Thomson NR, Crow MA, McGowan SJ, Cox A, Salmond GPC. Biosynthesis of carbapenem antibiotic and prodigiosin pigment in *Serratia* is under quorum sensing control. Mol Microbiol 2000; 36: 539-56.

[76]   Gerber NN. Prodigiosin like pigments from *Actinomadura (Nocardia) pelletieri* and *Actinomadura madurae.* Appl Microbiol 1969; 18: 1-3.

[77]   Gerber NN. Prodigiosin-like pigments. Crit Rev Microbiol 1975; 3: 469-85.

[78]   Williams RP. Biosynthesis of prodigiosin, a secondary metabolite of *Serratia marcescens.* Appl Microbiol 1973; 25: 396-02.

[79]   Perez-Tomas R Montaner B, Lagostera E, Soto-Cerrato V. The prodigiosins, proapoptotic drugs with anticancer properties. Biochem Pharmacol 2003; 66: 1447-52.

[80]   Okamoto H, Sato M, Sato Z, Isaka M. Biocontrol of *Phytophthora capsici* by *Serratia marcescens* F-1-1 and analysis of biocontrol mechanisms using transposon insertion mutants. *Ann Phytopathol Soc Japan* 1998; 64: 287-93.

[81]   Harris AKP, Williamson NR, Slater H, Cox A, *et al.* The *Serratia* gene cluster encoding biosynthesis of the red antibiotic, prodigiosin, shows species and strain-dependent genome context variation. Microbiology 2004; 150: 3547-60.

[82]   Coulthurst SJ, Williamson NR, Harris AKP, Spring DR, *et al.* Metabolic and regulatory engineering of *Serratia marcescens*: mimicking phage-mediated horizontal acquisition of antibiotic biosynthesis and quorum-sensing capacities. Microbiology 2006; 152: 1899-11.

[83]   Someya N, Nakajimai M, Hirayae K, Hibi T, *et al.* Synergistic antifungal activity of chitinolytic enzymes and prodigiosin produced by biocontrol bacterium *Serratia marcescens* strain B2 against grey mould pathogen *Botrytis cineria.* J Gen Plant Pathol 2001; 67: 312-17.

[84]   Someya N, Nakajima M, Watanabe K, Hibi T, *et al.* Influence of bacteria isolated from rice plants and rhizospheres on antibiotic production by the antagonistic bacterium *Serratia marcescens* strain B2. J Gen Plant Pathol 2003; 70: 367-70.

[85]   Hammer PE, Hill DS, Lam ST, van Pee KH, *et al.* Four genes from *Pseudomonas fluorescens* that encode the biosynthesis of pyrrolnitrin. *Appl Environ Microbiol* 1997; 63: 2147-54.

[86]   Hwang J, Chilton WS, Benson DM. Pyrrolnitrin production by *Burkholderia cepacia* and biocontrol of Rhizoctonia stem rot of poinsettia. Biol Control 2002; 25: 56-63.

[87]   Hill DS, Stein JI, Torkewitz NR, Morse AM, *et al.* Cloning of genes involved in the synthesis of pyrrolnitrin from *Pseudomonas fluorescens* and the role of pyrrolnitrin synthesis in biological control of plant disease. *Appl Environ Microbiol* 1994; 60: 78-85.

[88]   Kirner S, Hammer PE, Hill DS, Altmann A, *et al.* Functions encoded by pyrrolnitrin biosynthetic genes from *Pseudomonas fluorescens.* J Bacteriol 1998; 180: 1939-43.

[89]   Gavriel S, Ismailov Z, Ovadis M, Chet I, *et al.* Comparative study of chitinase and pyrrolnitrin biocontrol activity in *Serratia plymuthica* strain IC1270. IOBC/wprs Bulletin 2004; 27: 347-50.

[90]   Meziane H, Gavriel S, Ismailov Z, Chet I, *et al.* Control of green and blue mould on orange fruit by *Serratia plymuthica* strains IC14 and IC1270 and putative modes of action. Postharvest Biol Technol 2006; 39: 125-33.

[91]   Thaning C, Welch CJ, Borowicz JJ, Hedman R, *et al.* Suppression of *Sclerotinia sclerotiorum* apothecial formation by the soil bacterium *Serratia plymuthica*: identification of a chlorinated macrolide as one of the causal agents. Soil Biol Biochem 2001; 33: 1817-26.

[92]   Shen SS, Piao FZ, Lee BW, Park CS. Characterization of antibiotic substance produced by *Serratia plymuthica* A21-4 and the biological control activity against pepper Phytophthora blight. Plant Pathology J 2007; 23: 180-86

[93]   Kai M, Evmert U, Berg G, Piechulla B. Volatiles of bacterial antagonists inhibit mycelial growth of the plant pathogen *Rhizoctonia solani.* Arch Microbiol 2007; 187: 351-60.

[94]   Kim YC, Jung H, Kim KY, Park SK. An effective biocontrol bioformulation against *Phytophthora* blight of pepper using growth mixtures of combined chitinolytic bacteria under different field conditions Eur J Plant Pathol 2008; 120: 373-82.

[95]   Pleban S, Chernin L, Chet I. Chitinolytic activity of an endophytic strain of *Bacillus cereus.* Lett Appl Microbiol 1997; 25: 284-88.

[96]   Manjula K, Kishore GK, Girish AG, Singh SD. Combined application of *Pseudomonas fluorescens* and *Trichoderma viride* has an improved biocontrol activity against stem rot in groundnut. Plant Pathology J 2004; 20: 75-80.

[97]   Chernin L, Ismailov Z, Haran S, Chet I. Chitinolytic *Enterobacter agglomerans* antagonistic to fungal plant pathogens. Appl Environ Microbiol 1995; 61: 1720-26.

[98]   Nielsen MN, Sorensen J, Fels J, Pedersen HC. Secondary metabolite- and endochitinase- dependent antagonism toward plant-pathogenic microfungi of *Pseudomonas fluorescens* isolates from sugar beet rhizosphere. Appl Environ Microbiol 1998, 64: 3563-69.

[99]   Ajit NS, Verma R, Shanmugam V. Extracellular chitinases of fluorescent pseudomonads antifungal to *Fusarium oxysporum* f. sp. dianthi causing carnation wilt. Curr Microbiol 2006; 52: 310-16.

[100]   Nawani NN, Kapadnis BP. One-step purification of chitinase from *Serratia marcescens* NK1, a soil isolate. J Appl Microbiol 2001; 90: 803-08.

[101]   Mehmood MA, Xiao X, Hafeez FY, Gai Y, *et al*. Purification and characterization of a chitinase from *Serratia proteamaculans*. World J Microbiol Biotechnol 2009; 25: 1955-61.

[102]   Jones JDG, Grady KL, Suslow TV, Bedbrook JR. Isolation and characterization of genes encoding two chitinase enzymes from *Serratia marcescens*. EMBO J 1986; 5: 467-73.

[103]   Watanabe T, Kimura K, Sumiya T, Nikaidou N, *et al*. Genetic analysis of the chitinase system of *Serratia marcescens* 2170. J Bacteriol 1997; 179: 7111-17.

[104]   Suzuki K, Taiyoji M, Sugawara N, Nikaidou N, *et al*. The third chitinase gene (chiC) of *Serratia marcescens* 2170 and the relationship of its product to other bacterial chitinases. Biochem J 1999; 343: 587-96.

[105]   Frankowski F, Lorito M, Scala F, Schmid R, *et al*. Purification and properties of two chitinolytic enzymes of *Serratia plymuthica* HRO-C48. Arch Microbiol 2001; 1176: 421-26.

[106]   Kishore GK, Pande S, Podile AR. Chitin supplemented foliar application of *Serratia marcescens* GPS 5 improves control of late leaf spot disease of groundnut by activating defence- related enzymes. Journal of Phytopathology, 2005; 153: 169-73.

[107]   Kamensky M, Ovadis M, Chet I, Chernin L. Soil-borne strain IC14 of *Serratia plymuthica* with multiple mechanisms of antifungal activity provides biocontrol of *Botrytis cinerea* and *Sclerotinia sclerotiorum* diseases. Soil Biol Biochem 2003; 35: 323-31.

[108]   Mller H, Westendorf C, Leitner E, Chernin L, *et al*. Quorum-sensing effects in the antagonistic rhizosphere bacterium *Serratia plymuthica* HRO-C48. FEMS Microbiol Ecol 2009; 67: 468-78.

[109]   O'Sullivan DJ, O'Gara F. Traits of fluorescent *Pseudomonas* spp. involved in suppression of plant root pathogens. Microbiol Rev 1992; 56: 662-76.

[110]   Neilands JB. Siderophores: structure and function of microbial iron transport ompounds. J Biol Chem 1995; 270: 26723-26.

[111]   Dhar Purkayastha G, Saha A, Saha D. Characterization of antagonistic bacteria isolated from tea rhizosphere in sub-Himalayan West Bengal as potential biocontrol agents in tea. J Mycol Plant Pathol 2010; 40: 27-37.

[112]   Press CM, Wilson M, Tuzun S, Kloepper JW. Salicylic acid produced by *Serratia marcescens* 90-66 is not the primary determinant of induced systemic resistance in cucumber or tobacco. Mol Plant Microbe Int 1997; 10: 761-68.

[113]   Van Loon LC, Bakker PA, Pieterse CM. Systemic resistance induced by rhizosphere bacteria. Ann Rev Phytopathol 1998; 36: 453-83.

[114]   Someya N, Nakajimai M, Hibi T, Yamaguchi I, Akutsu K. Induced resistance to rice blast by antagonistic bacterium *Serratia marcescens* strain B2. J Gen Plant Pathol 2002; 68: 177-82.

[115]   [114] Pang Y, Liu X, Ma Y, Chernin L, *et al*. Induction of systemic resistance, root colonisation and biocontrol activities of the rhizospheric strain of *Serratia plymuthica* are dependent on N-acyl homoserine lactones. Eur J Plant Pathol 2009; 124: 261-68.

[116]   De Vleesschauwer D, Chernin L, Höfte MM. Differential effectiveness of *Serratia plymuthica* IC1270-induced systemic resistance against hemibiotrophic and necrotrophic leaf pathogens in rice. BMC Plant Biology 2009; 9: 9 Available from: http://www.biomedcentral.com/1471- 229/9/9.

[117]   Van Loon LC. Plant responses to plant growth-promoting rhizobacteria. Eur J Plant Pathol 2007; 119: 243-54.

[118]   Lipasova VA, Voloshina PV, Danilova NN, Chernin LS, *et al*. Effect of Mutations in Global Regulator Genes *grrS* and *rpoS* on Synthesis of Phosphatases in *Serratia plymuthica*. Russ J Genet 2007; 43: 1428-30.

[119]   Ryu Choong-Min, Hu Chia-Hui, Locy RD, Kloepper JW. Study of mechanisms for plant growth promotion elicited by rhizobacteria in *Arabidopsis thaliana*. Plant Soil 2005; 268: 285-2.

[120]   Berg G. Plant-microbe interactions promoting plant growth and health: perspectives for controlled use of microorganisms in agriculture. Appl Microbiol Biotechnol 2009; 84: 11-18.

[121]   Duffy BK, Defago G. Environmental factors modulating antibiotic and siderophore biosynthesis y *Pseudomonas fluorescens* biocontrol strains. Appl Environ Microbiol 1999; 65: 2429-38.

[122]   Someya N, Nakajima M, Hamamoto H, Yamaguchi I, *et al*. Effects of light conditions on rodigiosin stability in the biocontrol bacterium *Serratia marcescens* strain B2. J Gen Plant Pathol 2004; 70: 367-70.

[123]   Wang C, Knill E, Glick BR Défago G. Effect of transferring 1-aminocyclopropane-1-carboxylic acid (ACC) deaminase genes into *Pseudomonas fluorescens* strain CHA0 and its *gacA* derivative CHA96 on their growth-promoting and disease-suppressive capacities. Can J Microbiol 2000; 46: 898-07.

[124] Raupach GS, Kloepper JW. Mixtures of plant growthpromoting rhizobacteria enhance biological control of multiple cucumber pathogens. Phytopathology 1998; 88: 1158-64.

[125] Roberts DP, Lohrke SM, Meyer SLF, Buyer JS, *et al.* Biocontrol agents applied individually and in combination for suppression of soilborne diseases of cucumber. Crop Prot 2005; 24: 141- 5.

[126] Lutz MP, Wenger S, Maurhofer M, Défago G, *et al.* Signaling between bacterial and fungal biocontrol agents in a strain mixture. FEMS Microbiol Ecol 2004; 48: 447-55.

[127] Porras M, Barrau C, Romero F. Effects of soil solarization and *Trichoderma* on strawberry production. Crop Prot 2007; 26: 782-87.

[128] Sivan A, Chet I. Integrated control of fusarium crown and root rot of tomato with *Trichoderma harzianum* in combination with methyl bromide or soil solarization Crop Prot 1993; 12: 380-86.

[129] Stevens C, Khan VA, Rodriguez-Kabana R, Ploper LD, *et al.* Integration of soil solarisation with chemical, biological and cultural control for the management of soilborne diseases of vegetables. Plant Soil 2003; 253:493-06.

CHAPTER 8

Molecular Approaches for Detection of Plant Pathogens

Dipali Majumder[1,*], Thangaswamy Rajesh[2] and Thalhun Lhingkhanthem Kipgen[3]

[1]Associate Professor, Plant Bacteriology; [2]Assistant Professor, Plant Pathology and [3] M.Sc (Agri), Plant Pathology student, School of Crop Protection, College of Post Graduate Studies, Central Agricultural University, Umiam, Meghalaya- 793 103, India

Abstract: Accurate identification and early detection of pathogen is a crucial step in plant disease management programmes. Conventional methods were followed, which were time consuming, relied on the interpretation of visual symptoms, culturing, laboratory identification and required extensive taxonomic expertise. Rapid development of genomic techniques has greatly simplified and has revolutionarised research in the area of life sciences. Nucleic acid- based detection methods overcome various problem associated with microscopical and immunological detection methods. DNA probes are used for the precise and accurate detection of pathogen propagules in infected tissue. DNA microarray technology is currently a new and emerging diagnostic technology for plant pathogens, when coupled with PCR results in high level of sensitivity, specificity and throughput. Dot blot hybridization, microarray, polymerase chain reaction(PCR)-based methods *e.g.* PCR-restriction fragment length polymorphism (PCR-RFLP), nested-PCR, multiplex-PCR, reverse-transcription-PCR (RT-PCR) *etc.* are the techniques being employed for the detection of major pathogens *viz.* tobacco mosaic tobamovirus. Introduction of real-time PCR technique has improved and simplified the methods for PCR-based diagnosis of plant pathogens. Routine application of real-time PCR and metagenomic analysis may expedite entire process of diagnosis of plant pathogens.

Keywords: Plant diseases, molecular detection, PCR based, microarray.

INTRODUCTION

Bio-technological approaches in detection of plant pathogens can provide considerable information about the identification of specific plant pathogenic micro organism related to the disease. Plant diseases are causing significant economic losses worldwide which are due to various pathogens like fungi, bacteria, viruses, viroids, phytoplasma, *etc.* Losses due to diseases are estimated to be about 30% in developing countries, whereas European and North and Central American countries may lose about 15–25% of the produce [1, 2]. The first and foremost step in management of plant diseases is the detection of correct responsible pathogen. This step is critical and essential for making sustainable agriculture by maintaining plant health. Early information on crop health and disease detection can facilitate the control of diseases through proper management strategies such as application of pesticides and can improve productivity. It is also essential for development of disease free seeds and planting materials in certification programmes, identification and discard of infected seeds or any planting material to be imported or exported and identification of new pathogens rapidly to prevent further spread. A simple, cost effective, safe and rapid method of pathogen detection is a pre-requisite of all the disease management programmes [3]. Visual identification of plant diseases is difficult to perform by inexperienced personnel and is limited particularly to diseases affecting aerial parts of the plants. The microscopic examination of diseased tissues has often depended on identification of disease symptoms, isolation and culturing of the organisms and identification by morphology and biochemical tests. Tests to determine physiological and biochemical characteristics were the basic tools used to identify and differentiate between bacterial pathogens for several decades from 1930s. These tests were occasionally applied for the identification of filamentous fungi which in general exhibit greater phenotypic plasticity than bacteria [4]. The physiological and biochemical tests applied for the bacterial and fungal pathogens cannot be employed for the identification and differentiation of plant viruses, since they are extremely small and do not have any detectable physiological activity. The

Address correspondence to Dipali Majumder: School of Crop Protection, College of Post Graduate Studies, Central Agricultural University, Umiam, Meghalaya- 793103, India, Phone: +91-09436335882, 0364-2570031, E-mail: dipali_assam@yahoo.co.in

physiological and biochemical properties vary widely between different groups of bacterial pathogens. No single standard set of tests can be used for all bacteria. Hence, different sets of tests have to be employed to identify isolates of different bacteria. However, commercial kits have been developed for gram-positive and gram-negative bacteria based on assimilation tests by Biolog Inc., USA. As some metabolites like mycotoxins (aflatoxin, ochratoxin) are produced only by a narrow range of fungal species, this property may be of significance in the systematics of certain filamentous fungi like *Aspergillus* spp. and *Penicillium* spp. [5].

The major limitations of these conventional approaches are the reliance on the ability of the organism to be cultured, the time consuming nature, requirement of extensive taxonomic expertise and difficult for samples with hidden symptoms. Therefore, the conventional methods were replaced by immunological techniques. Pathogen detection based on antigen-antibody interaction is rapid method of disease diagnosis which involves monoclonal antibodies [6] and enzyme linked immunosorbant assays (ELISA) [7]. These techniques are known as immunological methods or serological methods. However, the techniques are predominantly used in the detection of viral diseases of plants. The introduction of antibody based serological detections such as the monoclonal antibodies and the ELISA developed during 1970s were significant turning point in virology and bacteriology which were followed by conventional methods. The serological techniques like ELISA which are very much reliable to detect and quantify the target pathogens but have some limitations like requirement of antibodies, specific antigen, *etc.* which are more expensive. However, the inherent limitations of the conventional methods necessitated to look for techniques with higher levels of precision and reliability [8, 9].

The failure to adequately identify and detect plant pathogens using conventional, culture based morphological techniques has led to the development of nucleic acid (NA) based molecular approaches [10, 11]. During the three decades rapid advances were made in the study of molecular biology of microbial pathogens. Molecular diagnostics began to develop a real momentum after the introduction of polymerase chain reaction (PCR) in the mid 1980s and the first PCR based detection of a pathogen in diseased plants was published in the beginning of 1990s [12]. To date an increasing numbers of diagnostic laboratories are adopting molecular methods for routine detection of pathogens. Molecular methods can provide accurate consistent and reproducible results rapidly [13]. Another advantage of these molecular methods lies in the identification of obligate parasites, which otherwise difficult to culture [14]. The use of these approaches can avoid many of these shortcomings. Biochemical and molecular methods of identification provide accurate, reliable diagnostic approaches for the identification of plant pathogens [15]. NA based detection methods overcome various problems associated with conventional microscopic and immunological detection of plant pathogens. Besides, these can be used at any developmental stages of the plants since all living cells contain entire set of the genome and not affected by the environment. DNA probes are basically used for the precise and accurate detection of pathogens or pathogen propagules in infected tissues.

The DNA based technologies like PCR revolutionized molecular diagnostics and biological sciences. These NA based techniques were developed to detect organism specific DNA/RNA sequence. They are also useful in studying systemic infections or in early detection of disease, even in hidden symptoms. Besides conventional PCR, other technologically advanced methodologies such as the second generation PCR known as the real time PCR and microarrays which allows unlimited multiplexing capability [15]. A wide range of techniques have been employed to suit the pathosystem [16, 17]. The details of different molecular approaches along with their advantages over the earlier two detection methods will be reviewed in this eBook review.

NUCLEIC ACID BASED METHODS

The NA based methods were developed in between mid to late 1980s, which detect organism specific DNA /RNA sequence. NA based detection methods currently applied in pathogen detection are based on NA hybridization or PCR [18]. These methods can be designed to detect either DNA or mRNA, whereas DNA based detection method is often more straightforward than that of mRNA, the stability of DNA leads to the possibility that DNA based methods yield positive results from non-viable or dead pathogens. This method is still followed till now due to their usefulness in studying systemic infections or in early detection of disease, before symptoms are visible. PCR based methods include Nested PCR, Multiplex PCR and Real-Time PCR methods. PCR is followed since last three decades for the detection of plant pathogens.

ADVANTAGES OF NA BASED METHODS OVER CONVENTIONAL AND SEROLOGICAL METHODS

➢ It requires very small quantity of sample.

➢ Highly sensitive and can detect minute quantities of pathogen DNA. *e.g.* from a single fungal spore.

➢ They are very specific, rapid and cheaper than serological techniques.

TECHNIQUES INVOLVED IN NUCLEIC ACID BASED METHODS

The development and use of NA-based diagnostics generally requires three basic steps:

i. Selection of specific NA sequences to be used to identify the pathogen.

ii. Extraction of DNA/RNA from the sample.

iii. Method for identifying the presence of target sequence(s) in the sample.

SELECTION OF SPECIFIC NUCLEIC ACID SEQUENCES TO BE USED TO IDENTIFY THE PATHOGENS

Two general approaches are used to select target DNA/RNA sequences for use in diagnostics: one is to develop a method using known conserved genes, common to all fungi, bacteria, viruses *etc.*, but which have useful sequence variation within them that can be exploited; the other is to screen random parts of the specific genome to find regions that show the required specificity. The main DNA region targeted for diagnostic development is ribosomal DNA. They are present in all organisms at high copy number and this allows very sensitive detection. These specific regions of DNA/RNA target could then be sequenced to design primers for PCR-based techniques.

EXTRACTION OF NUCLEIC ACID FROM THE SAMPLE

Various standard protocols are available for extracting NA from pathogenic organisms may be followed. The most common method of collection, extraction of the bacterium directly from leaf tissue has several limitations such as low cell numbers collected and high amounts of plant DNA and organic matter that can interfere with molecular diagnostics, hindering early detection and often resulting in false negatives.

METHOD FOR IDENTIFYING THE PRESENCE OF TARGET SEQUENCE(S) IN THE SAMPLE

Before the advent of PCR, DNA-based diagnostics generally involved the use of DNA probes. Probes are small fragments of DNA that have been labeled with a reporter molecule and are used to detect complementary sequences of DNA in test samples. Probes can be generated from double-stranded (ds) fragments of DNA (obtained by cloning or PCR, or from whole genomes) or short single-stranded oligonucleotides. DNA probes can be used as an alternative to PCR for detection, but are now more commonly used in conjunction with PCR. Reverse probing methods also involved for analysis of complex mixture of pathogens or other microbial communities and can screened 100s/1000s of different organisms simultaneously in the same sample.

NUCLEIC ACID BASED METHODS USED IN DETECTION OF PLANT PATHOGENS

Nucleic acid based methods used for detection of plant pathogens are of two types:

    i.   Hybridization based methods.

    ii.   PCR based methods.

Hybridization based methods are of mainly two types:

    i.   Dot-Blot Hybridization.

    ii.   Microarrays.

HYBRIDIZATION BASED METHODS

NA hybridization is a method for identifying closely related NA molecules within two populations, a complex target population and a comparatively homogeneous probe population. Detection of plant pathogen by hybridization is based on the production of NA by specific hybridization between the ss target NA sequence (denatured DNA or RNA) and a complementary ss NA probe. This method is having two steps *i.e.* denaturation and hybridization. Under denaturation step, dsDNA is denatured or separated into ssDNA by heating. During hybridization, ss pieces of RNA and DNA annealed to complementary strands to form ds hybrids. The conditions of denaturation must be such that specific binding between complementary strands is maintained while nonspecific matching are dissociated. Detection of a very small amount of microorganisms in the plant material is possible by using this method. Detection of two or more viruses simultaneously through the nonisotopic molecular hybridization technique was demonstrated by using a cocktail of specific single probes against viruses infecting vegetable crops [19], ornamental plants [20] and stone fruit crops [21].

DOT-BLOT HYBRIDIZATION

It is a NA hybridization technology which is extremely used for the detection of plant pathogens, especially viroids. Method involves immobilization of spot or dot of sap extract or test DNA/RNA of plant to be tested on a solid matrix membrane. The detection of microorganism specific NA in the spot by hybridization with NA probes is possible. The probe consists of radioactively labelled or biotin labelled NA sequences complementary to NA of microorganism. Intensity of dot corresponds to the DNA/RNA represented in the sample.

ADVANTAGES OF THE METHOD

> Very useful for qualitative detection, since method can discriminate between closely related but different target sequences.

> Useful to detect, differentiate and quantify non-culturable phytoplasmas infecting plants.

MICROARRAY

Plant pathogen detection by employing DNA array technology aims to miniaturize traditional bioanalytical detection system so that hundreds or even thousands of biomolecules with unique identity can be detected simultaneously in one single experiment by using a very small amount of test samples [22]. DNA arrays were earlier developed for the detection of human pathogens such as *Escherichia coli* 0.57: H7 [23, 24]. Microarray technology has been primarily developed to allow highly parallel examination of gene expression [25]. Later, the possibility of exploitation of microarray methodology was examined for its potential in diagnostics. As crops are infected by several pathogens, it is desirable to use assays that are able to detect multiple pathogens simultaneously [26, 22]. The DNA Microarray technology, originally designed to study gene expression and generate single nucleotide polymorphism (SNP) profiles, is currently a new and emerging pathogen diagnostic technology, which in theory, offers a platform for unlimited multiplexing capability. It is based on the principle of hybridization between a "probe" and target molecules in the experimental sample. They accommodate thousands/millions of microscopic spots of DNA oligonucleotides, called features, each containing picomoles of a specific DNA sequence, allowing all the genes present in the genome to be analyzed at once. This can be a short section of a gene or other DNA element that are used as probes to hybridize a cDNA or cRNA sample (called target) under high-stringency conditions. Probe-target hybridization is usually detected and quantified by detection of fluorophore-silver, or chemiluminescence-labeled targets to determine relative abundance of NA sequences in the target. It is used in detection of Oomycetes, nematode, bacterial and fungal DNA from pure culture and is very useful for quarantine purposes. Currently, DNA array technology is the most suitable technique for multiplex detection of plant pathogens [10]. The unique sequences from a wide range of pathogens and beneficial organisms could be used to develop micro-arrays for the simultaneous detection of large number of microorganisms. Depending on the size of the deposited sample spots, the DNA arrays may be designated as macoarrays (>300 microns in diameter) or microarrays (<200 microns in diameter).

A microarray protocol was developed to detect four different potato viruses *viz.*, *Potato virus X* (PVX), *Potato virus Y* (PVY), *Potato virus A* (PVA) and *Potato Virus S* (PVS) either individually or in mixtures in infected plants [27]. This technique was able to detect closely related viruses and strains and also to discriminate sequences with less than 80% sequence identity. It could select sequence variants with greater than 90% sequences identity. The technique was comparable to ELISA in the sensitivity of detection.

There are two types of DNA Microarrays:

CDNA MICROARRAY

cDNA microarrays type contains cDNA fragments of 600 to 2400 nucleotides in length. Here, each of the different probes must be chosen independently and made by PCR or traditional cloning. The pure cDNA probes are spotted onto the glass slide.

OLIGONUCLEOTIDE MICROARRAY

Oligonucleotide microarray contains oligonucleotides of 20 to 50 nucleotides in length. Here oligonucleotide arrays are created by synthesizing the DNA directly on the glass slide. Limitation of oligonucleotide microarray is it's expensiveness for routine application.

Polymerase Chain Reaction (PCR) based methods (Kary Mullis, 1985)

- -PCR-Restriction Fragment length Polymorphism (PCR-RFLP).

- -Multiplex PCR.

PCR BASED METHODS

PCR is the most important technique used in diagnostics. It exponentially amplifies a small amount of DNA into a large amount in a few hours. This method requires a ds DNA template, a DNA polymerase, nucleotides and primers. In the first stage, the DNA to be amplified is heated to 90-95°C for 30 sec. The two strands are thereby denatured (pulled apart). In the next step, primers are then added to the mixtures which allow DNA synthesis to begin. The mixture is cooled to 50-55°C so that the primers may anneal to the separated DNA strands and polymerization may begin. Taq DNA polymerase, isolated from *Thermus aquaticus*, bacteria which live in hot springs, is required for the polymerization step. The enzyme has highest activity at 75°C, so the reaction is then reheated. In the last step, The Taq polymerase binds to the 3' OH groups and each of the initial strands is used as a template to create complementary strands of DNA. In this manner, the quantity of DNA is doubled. The PCR cycle is then repeated for about 30 times, creating up to one billion molecules [28].

Recently, the sequence of *pthN* has been extensively exploited for the development of diagnostic tool for detection and differentiation of *Xanthomonas axonopodis* pv. *malvacearum* (*Xam*) strains [29]. A simple and rapid protocol was developed for the detection of *Xam* in axenic cultures or infected cotton by amplification of a 0.4 kb DNA fragment, without the need for extraction of DNA. Since the primer is designed from the critical region of the pathogenicity or avirulence genes present in *Xam*, any strain of this pathogen, irrespective of their geographic region or race designation can be detected [30].

The complete genome sequence of few important plant pathogens, *Xyllela fastidiosa* [31], *Ralstonia Solanacearum* [32] and *Magnaporthe grisea* (http://www-genome.wi.mit.edu/annotation/fungi/ magnaporthe/) have been deduced. The probe and primers could be designed for the differential detection of pathogens and their characterization at molecular level by using the unique sequence data of the pathogens DNA. The massive genome sequencing data being generated on different microorganisms can be used for the simultaneous detection of thousands of plant pathogens and beneficial microorganisms.

MODIFIED PCR BASED METHODS

Several modifications of PCR based methods have been exploited to detect various plant pathogens. Details of important modifications are as follows:

  ➢   PCR-Restriction Fragment Length Polymorphism (PCR-RFLP).

  ➢   Nested PCR.

  ➢   Multiplex PCR.

  ➢   Real-time PCR.

  ➢   Reverse-transcriptase PCR.

PCR-RESTRICTION FRAGMENT LENGTH POLYMORPHISM (PCR-RFLP)

 It is a modified method in which PCR method coupled with RFLP technique. In the first step, DNA of 2 or 3 individuals from the same species are extracted and they are amplified under normal PCR process. Restriction enzymes can be used to identify some differences in sequence. If sequence difference falls in a restriction enzyme recognition site, it gives a RFLP. If there is a difference in restriction sites between two related DNA molecules that results in production of fragments of different lengths under gel electrophoresis. The number and size of one or two fragments will be affected for each base difference that affects a cut site when the restriction enzyme patterns are compared. Detection and differentiation of *Xanthomonas axonopodis* pv. *citri* (*Xac*) and *X. oryzae* pv. *oryzae* (*Xoo*), causing citrus canker and rice bacterial leaf blight diseases respectively and their strains were performed by RFLP analysis [33, 34].

ADVANTAGES

  ➢   Highly sensitive.

  ➢   Highly reproducible (repeatable).

  ➢   Detection and strain variation.

Limitation of the method is that, it is expensive.

NESTED PCR

 Nested PCR is a variation of the PCR, in that two pairs of PCR primers are used to amplify a fragment instead of one pair. The sensitivity of detection of fungal pathogens has been enhanced by performing nested PCR assay. The first pair of PCR primers amplifies a fragment similar to a standard PCR. However, a second pair of primers called nested primers (as they lie / are nested within the first fragment) bind inside the first PCR product fragment to allow amplification of a second PCR product which is shorter than the first one.

STEPS INVOLVED IN NESTED PCR

Step One: The DNA target template is bound by the first set of primers shown in blue. The primers may bind to alternative, similar primer binding sites which give multiple products however only one of these PCR products give the intended sequence (multiple products not shown).

Step Two: PCR products from the first PCR reaction are subjected to a second PCR run however with a second new set of primers shown in red. As these primers are nested within the first PCR products, they make it very unlikely that non-specifically amplified PCR products would contain binding sites for both sets of primers. This nested PCR amplification ensures that the PCR products from the second PCR amplification have little or no contamination from non-specifically amplified PCR products from alternative primer target sequences.

Nested-PCR procedure was applied for the detection of *Phytophthora infestans* in commercial potato seed tuber stocks and soil. The primers DC6 and ITS4 amplified a 1.3 kb PCR product from the Oomycetes like *Phytophthora*, *Pythium* and downy mildew pathogens in the first-round PCR. *P. infestans*–specific primers were employed for the second-round amplification. The detection limit was 5 pg. This assay has practical application for testing the seed tubers for the presence of *P. infestans.* The test can be completed in less than 4 h, including time required for DNA extraction [35]. Based on the sequences of microsatellite regions of the genome of *Monilinia fructicola* infecting stone fruits, primers were designed and a nested-PCR protocol was developed for the detection of the pathogen [36].

Advantage of nested PCR is that, if the wrong PCR fragment was amplified, the probability is quite low that the region would be amplified a second time by the second set of primers. Thus, Nested PCR is a very specific PCR amplification.

MULTIPLEX PCR

It is a PCR based method which allows the simultaneous and sensitive detection of different DNA or RNA targets in a single reaction or is designed to amplify multiple target pathogens by using multiple primer sets in the same reaction at the same time. It helps in reducing the number of tests required, reagent usage, time for analysis and consequently the cost [1], but care is needed to optimise the conditions so that all of the different amplicons can be generated efficiently. It can utilize multiple primer sets, a pair for each amplicon to be generated or the same primer pair can be used to amplify two different amplicons. Two successful examples of multiplex system are the simultaneous detection of the six major characterized viruses described in olive trees *i.e.* CMV, CLRV, SLRV, *Arabis mosaic virus* (ArMV), *Olive talent virus-1 and Olive latent virus-2* [37] and the simultaneous detection of nine grapevine viruses *i.e.* ArMV, *Grapevine fanleaf virus* (GFLV), *Grape vine virus A, Grapevine viruses , Rupestris stem pitting-associated virus, Grape vine fleck virus, Grapevine leafroll- associated virus-1, 2 and 3* [38].

REAL-TIME PCR

A recent method of PCR quantification has been invented and is called "real-time PCR" because it allows the scientist to actually view the increase in the amount of DNA as it is amplified [39]. It is also known as second generation PCR. A new approach for the detection of pathogenic microbes using molecular beacons was attempted [40]. This novel fluorescence based NA detection involves the use of molecular beacon (the probe) consisting of a ss DNA with a stem-loop structure. The probe consists of two types of fluorophores which are the fluorescent parts of reporter proteins. Green Fluorescent Protein (GFP) has an often-used fluorophore. When the probe is attached or unattached to the template DNA and before the polymerase acts, the quencher (Q) fluorophore (usually a long-wavelength colored dye such as red) reduces the fluorescence from the reporter (R) fluorophore (usually a short-wavelength colored dye, such as green). It does this by the use of Fluorescence (or Förster) Resonance Energy Transfer (FRET), which is the inhibition of one dye caused by another without emission of a proton. The reporter dye is found on the 5' end of the probe and the quencher at the 3' end. The PCR portion of real-time PCR is standard. Two PCR primers are used to amplify a segment of DNA (PCR product) of interest. The two primers are shown as purple arrows and the base pairing between the two strands are shown as pink. When the beacon binds to the PCR product, it is able to fluoresce when excited by the appropriate wavelength of light. The amount of fluorescence is directly proportional to the amount of PCR product amplified. Then, it releases a signal and result is shown on the screen of computer system. Molecular beacons were designed specific to the RNA-dependent RNA polymerase (RdRP) and coat protein (CP) genes of two viruses *viz., Cymbidium mosaic virus* (CymMV) and *Odontoglossum ringspot virus* (ORSV). The molecular beacons were detected up to 0.5 ng of both CymMV and ORSV purified RNA. Only tubes containing total RNA isolated from CymMV- and ORSV-coinfected *Oncidium* leaves exhibited significant increases in fluorescence intensities following addition of both sets of molecular beacons specific for CymMV and ORSV [41]. The pathogen can be accurately and realiably detected in infected seed within 50 h [42]. This diagnostic technique was as effective as real-time RT-PCR procedure for the detection of GFLV in the nematodes [43].

REAL-TIME PCR ASSAY DESIGN

For all PCR assays, a unique DNA sequence (for a given pathogen) is required for the design of a real-time PCR assay. This can be obtained from sources such as GenBank (http://www.ncbi.nlm.nih.gov/gquery/gquery.fcgi), internet resources such as the Perlprimer freeware (http://perlprimer.sourceforge.net/) available to help assay design.

Molecular beacon technology offers several advantages over other NA-based diagnostic tests. The high specificity of molecular beacons can be useful to discriminate even with one nucleotide mismatch, because of the presence of the stem-loop structure as probes. There is no necessity of removing the unhybridized molecular beacons which do not fluoresce. Extra care must be taken when analyzing samples due to the highly sensitive real time PCR methodology, as it is possible that some samples may produce false positives resulting in a closely related bacterium being detected that does not cause disease. In this case, sequencing the PCR products is the best way to confirm the identity of the amplicon [44].

ADVANTAGES

o   Sensitive, specific, rapid assay and capable of quantitating PCR products.

o   Liable to detect minute amounts of inocula present.

o   No post-PCR processing (gel electrophoresis) is required.

DISADVANTAGE

The major disadvantage of real-time PCR is the initial cost of the equipment though significant reductions in unit cost have occurred over the past few years. Real-time PCR machines have recently been bought to the market which can be used on-site [45]. As such machines become smaller and less expensive their use in locations such as custom depots and docks, where rapid clearance of plant cargos is required which likely to become more prevalent.

REVERSE-TRANSCRIPTASE PCR (RT-PCR)

Detection of plant viruses with ss DNA and ds DNA genome can be carried out directly, while a reverse transcription step is necessary to generate the cDNA prior to PCR amplification. Specific PCR and RT-PCR protocols have been developed for most of the viruses causing economically important crop diseases. The presence of plant viruses in their respective insect vectors, nematodes and fungi have also been revealed by PCR assay [16, 46, 47]. In molecular biology, RT-PCR is a powerful tool used to amplify RNA molecule by two step reactions mediated by reverse transcriptase (RT) since it uses the enzyme RT to convert mRNA into ds DNA from an organism. Here, oligo-dT serves as primer, pairs with 3' poly A tail of mRNA and cDNA copy of RNA is produced. Two steps involved in the method are:

➢ In first step, reverse transcriptase recognizes the 3' end of primers containing repeated thymines and synthesizes a DNA strand that is complementary to the mRNA. Then the RNA strand is replaced with another DNA strand, leaving a ds DNA (cDNA).

➢ In second step, the cDNA is amplified using a normal PCR reaction containing appropriate primers, Taq polymerase and nucleotides. The PCR cycle is then repeated for 30 times to generate about millions of copies of the sequences of interest. The original RNA template is degraded by RNase H, leaving pure cDNA (plus spare primer).

OTHER MODIFICATIONS OF PCR BASED METHODS

Competitive PCR

This technique allows the simultaneous detection and quantification of target organisms and it relies on the construction of a competitor or internal DNA standard which is amplified using the same primers as designed for the target organism. The competitor DNA is generally cloned into a suitable vector, *Escherichia coli* and known amounts of whole cells or extracted DNA are introduced into the assay. As compared to immunofluorescence, this PCR was about ten folds more sensitive and could detect as few as 100 cells.

Bio-PCR

The term Bio-PCR was coined to describe a procedure that involves an enrichment technique prior to extraction and amplification of DNA of the target bacterium. The intent of the method is to achieve rapid and specific growth of the target bacterium and suppress the growth of non-target bacteria. However, quantification of bacterial populations cannot be readily done with Bio-PCR, and if an adequate selective medium is lacking, organisms which compete or antagonistic to the target bacterium during enrichment can result in decreased rather than increased sensitivity of detection. Bio-PCR not only increases the sensitivity of detection, but also avoids the possibility of detecting dead bacterial cells

Nucleic Acid Sequence- Based Amplification (NAS-BA)

NAS-BA is an isothermal amplification technique making use of a dual function reverse transcriptase/DNA polymerase, RNA polymerase, RNaseH and a T7 promoter-labelled target-specific primer. This combination of enzymes and primers targets a specific RNA transcript and initially produces RNA: DNA hybrid fragment with a T7 promoter. The RNA in the hybrid is degraded by the RNase and the DNA is extended to form a dsDNA fragment with a T7 promoter, forming a template for the production of more RNA transcripts by T7 RNA polymerase. These transcripts can then be used for the production of additional DNA fragments with T7 promoters.

Applications of each method for detection of plant pathogens are being tabulated in Table **1**.

Table 1: Crop wise pathogens detection using molecular methods

Crop	Target Pathogens	Molecular Method	References
Wheat	*Septoria tritici* *Puccinia striiformis* and *P. recondita*	PCR, Multiplex-PCR PCR, Multiplex-PCR	[48, 49]
	Fusarium spp	Competitive-PCR	[50]
	Tilletia controversa and *Tilletia caries*	PCR	[51]
	Phytophthora ramorum	Real-Time PCR	[45]

		Rust pathogens	Real-Time PCR	[52]
Rice		*Magnaporthe grisea*	PCR	[53]
		Rice tungro virus	Dot-Blot hybridization, RT-PCR	[54]
Maize		*Peronosclerospora sacchari*	Dot-Blot Hybridization	[55]
		Ustilago maydis	PCR	[56]
Barley		*Barley yellow dwarf virus*	Multiplex PCR	[57]
Soyban		*Diaporthe phaseolorum* and *Phomopsis longicolla*	PCR-RFLP	[58]
		Phakopsora pachyrhizi and *P. meibomiae*	PCR	[59]
Cotton		*Cotton leaf curl begomovirus*	PCR	[60]
Tobacco		*Tobacco Mosaic toba-movirus*	Dot-Blot Hybridization	[61]
Potato		*Potato Spindle Tuber Viroid*	Dot-Blot Hybridization	[62]
		Phytophthora . infestans	Multiplex PCR	[63]
		Verticillium spp.	Micro array	[27]
		Potato Leaf Roll virus	Multiplex real-time PCR	[27, 64]
		Phytophthora spp., Spiroplasma, Vascular bacteria	Nucleic acid based	[65]
		Ralstonia Solanacearum (Brown rot)	Multiplex PCR	[66]
		Phytoplasma associated potato diseases	Multiplex PCR	[67]
		Potato leaf roll virus	Immunocapture and fluorogenic 5 nuclease RT PCR	[68, 69]
		Potato poty virus	RT PCR	[70]
		R. solanacearum	Real-Time PCR	[71]
		Clavibacter michiganensis	RFLP and RT-PCR	[72]
Tomato		*Tomato spotted wilt virus*	RT PCR	[73]
		Tomato yellow leafcurl virus	Immunocapture PCR	[74]
		Phytophthora and *Pythium*	Real-Time PCR	[75]
		Phytoplasma (little Leaf pathogen)	Nested-PCR	[76]
		Xanthomonas vesicatoria	PCR	[77]
		X. campestris	PCR	[78]
Brinjal		Phytoplasma (little leaf pathogen)	PCR	[79]
Pepper		*Xanthomonas vesicatoria*	PCR	[77]
Cucumber		*Cucumber vein yellowing virus*	Real-Time PCR	[80]
Cauliflower		*Cauliflower Mosaic Virus*	Real-Time PCR	[81]
Lettuce		*Fusarium oxysporum* fsp. *lectucae*	Nested PCR	[82]
Sugar beet		*Beet Necrotic Yellow Vein Virus*	RT-PCR	[83]
Grape-vine		*Eutypa lata*	PCR-RFLP	[84]
		Xanthomonas campestris pv. *viticola*	PCR-RFLP	[85]
		Eutypa lata & *E. vitris*	Nested multiplex PCR	[86]
Citrus		*Citrus tristeza virus*	RT-PCR	[87]
		Phytophthora hibernali	PCR	[88]
		Citrus exocortis viroids	RT-PCR	[89]
		Multiple citrus virus	Multiplex PCR	[90]
Banana		*Bunchy top*	PCR	[91]
		Banana streak badna virus	IC-PCR	[92]
Apple		*Erwinia amylovora*	PCR	[93]

Stone fruits	*Prune Dwarf virus* and *Plum Pox virus*	Multiplex RT-PCR	[94]
Straw berry	*Verticillium dahliae*	PCR	[95]
Plum	*Plum pox virus*	Microarray Real time RT-PCR	[96, 97]
Faba bean	*Faba bean necrotic yellows virus* (FBNYV)	Real time PCR	[98]
Pea	*Pseudomonas syringae* pv *pisi*	PCR	[12]
Clover	*Subterranean clover stunt virus* (SCSV)	Real time PCR	[98]
Olive	*Pseudomonas savastanoi* pv. *savastanoi.*	PCR, Hybridization	[37]
Orchid	*Cymdidium mosaic potexvirus* & *odontoglossum ringspot tobamovirus*	RT-PCR	[41]
Gladiolus	*Fusarium oxysporum* f. sp. *Gladioli*	PCR	[99]
Anthurium	*X. axonopodis* pv. *dieffrenbachiae*	Immunocapture-PCR and Nested PCR	[100, 101]
Carnation	*F. oxysporum* f. sp. *Dianthi*	PCR	[102]

ANALYSIS OF MOLECULAR VARIABILITY IN PLANT PATHOGENS

In-depth knowledge of variability in pathogen population of particular region is very critical for employing viable disease management strategies. In the studies related to microbial pathogen variability, molecular markers have been applied primarily. Various methods like differential host reactions, culture characteristics, morphological markers and biochemical tests are employed for studying pathogen variability. These markers distinguish pathogens on the basis of their physiological characters *i.e.* pathogenicity and growth behaviours and are highly influenced by the host age, quality of inoculum and environmental conditions.

Use of differential hosts for pathotyping of plant pathogens is a time consuming and laborious process. Moreover, differential hosts are available only for a few host-pathogen systems, thus limiting the analysis of pathogenic variability. In such cases, use of molecular markers has been advocated for the characterization of genetic variability in phytopathogenic microbes [103]. Two types of marker systems have been followed: (i) dominant and (ii) co-dominant markers. The dominant markers are derived from random amplified polymorphic DNA (RAPD) and amplified fragment length polymorphism (AFLP). These markers give complex patterns of bands in which each band can be considered to represent a locus. If a band is amplified in some, but not in all individuals, it is polymorphic. In the case of co-dominant markers such as RFLP and simple sequence repeats (SSRs), each allele is revealed as a unique band (SSR) or number of bands (RFLPs). Co-dominant markers are considered to be better for population studies, since populations of pathogens can be distinguished and the extent of gene flow may be determined more precisely [22]. DNA markers like RFLP [104], RAPD [105], AFLP [106], DNA amplification fingerprinting (DAF) and sequence tagged sites (STS) being used commonly for the molecular characterization of plant pathogens. Molecular detection and characterization of plant pathogens would be better done by the use of single nucleotide polymorphisms (SNP's) by making unique sequence polymorphisms [107]. These protocols do not require prior knowledge of the pathogen's genome. These methods make use of PCR to amplify large amounts of specific DNA from small amounts of total genomic DNA [108].

DNA AMPLIFICATION FINGERPRINTING (DAF) AND RANDOM AMPLIFIED POLYMORPHIC DNA (RAPD)

DAF and RAPD utilize arbitrary primers of short length, often 12 bases or less, that anneal throughout the genome [109]. The sequence of the primers is random, and the probability of two primers annealing within 1.5 kb of each other can be calculated based on the size of the genome being assayed. However, genomes are not composed of random DNA sequences, so primer pairs must be empirically tested. Size separation of the PCR fragments can be done using either acrylamide or agarose gels. The banding patterns produced indicate genetic similarities and differences between samples. Same-sized PCR fragments produced under identical cycling conditions indicate genetic similarity, whereas fragments that are different between samples or polymorphic, suggest that a mutation disrupted the PCR amplification. Missing bands can be due to nucleotide changes in the primer annealing site, which would eliminate the production of the polymorphic fragment or an insertion/deletion mutation in the region to be amplified which would change the size of the fragment. PCR fragments that are unique to a group of samples suggest inheritance of that mutation and can be used to reconstruct the genetic relatedness of individuals and estimate genetic diversity within and between populations.

Reproducibility is often cited as a disadvantage of DAF and RAPD analyses. This is understandable because slight differences in temperature or reagents might bias the amplification process toward or away from certain PCR fragments. However, the overall conclusions from DAF and RAPD analyses have proven reliable as long as the necessary controls are observed [110]. Because of the low cost and virtually infinite combinations of primers and amplification conditions, researchers with enough dedication and time can compare an exhaustive number of loci between individuals. Randomly sampling genomic DNA includes comparisons between coding, noncoding, and mobile DNA, which increases the chances of finding a difference because some regions of the genome are more prone to accumulating mutations than others. Thus, DAF and RAPD potentially offer greater resolving power between highly related individuals.

AMPLIFIED FRAGMENT LENGTH POLYMORPHISM (AFLP)

AFLP is the technique to generate DNA fingerprinting of organisms [111]. AFLP is based on RFLP in which genomic DNA is cut into small chunks by restriction endonuclease enzymes. The resulting DNA fragments are visualized by radiolabeled probes made from known genes [106]. AFLP employs the same restriction endonuclease to restrict the genomic DNA, but then utilises PCR selectively amplify hundreds of discreet fragments. Unlike DAF and RAPD, which use random primers, AFLP primers are specific sequences designed to anneal to manmade linker DNA that is attached to the ends of the pieces of chopped-up genome. Linkers are anneal to the overhangs left behind by the restriction endonuclease enzymes and attached by litigating them to the sugar-phosphate backbone. The resulting pool of DNA contains small fragments, typically 500-2000 bp, which can be PCR amplified using primers designed from the linker DNA sequence. Primers are radioactive or fluorescently labelled so that amplified fragments can be visualized on acrylamide gels, which can separate fragments that differ by a single nucleotide.

Because there may be hundreds to thousands of amplifiable and detectable DNA, additional bases are sometimes included in the primer at the 3' end to further reduce the complexity of the amplified DNA. The resulting DNA fingerprint consists of size-separated fragments for each sample that can be compared side by side. Much like RAPD results, fragments that are present or absent in one sample but not the other suggest a genetic difference. Same-sized fragments produced by both samples indicate genetic similarity. Thus, AFLP, generates robust DNA fingerprints from loci across the entire genome, but uses specific PCR primers that increase reliability and repeatability. AFLP has been adapted to run on automated capillary array sequencing instruments using fluorescent labels for greater throughput and automated data analysis.

AFLP, DAF and RAPD produce dominant markers. Data is tabulated as DNA fragments that are present (1) or absent (0). The exact nature of the genetic mutation creating differences in the DNA fingerprints is unknown. Genetic similarity between samples and phylogenetic inference regarding shared ancestry are based solely on the mathematical frequency of DNA fragments, not biological models for DNA evolution. Amplified DNA fragments that are unique to a specific individual or population can be purified, ligated into plasmid vectors, and cloned into *E.coli*. The plasmid DNA can then be sequenced to identify the underlying nature of the polymorphism. These sequenced regions are referred to as sequence characterized regions (SCARs) which once described, can be exploited as co-dominant markers. In a sexually reproducing diploid organism, researchers expect to see two alleles per sample, one corresponding to the paternal chromosome, and another allele from the maternally contributed chromosome. Co-dominant data is not scored as a binary (absence or presence) but as allele variation. Identical alleles suggest shared ancestry, whereas different alleles indicate genetic divergence. Co-dominant data is considered more informative because every PCR amplification produces DNA fragments and a technical failure during PCR amplification cannot be mistakenly scored as absence of a fragment (0).

SIMPLE SEQUENCE REPEATS (SSR)

One technique that generates co-dominant data is SSR microsatellites or SSR markers [112]. Eukaryotic genomes contain many short regions of repeated DNA. Generally, the repeat unit is 1 to 4 bp long and repeated 10 to 100 or more times. These repeats have a tendency to change in number when DNA is replicated due to a phenomenon known as DNA polymerase slippage. PCR primers adjacent to the SSR region can amplify the repeat. Size differences in the repeat length can be visualized by radiolabel or fluorescent molecules incorporated into the PCR products during amplification. Because they are uniformly spread around the genome, and some mutate faster than others. SSRs are robust molecular markers.

Polyploidy and multinucleonic conditions found in certain pathogens can produce numerous SSR alleles per amplification, making analysis of the data somewhat complex. Because SSR markers are specific types of DNA, they are not present in all genomes. Therefore, viral and bacterial plant pathogens are less likely to produce results. SSR markers are particularly suited for diploid organisms that reproduce sexually and for evaluating genetic differences between species, genera, and higher-order taxa. The main disadvantage of the SSR technique is its expensiveness. To develop SSR markers, researchers must locate and sequence SSR regions before they can develop specific primers. There are also specific models for the evolution of SSR regions that can be invoked during data analysis for more accurate conclusions. For example, a trinucleotide repeat consists of three base pair (bp) units. Changes in allele size can be weighted during analysis such that a 15 bp change is weighted more than 3 bp change because it is likely that multiple slippage events, or more than one mutation, contributed to the 15 bp variant.

CONSERVED DNA SEQUENCES

Most DNA sequencing methods focus on only a few loci. DNA sequence variation can range from highly conserved to highly variable, and researchers often use different regions of the genome to answer different questions. Conserved DNA sequences are more appropriate for evaluating higher-level relationships such as comparing genera or families whereas more variable regions are appropriate for comparing individuals and populations. Public databases such as GenBank (www.ncbi.nih.nlm.gov) contain more than 100 gigabases of DNA sequence data and computational tools to search for analogous DNA sequences that share a high level of similarity [113]. Gene sequences, particularly conserved gene sequences from related taxa can be used to design PCR primers to amplify the same DNA regions in unknown plant pathogens. DNA sequences from the unknown pathogens can then be compared to related organisms in order to estimate genetic diversity and relatedness. Studies using conserved loci for species identification are also catalogued in GenBank, making it possible to classify unknown plant pathogens based solely on DNA sequence comparisons to previously classified organisms [114]. Genes commonly sequenced include elongation factor genes, tublin genes and other universally conserved eukaryotic sequences.

The optimum scenario is a short hyper variable region sandwiched between conserved gene sequences such that PCR primers can be designed to anneal to the conserved regions and amplify the more variable internal DNA.

DETECTION SPECIFICITY AND SENSITIVITY

Sensitivity and specificity are measures of effectiveness of a detection method. Diagnostic specificity is defined as a measure of the degree to which the method is affected by non target components present in a sample, which may result in false positive responses. Too low sensitivity often leads to false negatives. High degree of diagnostic accuracy is characterized by the ability to detect, true and precisely the target micro organism from a sample without interference from non target components.

Most important advantages that molecular based detection over conventional diagnostic detection methods is the high specificity, the ability to distinguish closely related organisms. The specificity of PCR, be it conventional or real-time, depends upon the designing of proper PCR primers that are unique to the target organism. Highly conserved gene regions are often the target for designing primers. Closely related microbial species often differ in a single (SNPs) to few bases in such genes. PCR allows detection of such

SNPs [115]. With the advancements in high throughput DNA sequencing more and more genomes of plant pathogens are sequenced and nucleotide sequence data will be available increasing the possibility for designing unique primers and probes for specific detection of pathogens.

Sensitivity of PCR is greatly affected by the presence of inhibitors which prevent or reduce amplification. A wide range of inhibitors are reported [116]. Although their mode of action is not clear, these inhibitors are believed to interfere with the polymerase activity for amplification of the target DNA. On the other hand, it is worth mentioning that the high sensitivity of PCR also causes one of the limitations of PCR, that is detection sensitivity exceeding threshold levels or clinical significance and false positive results from slight DNA contamination [116]. Hence, stringent conditions are necessary in conducting the assay and proper negative controls must be included in the test.

PATHOGEN QUANTIFICATION

Quantification of a pathogen upon its detection and identification is an important aspect as it can be used to estimate its potential risk regarding epidemic outbreak, spread of inoculam and economic losses. PCR is the ideal technique for detection of small amount of the target. But problem lies with quantification. Three PCR variants *viz.,* limiting dilution PCR, kinetic PCR and competitive PCR [117] have been used for quantitative analysis of DNA. However, all are based on end point measurements of the amount of DNA produced which makes estimation of initial concentration of DNA and quantification problematic. Even the emerging microarray technology has limitations with respect to microbial quantification in complex environmental samples due to the fact that microarray hybridization signals could vary depending on target abundance and hybridization efficiency [118]. In other words, a low abundance target with high genetic similarity to a microarray probe might produce a stronger hybridization signal compared with a higher abundance target that has low similarity to the same microarray probe. Efforts are underway towards adding a quantitative aspect in the array technology [10, 119].

FUTURE PROSPECTS

Routine use of real-time PCR and micoarrays for detection and identification of plant pathogens in quarantine programmes, seed certification programmes and other detection programmes since the methods are very much specific, sensitive and very rapid, can detect few to multiple pathogens at the same time within a few hour. Metagenomic analysis is a universal diagnostic tool in plant virology. Next-generation sequencing coupled with metagenomic analysis can expedite the entire process of novel virus discovery, identification, viral genome sequencing and subsequently, the development of more routine assays for new viral pathogens [120].

CONCLUSION

Modern molecular diagnostics have the potential to greatly improve early disease detection. The fast growing databases generated by genomics provide unique opportunity to design more versatile, high throughput, sensitive and specific molecular assays which will help in diagnosis of plant pathogens and will address the major limitations of the current technologies and benefit Plant Pathology in future. PCR and especially real-time PCR are the methods of choice for rapid and accurate diagnosis of plant pathogenic organisms, but conventional serological methods, such as immunofluorescence are still widely used and ELISA is the most frequently applied method for virus detection. Molecular beacon techniques can provide a rapid and specific NA-based detection method for screening plants for disease free certification, quarantine verification, germplasm collection and selection of disease resistant plants. The use of these new approaches like real-time PCR, Multiplex-PCR and DNA microarray coupled with PCR also can be extended to the detection of multiple plant pathogens in various crops.

Research into biosensors or 'lab-on-a-chip' devices may generate further benefit by elimination of sample preparation, enhancement in specificity and reduced duration of analysis. The increasing use of microarray technologies will also allow for a wide range of potential pests and pathogens to be screened simultaneously, with a high degree of speed and accuracy.

With the advancement in biotechnological research, there has been a rapid development on various molecular biology techniques for handling plant pathological problems in a better way. Specificity and sensitivity of detection of pathogens are greatly improved and pathogen detection is becoming simple and fast. These new molecular techniques are effective management tools to be used in parallel with knowledge of the crop, understanding the population biology of the pathogen and the ecology of the disease. The future will bring more novel tools to detect plant pathogens, probably based on the new sequences available. However, only some of these will be accepted by phytopathologists, keeping in mind not only the quality of test results, but also their sensitivity and specificity, selecting the best cost-effective diagnostic strategies [75].

ABBREVIATIONS USED

ELISA: Enzyme linked immunosorbant assays

PCR: Polymerase chain reaction

NA: Nucleic acid

MO: Microorganism

SNP: Single nucleotide polymorphism

PVX: *Potato Virus X*

PVY: *Potato Virus Y*

PVA: *Potato Virus A*

PVS: *Potato Virus S*

Xoo*: Xanthomonas oryzae* pv. *oryzae*

Xav*: Xanthomonas axonopodis* pv. *citri*

CMV: Cauliflower mosaic virus

CLRV: Cherry leaf roll virus

SLRV: Strawberry latent ringspot virus

ArMV: Arabis mosaic virus

GFP: Green fluorescent protein

FRET: Fluorescence resonance energy transfer

RdRP: RNA development RNA polymerase

CP: Coat protein

CymMV:*Cymbidium Mosaic Virus*

ORSV:*Odontoglossum Ringspot Virus*

RT-PCR: Reverse transcription PCR

GFLV: Grape fan leaf virus

NASBA: Nucleic acid sequence based amplification

RFLP: Restriction fragment length polymorphism

AFLP: Amplified fragment length polymorphism

RAPD: Random amplified polymorphic DNA

SSR: Simple sequence repeats

DAF: DNA amplification fingerprint

STS: Sequence tagged sites

SCAR: Sequence

REFERENCES

[1]   James D, Varga A, Pallas V, Candresse T. Strategies for simultaneous detection of multiple plant viruses. Can J Pl Pathol 2006; 28: 16- 29.

[2]   Pimentel D. Pest management in agriculture. In: Pimentel D, Ed. Techniques for reducing pesticide use. John Wiley & Sons, Chichester 1997; pp 1–11

[3]   Weller SA, Elphinstone JG, Parkinson N, Thwaites R. Molecular Diagnosis of Plant Pathogenic Bacteria. Arab J Pl Prot 2006; 24: 143-6.

[4]   Bridge P. Biochemical molecular techniques. In: Waller JM, Lenn`e JM, Wallter SJ, Ed. Plant Pathologis'ts Pocketbook, CAB International, Wallingford, UK 2002.

[5]   Frisvad JC, Thrane U, Filtenborg O. Role and use of secondary metabolites in fungal taxonomy. In: Frisvad JC, Bridge PD, Arora DK, Ed. Chemical fungal taxonomy, Marcel Dekker, New York, USA 1998; pp. 289–319.

[6]   Kohler G, Milstein C. Continuous culture of fused cells secreting antibody of predefined specificity. Nature 1975; 256: 495-7.

[7]   Clark MF, Adams, AN. Characteristics of the microplate method of enzyme-linked immunosorbant assay (ELISA) for the detection of plant viruses. J Gen Virol 1977; 34: 475-83.

[8]   Narayanasamy P. Microbial plant pathogens and crop disease management. Science Publishers, Enfield, USA 2002.

[9]   Narayanasamy P. Postharvest pathogens and disease management. Wiley-InterScience, John Wiley & Sons, Hoboken, NJ, USA 2006.

[10]   Lievens B, Bart P, Thomma PHJ. Recent developments in pathogen detection arrays: Implications for fungal plant pathogens and use in practice. Phytopathol 2005; 95: 1374-80.

[11]   Lopez ML, Bertolini E, Olmos A, Caruso P, Goris MT, Llop P, Penyalver R, Cambra M. Innovative tools for detection of plant pathogenic viruses and bacteria. Int Microbiol 2003; 6: 233-43.

[12]   Rasmussen OF, Wulff BS. Detection of *Pseudomonas syringae* pv *pisi* using PCR. In Proceedings 4th International working group on *Pseudomonas syringae* pathovars, 1991: KLuwer Academic Publishers, Dordrecht, The Netherlands: pp 369-76.

[13]   McCartney HA., Foster SJ, Fraaije BA, Ward E. Molecular diagnostics for fungal plant pathogens. Pest Management Science 2008; 59(2):129-34.

[14]   Kahekashan T, Lalani SS, Baig MMV. Molecular methods for detection of plant pathogen with special reference to DNA based methods. In: Gangawane, LV, Khilare VC, Ed. Molecular Biology of Plant Pathogens 2010; pp. 360-76.

[15]   Mumford RA, Walsh K, Boonham N. A comparison of molecular method for the routine detection of viroid. EPPO Bulletin 2000; 30: 431-5.

[16] Narayanasamy P. Plant pathogen detection and disease diagnosis. 2nd edn. Marcel Dekker, Inc, New York 2001.

[17] Narayanasamy P. Immunology in plant health and its impact on food safety. The Haworth Press, Inc, New York 2005.

[18] Gans JD, Wolinsky M. Improved assay-dependent searching of nucleic acid sequence databases. Nucleic Acids Res 2008; 36 (12): 74.

[19] Saito N, Ohara T, Sugimoto T, Hayashi Y, Hataya T, Shikata E. Detection of *Citrus exocortis viroid* by PCR-microplate hybridization. Res Bull Plant Protec Serv Jpn 1995; 31: 47–55.

[20] Sanchez-Navarro JA, Canizares MC, Cano EA, d Pall′as V. Simultaneous detection of five carnation viruses by non-isotopic molecular hybridization. J Virol Meth 1999; 82: 167–75.

[21] Saade M, Aparicio F, S′anchez-Navarro JA, Herranz MC, Myrta A, Di Terlizzi B, *et al.* Simultaneous detection of the three ilarviruses affecting stone fruit trees by nonisotopic molecular hybridization and multiplex reverse-transcription polymerase chain reaction. Phytopathology 2000; 90: 1330–6.

[22] Narayanasamy P. Microbial Plant Pathogens. In: Narayanasamy P Ed. Molecular Biology in Plant Pathogenesis and Disease Management .Volume 1. Springer Publication, 2008; pp. 109.

[23] Schena M, Shalon D, Heller R, Chai A, Brown PO, Davies RW. Parallel human genomic analysis: microarray-based expression monitoring of 1000 genes. Proc Natl Acad Sci USA 1996; 93: 10614–9.

[24] Call DR, Brockman FJ, Chandler DP. Detecting and genotyping *Escherichia coli* 0157:H7 using multiplexed PCR and nucleic acid microarrays. Internat J Food Microbiol 2001; 67: 71–80.

[25] Schena L, Hughes KJD, Cooke EL. Detection and quantification of *Phytophthora ramorum, P kernoviae, P citricola* and *P quercina* in symptomatic leaves by multpliex real-time PCR. Mol Plant Pathol 2006; 7: 365–79.

[26] Ahmad YA, Royer M. Geographical distribution of four *Sugarcane yellow leaf virus* genotypes. Plant Dis 2006; 90: 1156–60.

[27] Boonham N, Walsh K, Smith P, Madagan K, Graham I, Barker I. Detection of potato viruses using micro array technology: towards a generic method for plant viral disease diagnosis. J of Virol Meth 2003; 108: 181- 7.

[28] Mullis KB. Specific synthesis of DNA *in vitro via* a polymerase-catalyzed chain reaction. Methods Enzymol 1984; 155: 335-350.

[29] Chakrabarty PK. Development of DNA based diagnostic tools for detection of strains of *Xanthomonas axonopodis* pv. *malvacearum*. CICR Newslwtter 2003; XIX: 8-9.

[30] De Boer SH, Elphinstone JG, Saddler GS. Molecular detection strategies for phythopathogenic bacteria. Biotechnology and plant disease management 2007; 1: 165- 94.

[31] Simpson AJG, Feinach FC, Arruda P. The genome sequence of plant pathogen *Xylella fastidiosa*. *Nature* 2000; 406: 151-7.

[32] Salanoubat M, Genin S, Artiguinave F. Genome sequence of plant pathogens *Ralstonia solancearum*. Nature 2002; 415: 497-502.

[33] Leach JE, White FE. Molecular probes for disease diagnosis and monitoring. In: Khush GS, Toenniessen GH Ed. Rice biotechnology. CAB International, UK and International Rice Research Institute, Philippines, 1991; pp. 281–307.

[34] Kanamori H, Sugimoto H, Ochiar K, Kaku H, Tsugumu S. Isolation of hrp cluster from *Xanthomonas campestris* pv *citri* and its application for RFLP analyses of Xanthomonads. Ann Phytopathol Soc Jpn 1991; 65: 110–5.

[35] Hussain S, Lees AK, Duncan JM, Cooke DEL. Development of a species-specific and sensitive detection assay for *Phytophthora infestans* and its application for monitoring of inoclum in tubers and soil. Plant Pathol 2005; 54: 373–82.

[36] Ma Z, Luo Y, Michailides TJ. Nested PCR assays for detection of *Monilinia fructicola* in stone fruit orchards and *Botryosphaeria dothidea* from Pistachios in California. J Phytopathol 2003; 151: 312–22.

[37] Bertolini E, Olmos A, Martinez MC, Gorris MT, Camra M. Single step multiplex RT-PCR for simultaneous and colourimetrix detection of six RNA viruses in Olive tree. J of Virol meth 2001; 96: 33-41.

[38] Gamino G, Gribaudo I. Simultaneous detection of nine grapevine viruses y multiplex RT-PCR with coamplification of a plant RNA as internal control. Phythopathol 2006; 96: 1223-9.

[39] Schaad NW, Frederick RD. Real-Time PCR and its application for rapid plant disease diagnostics. Can. J. Plant Pathol 2002; 24: 250-8.

[40] Tyagi S, Kramer FR. Molecular beacons: Probes that fluroesce upon hybridization. Nature Biotechnol 1996; 14: 303–8.

[41] Eun AJC, Wong S. Molecular Beacons: A new approach to plant virus detection. Mol Plant Pathol 2000; 90(3): 269-75.

[42]   Guillemette T, Iacomi-Vasilescu B, Simoneau P. Conventional and real-time PCR-based assay for detecting pathogenic *Alternaria brassicae* in crucifers seed. Plant Dis 2004; 88: 490–6.

[43]   Finetti-Sialer MM, Ciancio A. Isolate-specific detection of *Grapevine fanleaf virus*. In: Fouad A. and Czosnek H. Ed. Virus detection: Potato Virus Y (PVY), Method: RT-PCRThe Herbrew University, 2005: Technical sheet No.1.

[44]   Shatters JR, Hunter RG, Hall DW, McKenzie CL, Duan YP. Optimizing a 16S rRNABased Double Stranded DNA Dye Binding Q-PCR Method for Detection of *Candidatus* Liberibacter Species Associated with Plants and Psyllids. Molecular and Cellular Probes 2008.

[45]   Tomlinson JA, Boonham N, Hughes KJD, Griffen RL, Barker I. On-situ DNA extraction and real-time PCR for detection of *Phytophthora ramorum* in the field. Appl Environ Microbiol 2005; 71: 6702-10.

[46]   Dietzgen RG. Application of PCR in Plant Virology. In: Khan JA, Dijkstra J Ed. Plant viruses as molecular pathogens. The Haworth Press, Inc, New York, USA, 2002; pp. 471–500.

[47]   Chomic A, Pearson MN, Clover GRG, Farreyrol K, Saul D, Hampton JG, Armstrong KF. A generic RT-PCR assay for the detection of *Luteoviridae*. Plant Pathol 2010; 59(3): 429-42.

[48]   Fraaije B, Lovell D, Coelho JM, Baldwin S, Holloman DW. PCR-based assay to asses wheat varietal resistance to blotch (*Septoria tritici* and *Stagonospora nodorum*) and rust (*Puccinia striiformis* and *P. recondita*) diseases. Eur J Plant Pathol 2001; 107: 905-17.

[49]   Cartney HA, Foster SJ, Fraaije BA, Ward E. Molecular diagnostics for fungal plant pathogens. Pest Management Science 2003; 59: 129-42.

[50]   Edwards SG, Pirgozliev SR, Hare MC, Jenkinson P. Quantification of trichothecene-producing *Fusarium* Species in harvested grain by competitive PCR to determine efficacies of fungicides against *Fusarium* head blight of winter wheat. Appl Environ Microbiol 2001; 67: 1575-80.

[51]   Kochanova M, Zouhar M, Prokinova E, Rysanek P. Detection of *Tilletia controversa* and *Tilletia cariesin* wheat by PCR method. Plant Soil Environ 2004; 50 (2): 75-7.

[52]   Barnes CW, Szabo LJ. Detection and identification of four common rust pathogens of cereals and grasses using Real-Time PCR. Phytopathol 2007; 97 (6): 717-27.

[53]   Chadha S, Gopalakrishna T. Detection of *Magnaporthe grisea* in infested rice seeds using polymerase chain reaction. J Appl Microbiol 2006; 100: 1147–53.

[54]   Mangrauthia SK. Rapid detection of Rice Tungro spherical virus by RT-PCR and Dot Blot Hybridization. J Mycol Pl Pathol 2010; 40 (3): 445- 9.

[55]   Yao CL, Frederiksen RA, Magill CW. Seed transmission of sorghum downy mildew: Detection by DNA hybridization . Seed Sci Technol 1990; 18: 201-7.

[56]   Xu ML, Melchinger AE, L¨ubberstedt T. Species-specific detection of the maize pathogens *Sporisorium reiliana* and *Ustilago maydis* by dot-blot hybridization and PCR-based assays. Plant Dis 1999; 83: 390–5.

[57]   Deb M, Anderson JM. Development of a multiplexed PCR detection method for Barley and *Cereal yellow dwarf viruses, Wheat spindle streak virus, Wheat streak mosaic virus*. J Virol Methods 2008; 148: 17-24.

[58]   Zhang AW, Hartman GL, Curio-Penny B, Pedersen WL, Becker KB. Molecular detection of *Diaporthe phaseolorum and Phomopsis longicolla* from soybean seeds. Phytopathol 1999; 89 (9): 797-804.

[59]   Frederick RD, Snyder CL, Peterson GL, Bonde MR. Polymerase Chain Reaction assays for the detection and discrimination of the soybean rust pathogens *Phakopsora pachyrhizi and P. meibomiae*. Phytopathol 92: 217-27.

[60]   Sivalingam PN, Padmalatha KV, Monga D, Ajmera BD, Malathi VG. Detection of begomoviruses by PCR in weeds and crop plants in and around cotton field infected with cotton leaf curl disease. Indian Phytopath 2007; 60(3): 356-61.

[61]   Ogras TT, El-Fadly G, Balaglu S, Yilmaz MA, Cirakoglu B, Bermek E. Development of a scientific detection technique for tobacco mosaic virus based on dot-blot hybridization using an oligonucleotide probe. Turkish J Biol 1994; 18: 91-8.

[62]   Owens RA, Diener TO. Sensitive and rapid diagnosis of potato spindle tuber viroid disease by nucleic acid hyridisation. Science 1981; 213: 670-2.

[63]   Tooley PW, Carras MM, Lambert DH. Application of a PCR-based test for detection of potato late blight and pink rot in tubers. Amer J Potato Res 1998; 75:187–94.

[64]   Boonham N, Walsh K, Mumford RA, Barker I. Use of Multiplex real-time PCR for the detection of potato viruses. EPPO Bull 2000; 30: 427-30.

[65]   Davis RE, Lee I, Zhao Y. Genome-based detection and identification of plant pathogenic Phytoplasmas, Spiroplasmas and Vascular bacteria. DNA and Cell Biol. 2005; 24: 832-40.

[66]   Pastrik KH, Elphinstone JG, Pukall R. Sequence analysis and detection of *Ralstonia Solanacearum* by multiplex PCR amplification of 16S-23S ribosomal intergenic spacer region with internal positive control. Eur J Plant Pathol 2002; 108: 831-42.

[67]   Leyva-Lopez NE, Ochoa-Sanchez JC, Leal-Klevezas DS, Martinez-Soriano JP. Multiplex Phytoplasma associated potato diseases in Mexico. Can J Microbiol 2002; 48(12): 1062-8.

[68]   Schoen CD, Knorr D, Leone G. Detection of potato leaf-roll virus in dormant potato tubers by immunocapture and a fluorogenic 5' nuclease RT-PCR assay. Phytopathol 2006; 86: 993-9.

[69]   Singh RP, Kurz J, Boiteau G, Moore LM. *Potato leafroll virus* detection by RT-PCR in field-collected aphids. Amer Potato J 1997; 74: 305.

[70]   Mirmomeni MH, Sharifi A, Sisakhtnezhad S. Rapid detection of *potato Y potyvirus* in potato farms of Kermanshah using RT-PCR amplification of the P1-protease gene and its cloning. Pak J Biol Sci 2008; 11: 1482-6.

[71]   Ozakman M, Schaad NW. A Real-Time Bio-PCR assay for detection of *Ralstonia solanacearum* race 3, biovar 2 in asymptomatic potato tubers. Can J Plant Pathol 2003; 25: 232-9.

[72]   Lopez MM, Bertolini E, Marco-Noales E, Llop P, Cambra M. Update on molecular tools for detection of plant pathogenic bacteria and viruses. In: Rao JR, Fleming CC, Moore JE Ed. Molecular diagnostics: current technology and applications. Horizon Bioscience, Wymondham, UK, 2006; pp. 1-46.

[73]   Tsuda S, Fujisawa I, Hanada K, Hidaka S, Higo K, Kameya-Iwaki M, Tomaru K. Detection of *Tomato spotted wilt virus* SRNA in individual thrips by reverse transcription and polymerase. Ann Phytopathol Soc Jpn 1994; 60: 99–103.

[74]   Rampersad SN, Umaheran P. Detection of Begomoviruses in clarified plant extracts: A comparison of standard, direct-binding and immunocapture polymerase chain reaction techniques. Phytopathol 2003; 93: 1153–7.

[75]   Lievens, B, Brouwer M, Cammue BPA, Thomma PHJ. Real-time PCR for detection and quantification of oomycetes fungal pathogens of tomatoin plant and soil samples. *Plant Science* 2006; 16:155–71.

[76]   Cervantes. Detection and Molecular Characterization of Two Little Leaf Phytoplasma Strains Associated with Pepper and Tomato Diseases in Guanajuato and Sinaloa, Mexico. Plant Dis 2008; 92 (7): 1007-11.

[77]   Moretti C, Amatulli MT, Buonaurio R. PCR-based assay for the detection of *Xanthomonas euvesicatoria* causing pepper and tomato bacterial spot. Letters in Applied Microbiol 2009; 49: 466-79.

[78]   Leite RP, Jones JB, Somodi GC, Minsavage GV, Stall RE. Detection of *Xanthomonas campestris* pv. *vesicatoria* associated with pepper and tomato seed by DNA amplification. Plant Dis 1995; 79 (9): 917-22.

[79]   Siddique AB, Agarwal GK, Alam N, Krishan RM. Electron microscopy and molecular characterization of *phytoplasma* associated with the little leaf disease of brinjal (*Solanum melongena*) and Periwinkle (*Cathrantus roseus*) in Banglaladesh. J Phytopathol 2001; 149: 237-44.

[80]   Pico B, Sifres A, Nuez F. Quantitative detection of cucumber vein yellowing virus in succeptible and partially resistant plants using Real time PCR. Journal of Virological Methods 2005; 128: 14-20.

[81]   Cankar K, Ravnikar M, Zel J, Gruden K, Toplak N. Real time PCR detection of cauliflower mosaic virus to compliment the 35S screening assay for genetically modified organisms. Journal of AOAC International 2005; 88: 814-22.

[82]   Mbofung G, Pryor BM. A PCR based assay for detection of *Fusarium oxysporum* fsp. *lactucae* in lettuce seeds. J Plant Pathol 2010; 94 (7): 860-6.

[83]   Shahnejat-Bushehri A, Adel J, Dehkordi MKH. 2006. Detection of beet necrotic yellow vein virus with reverse transcription-polymerase chain reaction. Int J Agri Biol 2006; 8 (2): 280-5.

[84]   Rolshausen PE, Trouillas FP, Gubler WD. Identification of *Eutypa lata* by PCR-RFLP. Plant Dis. 2004; 88: 925-9.

[85]   da Trindade LC, Marques E, Lopes DB, Ferreira MA. Development of a molecular method for detection of *Xanthomonas campestris* pv. *viticola. Summa Phytopathol.* 2007; 33 (1): 84-90.

[86]   Catal M, Jordan SA, Butterworth SC, Schilder AMC. Detection of *Eutypa lata* and *Eutypella vitis* in Grapevine by Nested Multiplex Polymerase Chain Reaction. Phytopathol 2007; 97 (6): 737-47.

[87]   Roy A, Ramachandran P. Bi-directional PCR- a tool for identifying strains of Citrus tristeza virus. Indian Phytopath. 2002; 55 (2): 182-6.

[88]   Montenegro ID, Aguín IO, Pintos IC, Sainz MJ, Mansilla JP. A selective PCR-based method for the identification of *Phytophthora hibernalis* Carne. Span J Agric Res. 2008; 6 (1): 78-84.

[89]   Singh RP, Dilworth AD, Ao X, Singh M, Baranwal VK. Citrus exocortis viroid transmission through commercially-distributed seeds of Impatients and Verbena plants. Eur J Plant Pathol 2009; 124: 691-4.

[90]   Roy A, Fayad A, Barthe G, Brlansky RH. A multiplex polymerase chain reaction method for reliable, sensitive and simultaneous detection of multiple viruses in citrus trees. J Virol Meth 2005; 128: 176–82.

[91] Manickam K. Molecular detection of *Banana bunchy top nanavirus* and production of disease-free plantlets. Doctoral Thesis Tamil Nadu Agricultural University, Coimbatore, India 2000.

[92] Harper G, Osuji JO, Heslop-Harrison JS, Hull R. Integration of Bananavirus into the MusaGenome: Molecular and Cytogenetic Evidence. Virology 1999; 255 (2): 207-13.

[93] Kokoskova B, Mraz I, Hyblova J. Comparison of specificity and sensitivity of immunochemical and molecular techniques for reliable detection of *Erwinia amylovora*. Folia Microbiol 2007; 52 (2): 175–82.

[94] Youssef SA, Shalaby AA, Mazyad HM, Hadidi A. Detection and identification of *prune dwarf virus* and *plum pox virus* by standard and multiplex RT-PCR probe capture hybridisation (RT-PCR-ELISA). Journal of Plant Pathology 2002; 84: 113-9.

[95] Kuchta P, Jęcz T, Korbin M. The suitability of PCR-based techniques for detecting *verticillium dahliae* in strawberry plants and soil. J Fruit Ornam. Plant Res 2008; 16: 295-304.

[96] Pasquinin G, Barba G, Hadidi A, Faggioli F, Negri R. Microarray based detection and genotyping of plum pox virus. J Virol Methods 2008; 147: 118-26.

[97] Capote N, Bertolini E, Olmos A, Vidal E, Martínez MC, Cambra M. Direct sample preparation methods for the detection of Plum pox virus by real-time RT-PCR. Int Microbiol 2009; 12: 1-6.

[98] Makkouk K, Kumari S. Molecular diagnosis of plant viruses. Arab J Pl Prot 2006; 24: 135-138.

[99] de Haan LAM, Numansen A, Roebroeck EJA, Van Doorn J. PCR detection of *Fussarium oxysporum* f. sp. *gladioli* race 1, causal agent of Gladiolus yellows diseases from infected corms. Plant Pathol 2000; 49: 89-100.

[100] Khoodoo MHR, Sahin F, Jaufeerally-Fakim Y. Sensitive detection of *Xanthomonas axonopodis* pv. *dieffenbachiae* on *Anthurium andrenum* by immunocapture- PCR using primers design from sequence characterised amplified region of the blight bacterium. Eur J Plant Pathol 2005; 112: 379-90.

[101] Robene-Soustrade I, Laurent P, Gangnevin L, Joen E, Pruvost O. Specific detection of *Xanthomonas axonopodis* pv. *dieffenbachiae* in *anthurium andrenum* tissues by nested PCR. Appl Envion Microbiol 2006; 72: 1072-8.

[102] Chiocchetti A, Bernardo I, Daboussi MJ, Garibaldi A, Gullino ML, Langin T, Migheli Q. Detection of *Fussarium oxysporum* f. sp. *dianthi* in carnation tissue by PCR amplification of transposon insertions. Phytopathol 1999; 89: 1169-75.

[103] Santos-Cervantes ME, Sinaloa U, de Dios J, Culiacán S, Chávez-Medina JA. Detection and molecular characterization of two little leaf phytoplasma strains associated with pepper and tomato diseases in Guanajuato and Sinaloa, Mexico. Plant Dis 2008; 92 (7): 1007-11.

[104] Botstein D, White RL, Skolnick M, davis RW. Construction of genetic linkage map in human using restriction fragment length polymorphism. Am J Hum Genet 1980; 32 (3): 314-31.

[105] Williams JGK, Kubelik AR, Livak KJ, Rafalski JA, Tingey SV. DNA polymorphisms amplified by arbitrary primers useful as genetic markers. Nucl acids Res 1991; 18: 6531-5.

[106] Vos P, Hogers R, Bleeker M, Reijans M, Van de Lee T, Hornes M, *et al.* AFLP: a new technique for DNA finger printing. Nucleic Acids Res 1995; 23: 4407–14.

[107] Sharma TR, Singh K, Sridhar R. Application of biotechnological approaches in Plant Pathological research in India. In: One Hundred Years of Plant Pathology in India: An Overview. Scientific Publishers (India), Jodhpur, 2006; Pp 349-76.

[108] Mullis K. Specific synthesis of DNA *in vitro via* a polymerase-catalyzed chain reaction. Methods Enzymol 1987; 155: 335-50.

[109] Welsh J, Mcclelland M. Fingerprinting genomes using PCR with arbitrary primers. Nucl Acids Res 1990; 18: 7213-8.

[110] Brown JKM. The choice of molecular marker methods for population genetic studies of plant pathogens. New Phytologist 1996; 133: 183-95.

[111] Hogers PR, Bleeker M, Reijans M, Van de lee T, Hornes M, Frijters A, *et al.* AFLP: a new technique for DNA fingerprinting. Nucl Acids Res 1995; 23: 4407-14.

[112] Tautz D. Hypervariability of simple sequences as a general source for polymorphic DNA. Nucl Acids Res 1989; 17: 6463-71.

[113] Benson DA, Karsch-Mizrachi I, Lipman DJ, Ostell J, Wheeler DL. GenBank. Nucl Acids Res 2006; 34: 16-20.

[114] Rinehart TA, Copes C, Toda T, Cubeta M. Genetic characterization of binucleate *Rhizoctonia* species causing web blight on azalea in Mississippi and Alabama. Plant Dis 2006; 91: 616-23.

[115] Papp AC, Pinsonneault JK, Cooke G, Sadee W. Single nucleotide polymorphism genotyping using allele-specific PCR and fluorescence melting curves. Biotechniques 2003; 34: 1068-72.

[116] Yang S, Rothman RE. PCR-based diagnostics for infectious diseases: uses, limitations, and future applications. Infectious Dis 2004; 4: 337-48.

[117] Hermansson A, Lindgren PE. Quantification of Ammonia-oxidizing bacteria in arable soil by real-time PCR. Appl Envion Microbiol 2001; 67: 972-6.

[118] Wu L, Thompson DK, Li G, Hurt RA, Tiedje JM, Zhou J. Development and evaluation of functional gene array for detection of selected genes in the environment. Appl Environ Microbiol 2001; 67: 5780-90.

[119] Rudi K, Rudi I, Holck A. A novel multiplex quantitative DNA array based PCR (MQDA-PCR) for quantification of transgenic maize in food and feed. *Nucleic Acids Res.* 2003; 31: 60-2.

[120] Adams IP, Glover RH, Monger W, Mumford R, Jackeviciene E, Navalin-skiene M, Samuitiene M, Boonham N. Next generation sequencing and metagenomic analysis: a universal diagnostic tool in plant virology. Mol Plant Pathol 2009; 10 (4): 537-45.

CHAPTER 9

Role of NACs in Regulation of Abiotic Stress Responses in Plants

S. Puranik[1,2] and M. Prasad[1,*]

[1]National Institute of Plant Genome Research, Aruna Asaf Ali Marg, New Delhi, India and [2]Department of Biotechnology, Faculty of Science, Jamia Hamdard, New Delhi, India

Abstract: Abiotic stresses such as drought, high salinity and cold are common adverse environmental conditions that significantly influence plant growth and productivity worldwide. NAC domain proteins are important plant-specific transcription factors (TFs) that regulate the expression of many stress-inducible genes. They act both by an ABA-dependent or independent manner and play a critical role in improving abiotic stress tolerance of plants by interacting with *cis-* element present in the promoter region of various abiotic stress-responsive genes. We summarize recent studies highlighting the structural and functional characters of specific members of this family, the current knowledge on the relation between NACs and their *cis*-elements, with emphasis on the expression and regulation of NACs in the adaptive responses to abiotic stresses. The progress of the practical and application value of NACs in crop improvement engineering has also been discussed.

Keywords: Abiotic stresses, NAC proteins, DNA-binding domains.

INTRODUCTION

Plants being sessile are frequently exposed to myriad factors in both natural and agricultural conditions throughout their life time. Such adverse conditions that mainly include abiotic stresses, lead to decreased agricultural productivity. Drought and salinity alone decline 50% of the average yields of major crops worldwide (Bray ASPB 2000) [1]. Hence, increasing the yield of crop plants in normal soils as well as in less productive lands is an absolute requirement for feeding the world. Crop improvement strategies to develop and use crops that can tolerate relatively more stress have to be employed. The complexity of stress tolerance traits along with barriers posed by incompatibility of gene transfer from wild species to cultivated crops limits the competence of traditional breeding approaches. Therefore, newer strategies like genetic engineering need to be used for developing stress tolerant crop plants.

Stress tolerance or susceptibility is a complex phenomenon as stress may occur at multiple stages of plant development and often more than one stress simultaneously affects the plant [2]. Thus, through understanding of the stress adaptation mechanisms has long intrigued the researchers. Such responses include several modifications at the physiological and biochemical levels, such as wilting and abscission of leaves, root growth stimulation, fluctuation in relative water content (RWC), formation of reactive oxygen species (ROS) which disrupt cellular homeostasis by reacting with lipids, proteins, pigments and nucleic acids resulting in lipid peroxidation, membrane damage, and inactivation of enzymes thus affecting cell viability [3]. Additionally, the stress phytohormone abscisic acid (ABA), promotes stomatal closure resulting in decreased transpirational water loss and reduced photosynthetic rate, thereby improving the water-use efficiency of plant. The early events of plant adaptation to environmental stresses are the stress signal perception and subsequent signal transduction through either ABA-dependent or ABA-independent pathways, leading to the activation of various physiological and metabolic responses.

A crucial part of such response involves transcriptional control of the expression of stress-responsive genes and regulation of their temporal and spatial expression patterns [4]. Transcription factors (TFs) are proteins that bind specific sequences of DNA (*cis*-regulatory sequences) in the promoter regions of various genes and thus are capable of activating and/or repressing transcription of many secondary responsive genes. In addition to their role as molecular switches for gene expression these TFs and *cis*-regulatory sequences also function as end points of signal transduction processes [10].

*Address correspondence to M. Prasad:** National Institute of Plant Genome Research, Aruna Asaf Ali Marg, New Delhi, India, Phone: +91-1126735160 (0), Fax: +91-1126741658, E-mail : manoj_prasad@nipgr.res.in

Aakash Goyal and Priti Maheshwari (Eds)

A large portion of the plant genome is dedicated to transcription factors, with the *Arabidopsis* genome coding for more than 1500 of them [5]. Often, members of the same family highly differ in their response to various stress stimuli. On the other hand, the same TF can regulate many genes, as indicated by the significant overlap of the gene expression profiles that are induced in response to different stresses [6, 7]. Often, these TFs are members of large gene families, which may sometimes be exclusive to plants. This chapter specifically focuses on one of the largest plant-specific family of transcription factors: the NAC superfamily. We provide an overview of research on NAC TFs and discuss their role in regulating abiotic stress responses in plants with an emphasis on their potential in improvement of plant stress tolerance.

THE NAC FAMILY OF TRANSCRIPTION REGULATORS

NAC is actually an ellipsis taken from names of three earliest characterized proteins from petunia <u>N</u>AM (no apical meristem), and from *Arabidopsis* AT<u>A</u>F1/2 and <u>C</u>UC2 (cup-shaped cotyledon) [8, 9]. Since then, a large number of NAC members have been identified and functionally characterized in both model and crop plants such as *Arabidopsis*, rice (*Oryza sativa*), soybean (*Glycine max*) and wheat (*T. turgidum* spp. durum) [10-14]. The availability of complete genomic sequences for several model and crop plants has helped to comprehensively analyse the NAC TF family members. There are about 106 members in *Arabidopsis*, 149 in rice, 205 in soybean and 152 in tobacco (*Nicotiana tabacum*), respectively [15-20]. The NAC TFs are multi-functional proteins and have been implicated in diverse processes including various developmental programmes [8, 9, 21-25], senescence [14], biotic [26-29] and abiotic stress responses [12, 28-30]. The importance of this protein family in the biology of plants has been demonstrated by extensive research into NAC transcription factors. Despite recent acceleration to investigate functions of NAC genes, efforts are still needed for comprehensive understanding of this important plant-specific TF family.

STRUCTURE OF THE NAC PROTEINS

DNA-Binding Domains

These proteins share a highly conserved N-terminal NAC domain and a diversified C-terminal (Fig. **1**) [11, 31]. The highly conserved region at the N-terminal, where the DNA-binding ability of NAC domain proteins may be localized, is generally divided into five conserved subdomains (A-E) on the basis of sequence similarities.

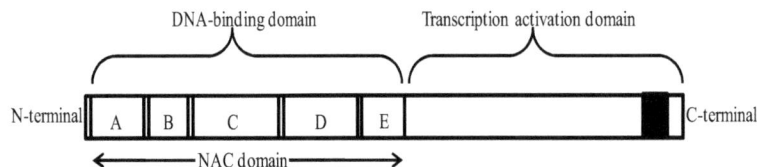

Figure 1: Schematic representation of NAC proteins. The highly conserved region at the N-terminal is divided into five conserved subdomains (A-E; shown as boxed). The C-terminal region of NAC domain proteins is more diverged and serves as a potential transcriptional activator or repressor. In some cases, it may contain a transmembrane motif (shaded black).

The B and E subdomains are highly diversified among members of different species while A, C and D are the most conserved. Apart from binding DNAs, the NAC TFs are reported to bind other proteins through their NAC domains. *SINAT5*, an *Arabidopsis* RING-finger protein, promoted ubiquitin-related degradation of *NAC1* by interacting with its NAC domain to attenuate auxin signaling in *Arabidopsis* [32]. Three stress responsive TFs from *Arabidopsis ANAC019, ANAC055* and *ANAC072* were shown to interact with a zinc-finger homeodomain *ZFHD*1 by their NAC domains. This interaction significantly improved activity of these NAC as well as the ZFHD proteins and enhanced the expression of *ERD1* gene [33]. X-ray crystallographic analysis of NAC domains further highlights their structure and conformation. Ernst and co-workers reported that the NAC domain of *Arabidopsis* ABA-responsive NAC (*ANAC*) consists of twisted antiparallel β-sheet that packs against an N-terminal α-helix on one side and a short helix on the other side [31]. One face of the NAC domain is rich in positive charges and is probably involved in DNA binding.

Similar findings have been recently reported from our laboratory during three-dimensional structure prediction of a foxtail NAC protein, SiNAC [34].

Transactivation Domains

The C-terminal region of NAC domain proteins is more diverged and serves as a potential transcriptional activator (TAR) [22, 29, 35, 36]. Ooka and coworkers found 13 common TAR motifs in the C-terminal regions of predicted NAC proteins from rice and *Arabidopsis* [19]. They suggested the conserved nature of TARs were in parallel with NAC domain structures. Generally the TAR contains simple amino acid repeats and regions rich in serine and threonine, proline and glutamine, or acidic residues [9, 26, 29, 35, 37, 38]. Some members of NAC family also act as transcription repressors. An *Arabidopsis* calmodulin-binding NAC protein (CBNAC), a transcriptional repressor was shown to specifically interact with calmodulin through its the C-terminal domain [39]. However, recently, a 35 residue long NAC repression domain (NARD) was identified in DNA-binding sub-domain D of soybean GmNAC20 [40].

The C-terminal regions of some NAC TFs also contain transmembrane (TM) motifs which are responsible for the membrane anchor. A NAC member, NTM1 (NAC with transmembrane motif1), was attached to the intracellular membranes as a dormant form and it's transcriptionally active form was released from the membranes upon proteolysis [41]. Several members of the NAC transcription factors, since then, have been confirmed to be intimately membrane associated and collectively designated NTLs for NTM1-Like's [42]. These membrane associated NAC proteins have amino acids ranging from 335-652 and are larger in molecular weight than other nuclear NAC proteins because of their C-terminal extensions, which contain the TM motif [43]. Recently, Kim and coworkers predicted that 13 members of the NAC domain proteins from *Arabidopsis thaliana* and 6 from rice were membrane associated TFs [44]. However, physiological roles of only few NAC membrane TFs like NTL8, NTL11, NTL12/NTM1) have been studied in detail [41, 43-45]. The presence of a membrane anchor causes cytoplasmic sequestration of the protein and nuclear localization is allowed only on their proteolytic cleavage by specific cues. For example, salt stress activates NTL8 by proteolysis of its membrane anchor thereby liberating it from its TM domain [46].

NAC PROTEIN DIMERIZATION

NAC proteins have been shown to homo- and heterodimerize with the dimerization ability localized to the NAC domain [11, 22, 31, 34, 47]. *In silico* analysis of *SiNAC* and *ANAC* revealed that both of them homodimerized and their dimeric interfaces were stabilized by hydrophobic effects. Several NACs are also known to heterodimerization. One such example is the *Brassica napus* NAC BnNAC14 which dimerizes with other NAC proteins like BnNAC5-8, BnNAC485, BnNAC5-11 and BnNAC3 [29]. OsNAC5, a rice NAC domain protein was shown to interact with other OsNACs, including itself [47].

INVOLVEMENT OF NAC DNA-BINDING SEQUENCES IN ABIOTIC STRESS

The ability of NAC proteins to bind DNA was first suggested by the activation of the cauliflower mosaic virus (CaMV) 35S promoter [8]. This was substantiated by Duval and coworkers who showed that the NAC domain of *A. thaliana NAM* recognized CaMV 35S *as-1* (activation sequence 1) element [35]. Since then, several NAC proteins have been shown to bind fragments of this promoter [11, 22, 29]. The consensus binding sequence was determined to be CGT(G/A) with CACG core motif. Later, NAC recognition sequence (NACRS) was also identified in promoter of drought inducible *EARLY RESPONSE TO DEHYDRATION1* (*ERD1*) gene from *Arabidopsis* [12]. Similarly, a drought-inducible rice ONAC TF bound to the NACRS [13, 15]. Several other NAC binding sites have also been identified by many groups. The *Arabidopsis* NAC1 protein has been shown to bind a fragment (CTGACGTAAGGGATGACGCAC) within the 35S promoter [22]. Kim *et al.,* identified a novel DNA binding sequence, CBNACBS, for an *Arabidopsis* calmodulin binding NAC. It consisted of a GCTT core sequence flanked on both sides by other repeated sequences [39]. Another NAC TF, IDEF2 was named after the *cis*-acting element it specifically recognized (iron deficiency-responsive *cis*-acting element 2; IDE2) with CA(A/C)G(T/C)(T/C/A)(T/C/A) as the core-binding site [48]. A jasmonic acid responsive ANAC019 transcription factor bound directly to

10-bp DNA segment (CATGTCCACG) of the Vegetative Storage Protein (VSP1) promoter at positions –369 to –360 relative to the predicted translation start codon [49]. Recently, *Arabidopsis* ANAC078 was found to recognize T[A/T/C] [A/T/G/C] C[T/G] TG[T/G]G sequence as a DNA-binding site by cyclic amplification and selection of targets [50]. Thus, the NAC TF family constitutes one of the largest TF groups with such vast array of DNA-binding sequences.

PHYLOGENETIC ANALYSIS OF NAC PROTEINS

The NAC phylogeny has been extensively studied in two model plants, *Arabidosis* and rice. Ooka *et al.,* performed phylogenetic analysis of 105 *Arabidopsis* and 75 rice NAC proteins and classified the NAC domains into two groups and several subgroups, each containing sequences from both species [19]. Another phylogenetic analysis classified 140 rice NAC genes into five groups (I-V), classifying all the published development-related genes under group I while all the published stress-related NACs into group III [15]. We have carried out a systematic phylogenetic analysis based on the similarities of NAC proteins isolated from various plant species using MEGA 4.0 by Neighbour-Joining method (Fig. **2**).

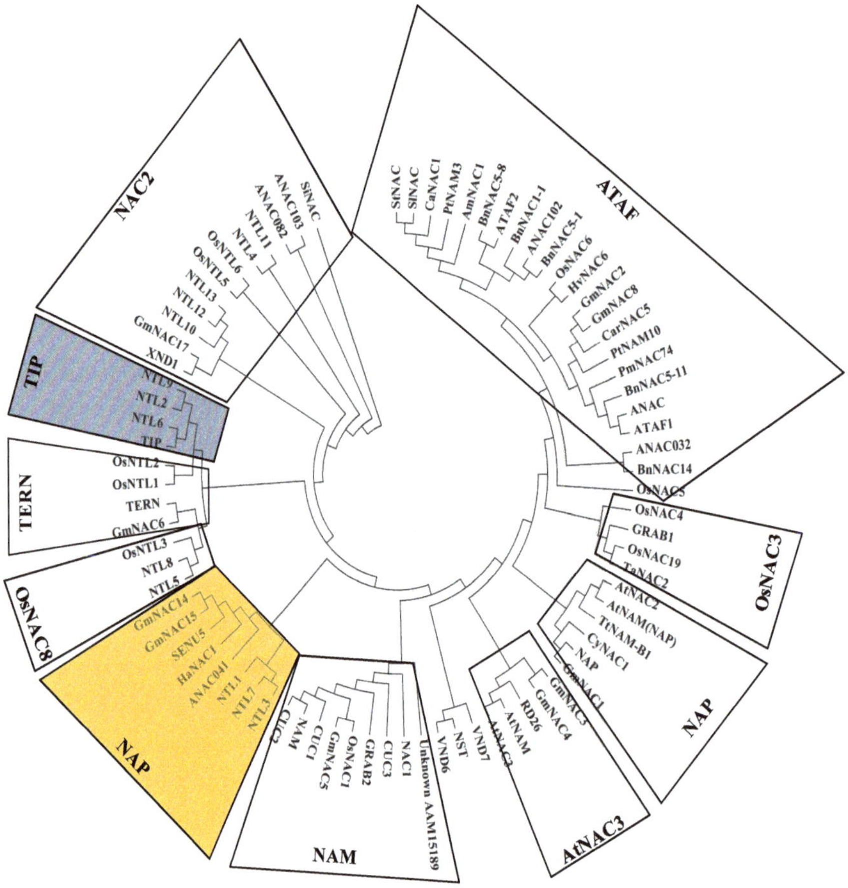

Figure 2: Phylogenetic analysis of NAC proteins from various organisms. Published NAC proteins in NCBI database were extracted and the tree was generated by MEGA 4.0 software. The NAC subgroup names are shown at the right of brackets.

The phylogenetic analysis showed that all the groups within the NAC subfamily could be easily classified as suggested by Ooka *et al.,* [19]. Interestingly, apart from being present in monocots and dicots, many NAC genes are also present in conifers and in the moss *Physcomitrella patens* as revealed by expressed sequence tag database searches. This points towards their importance for the evolution of the plant lineage as NAC proteins constitute one of the largest plant-specific families of transcription factors. This conservation ranging from mosses to eudicots makes them interesting candidates to study evolution of plant development.

FUNCTIONS OF NAC PROTEINS IN ABIOTIC STRESS RESPONSE

The wide range of stresses that this family encounters may be responsible for their such large size. The earliest NAC genes were shown to be involved mainly in plant development, such as maintenance of the shoot apical meristem) [8], cotyledon development [9], formation of flower organ primordia [21], lateral root development [22] and leaf senescence [51] and various stress conditions [11]. NAC genes have also been implicated in light responses [52, 53], programmed cell death [54, 55]. Quite a number of NAC TFs have been reported as responsive to abiotic stresses like drought, high salt, low temperatures and associated oxidative or osmotic stresses (Table **1**).

Table 1: NAC genes isolated from different plants and their transcript response to various abiotic stresses

Species	NAC Gene	Abiotic Stress	References[*]
Arabidopsis thaliana	AtNAC2	Salinity, ABA, Ethylene, Auxin	[56, 72]
A. thaliana	ANAC019	Dehydration, Salinity, ABA, MeJA	[12]
A. thaliana	ANAC055	Dehydration, Salinity, ABA, MeJA	[12]
A. thaliana	ANAC072	Dehydration, Salinity, Cold, ABA	[12]
A. thaliana	RD26	Dehydration, Salinity, Cold, ABA, MeJA, H_2O_2	[30]
A. thaliana	ATAF2	Drought	[73]
A. thaliana	NTL8	Salinity, Down regulation by GA	[45, 46]
A. thaliana	ANAC019	Drought, Salinity, ABA	[68]
A. thaliana	ATAF1	Dehydration, ABA	[57]
Oryza sativa	OsNAC5	Drought, Salinity, Cold, ABA, MeJA	[71]
O. sativa	OsNAC10	Drought, Salinity, ABA	[69]
O. sativa	SNAC2	Drought, Salinity, Cold, ABA	[59]
O. sativa	SNAC1	Drought, Salinity, Cold, ABA	[13]
O. sativa	OsNAC6	Dehydration, Salinity, Cold, ABA, H_2O_2	[36]
O. sativa	OsNAC6	Drought, Salinity, Cold, ABA,	[58]
O. sativa	ONAC063	Salinity, Mannitol, H_2O_2, Paraquat	[66]
O. sativa	ONAC045	Drought, Salinity, Cold, ABA	[70]
Gossypium hirsutum	GhNAC2	Cold, ABA, Drought	[74]
G. hirsutum	GhNAC3	Cold, ABA	[74]
G. hirsutum	GhNAC4	Drought, Salinity, Cold, ABA	[74]
G. hirsutum	GhNAC5	Drought, ABA, Cold	[74]
G. hirsutum	GhNAC6	Drought, Salinity, Cold, ABA	[74]
Cicer arietinum	CarNAC5	Drought, Heat, Wounding, SA, IAA	[76]
C. arietinum	CarNAC1	Drought, Salt, Cold, Ethephon, IAA, H_2O_2, GA	[62]
C. arietinum	CarNAC3	Drought, ABA, Ethephon, IAA, H_2O_2	[75]
Triticum aestivum	TaNAC4	ABA, Ethylene, Salinity, Low-temperature	[60]
T. aestivum	TaNAC8	Salinity, PEG, Low-temperature	[60]
T. aestivum	TaNAC69	Drought, Cold	[80]
Glycine max	GmNAC1	ABA	[61]
G. max	GmNAC2	PEG	[61]
G. max	GmNAC3	PEG, ABA, Salinity, Cold	[61]
G. max	GmNAC4	PEG, ABA, Salinity	[61]
G. max	GmNAC46	ABA	[61]
Brassica napus	BnNAC5-1	Dehydration, Cold	[29]
B. napus	BnNAC5-8	Cold	[29]
B. napus	BnNAC5-11	Dehydration, Cold	[29]

Table 1: cont....

Solanum lycopersicum	SlNAC1	Salt	[77]
S. lycopersicum	SlNAM1	Salt	[77]
Avicennia marina	AmNAC1	Salinity, ABA	[63]
Arachis hypogaea	AhNAC2	Dehydration, Salinity, ABA	[67]
Citrus sinensis	CsNAC	Cold, Ethylene, Low oxygen	[78]
Setaria italica	SiNAC	Drought, Salt stress, Ethephon, MeJA	[34]
Helianthus annuus	HaNAC	Salt Tolerance	[64]
Saccharum sp.	SsNAC23	Low temperature, Water stress	[79]

*The numbered references have been mentioned in the text.

In *Arabidopsis*, three NAC genes, *ANAC019*, *ANAC055*, and *ANAC072*, were induced by drought, salinity, and/or low temperature [12]. Expression of *AtNAC2*, an *Arabidopsis* NAC gene, elevated following high salinity, ABA, synthetic auxin (naphthalene acetic acid; NAA) and ethylene precursor (aminocyclopropane carboxylic acid; ACC) [56]. *RD26* which is a NAC domain TF accumulated following dehydration, cold, salinity as well as by hormones like abscisic acid (ABA) and Methyl Jasmonate (MeJA). ATAF1 which was one of the first identified NAC proteins in *Arabidopsis* was strongly induced by dehydration and ABA treatment, but inhibited by water treatment, suggesting a general role in drought stress responses [57]. In rice, a stress-responsive NAC gene, OsNAC6, has been reported to be induced by stresses like dehydration, salinity, cold, ABA and also by H_2O_2 [36, 58]. The mRNA levels of a stress induced NAC, *SNAC1* and *SNAC2*, were induced predominantly by dehydration, salinity, cold, ABA [13, 59]. Expression of a salt-responsive ONAC063 was induced by high-osmotic pressure and reactive oxygen species levels but not by high-temperature. Expression analysis of ONAC045 revealed that it was induced in leaves and roots by drought, high salt, low temperature stresses and ABA treatment. Recently, novel wheat NAC TF, *TaNAC8* was found to be induced in response to low temperature, salt and PEG [60].

Numerous reports for abiotic stress expression of NAC TFs in other plants are now available. *GmNAC2*, *GmNAC3* and *GmNAC4* were strongly induced by PEG-mediated osmotic stress, ABA, and salinity but differed in their response to cold [61]. BnNACs from *Brassica* were involved in the response to cold and dehydration stress [29]. Peng and coworkers have isolated three members of this family from chickpea namely *CarNAC1*, *CarNAC3* and *CarNAC5* [62-64]. These responded to various stress treatments as well as hormones. *AmNAC1*, a mangrove homologue of a potato NAC (*StNAC*) was induced by ABA and by both tolerable and severe salt concentrations [65]. Lately, a salt responsive NAC was identified from sunflower by Giordani and coworkers [66]. We recently showed up-regulation of a membrane associated NAC TF, *SiNAC* from foxtail millet by drought and salinity but not by ABA treatment [34]. Such stress-inducible NAC TFs are promising candidates for generation of stress tolerant transgenic plants possessing traits best suited for survival and/or decreased yield loss under stressed conditions.

ENGINEERING PLANT STRESS TOLERANCE BY NACs

Functional *in vivo* analyses are vital for understanding molecular mechanisms regulating stress tolerance in plants. It offers genomic tools that might improve crop productivity. A significant approach to achieve multiple stress tolerance is to overexpress transcription factors appear to control of biochemical and molecular pathways that can save plants under different stress conditions. Given their potential, the NAC TFs have been considered as potential targets for engineering plant tolerance to different stresses (Table **2**).

Using yeast one-hybrid system, three NAC-domain cDNA clones were isolated by their binding to *erd1* promoter containing the CATGTG motif. Transgenic plants overexpressing these three genes, ANAC019, ANAC055, and ANAC072 showed improved drought tolerance compared to the wild type [12]. A cDNA microarray revealed that many stress-inducible genes were upregulated in the transgenic plants except *erd1*. Thus the increased drought tolerance may be because of the elevated expression stress-inducible genes one of which encodes a glyoxalase I family protein. These enzymes are involved in glutathione-based detoxification of methylglyoxal, which is a by-product of carbohydrate and lipid metabolism. Therefore, it was postulated the glyoxalase pathway through reduction and, thus, detoxification of toxic aldehydes improved tolerance of the transgenics. Singla-Pareek *et al.,* have reported that overexpression of *Brassica juncea* glyoxalase I and *Oryza sativa* glyoxalase II, either alone or together, confers improved salinity tolerance in *Nicotiana tabacum* [67]. When a salt responsive NAC gene from rice, ONAC63 was overexpressed in *Arabidopsis* under the control of CaMV35S promoter, tolerance to high salinity and osmotic pressure was enhanced by a similar mechanism [68]. Microarray and quantitative real-time polymerase chain reaction analyses of *ONAC063-* expressing transgenic *Arabidopsis* showed upregulated expression of some salinity-inducible genes, including the amylase gene *AMY1*. Moreover, the seedlings of transgenic *Arabidopsis* showed a higher germination rate under high-salinity and mannitol-mediated osmotic stress. The survival of young seedlings after transient high-salinity stress was not significantly improved as compared to wild-type plants. Overexpression of a peanut *AhNAC2* in *Arabidopsis* increased its drought and salinity tolerance [69]. The transgenic lines were rendered ABA hypersensitive with root growth inhibition, lower seed germination rate and reduced stomatal aperture as compared to the wild type (WT) seedling. Apart from this, expression levels of 12 stress-regulated genes were higher in the transformed plants. Overexpression of *NAC1* and *AtNAC2* amended salt tolerance and enhanced lateral root development [22, 56]. Root growth, increased development of lateral roots and proliferation of root hairs are critical aspects of principal importance in plant adaptation to abiotic stress, particularly drought stress. Therefore genetic engineering by application of these NAC TFs for influencing root development is a good approach to provide stress tolerant plants. Transgenic *Arabidopsis* plants overexpressing *RD26* which codes for a NAC protein were highly sensitive to ABA, whereas *RD26*-repressed plants were ABA-insensitive. Microarray analysis illustrated that ABA- and stress-inducible genes were induced in the *RD26*-overexpressing plants and repressed in the *RD26*-repressed plants. Furthermore, *RD26* activated a promoter of its target gene encoding glyoxalase in *Arabidopsis* protoplasts. These results collectively indicate that *RD26* functions as a transcriptional activator in abiotic stress in plants *via* ABA-inducible gene expression [30].

Table 2: Functional analysis of NAC domain proteins in transgenic plants

NAC Gene	Donor Species	Transgenic Species	Stress Response	References
ANAC019, ANAC055, ANAC072	*A. thaliana*	*A. thaliana*	Drought tolerance	[12]
NAC1, AtNAC2	*A. thaliana*	*A. thaliana*	Enhanced later root development	[22, 56]
RD26	*A. thaliana*	*A. thaliana*	ABA hypersensitivity	[30]
ANAC019	*A. thaliana*	*A. thaliana*	ABA hypersensitivity	[68]
NTL8	*A. thaliana*	*A. thaliana*	Seed germination, Salt stress-mediated flowering	[44]
ONAC063	*O. sativa*	*A. thaliana*	Osmotic and Salinity tolerance	[66]
AhNAC2	*A. hypogeae*	*A. thaliana*	Drought and Salt tolerance	[67]
ONAC10	*O. sativa*	*O. sativa*	Drought, High salinity, Low temperature tolerance	[69]
ONAC045	*O. sativa*	*O. sativa*	Drought and Salt tolerance	[70]
SNAC1	*O. sativa*	*O. sativa*	Drought and Salt tolerance, ABA hypersensitivity	[13]
OsNAC6	*O. sativa*	*O. sativa*	Drought and Salt tolerance, Blast disease tolerance	[36]
SNAC2	*O. sativa*	*O. sativa*	Cold tolerance and ABA hypersensitivity	[59]
OsNAC5	*O. sativa*	*O. sativa*	Salt tolerance	[71]

*The numbered references have been mentioned in the text

Apart from *Arabidopsis*, several success stories have been reported for enhanced tolerance against abiotic stresses in cultivated crops like rice through application of NAC TFs. Recently, a stress-responsive NAC gene, *SNAC1*, was characterized in rice [13]. As it was predominantly induced in guard cells during stress, its overexpression resulted in significantly increased stomata closure thus preventing water loss from the plant under dehydration. The transgenic rice showed higher drought resistance in drought-stressed field conditions. More importantly, the transgenic plants had 22% to 34% greater seed setting than in the negative control population upon exposure to severe drought during flowering. However, the photosynthetic rate and yield of transgenic plants was not affected under normal growth conditions. So far, this is the only record of drought tolerance by *SNAC1* transgenic plants with positive results in a field trial. DNA chip analysis revealed that a large number of stress-related genes such as those involved in osmolyte production, detoxification and protection of macromolecules were up-regulated in the *SNAC1*-overexpressing rice plants. A rice *R2R3-MYB* gene (*UGS5*) containing core binding sequence of putative NACRS in the promoter region was also up-regulated in the *SNAC1*-overexpressing plants. Transgenic rice plants over-expressing *OsNAC6* showed an improved tolerance to dehydration and high-salt stresses though, they constitutively exhibited growth retardation and low reproductive yield [36]. Many stress-responsive genes were upregulated in the transgenic plants which was in good accordance with significant increase in stress tolerance. In an independent study, Hu *et al.,* established that *SNAC2/OsNAC6* transgenic rice plants had significantly higher germination and growth rate than WT under high salinity conditions [59]. Despite their improved tolerance to drought or PEG-induced osmotic stress in greenhouse conditions, the transgenics did not show any significant effect on drought tolerance in field conditions. Additionally, 50% of the transgenic plants remained vigorous after severe cold stress (4-8°C for 5 days) when all WT plants died. Cold tolerance of *SNAC2/OsNAC6* overexpressing plants was perhaps due to higher cell membrane stability in transgenic plants than in wild type during the cold stress. Transgenic rice plants overexpressing *ONAC045* showed enhanced tolerance to drought and salt treatments and expression level of two stress-responsive genes, *OsLEA3-1* and *OsPM1*, were highly elevated. Nevertheless, their expression was not regulated in *SNAC1* or *SNAC2* overexpressing plants described previously, signifying functional non-redundancy among different NAC genes inspite of their involved in stress responses. Hence NAC TFs play an indispensable role in physiological adaptation for successful plant propagation under abiotic stress conditions.

CONCLUSIONS AND FUTURE PERSPECTIVES

There is increasing evidence for involvement of transcription factors in response to abiotic stresses such as, drought, salinity, heat and cold. The physiological and molecular functions of several NAC proteins in such stresses have been established. They play a crucial role in providing tolerance to multiple stresses generally in both ABA-dependent and -independent manner and through respective *cis*-elements and DNA binding domains. The NAC TFs are capable of integrating the multiple stresses response as they tend to be regulated by diverse stimuli and are able to form multiple protein complexes. This area of research is however, still in its infancy. The complete knowledge about NAC functions becomes a daunting task given the large NAC gene families in plants. Further characterization of other members of this family will therefore undoubtedly shed light on integration and response of plants to different stresses benefit. Future NAC research by large-scale approaches to functional analysis may therefore prove to be a practical way of engineering multiple stress tolerance into crops.

ACKNOWLEDGEMENTS

This study was supported by the Department of Biotechnology, Govt. of India, New Delhi and core grant from the National Institute of Plant Genome Research (NIPGR). Ms Swati Puranik acknowledges the award of SRF from the CSIR, New Delhi. We appreciate the support and encouragement from Prof. Akhilesh Kumar Tyagi (Director, NIPGR)

ABBREVIATIONS USED

TFs: Transcription factors

ROS: Reactive oxygen species

RWC: Relative water content

ABA: Abscisic acid

NACRS: NAC recognition sequence

ERD1: Early response to dehydration1

ZFHD1: Zinc-finger homeodomain

ANAC: ABA-responsive NAC

CBNAC: Calmodulin-binding NAC protein

TM: Transmembrane motifs

NACRS: NAC recognition sequence

VSP1: Vegetative storage protein

MeJA: Methyl jasmonate

REFERENCES

[1]   Bray EA, Bailey-Serres J, Weretilnyk E. Responses to abiotic stresses. In: Gruissem W, Buchannan B, Jones R, Ed. Biochem Mol Biol Plants. American Society of Plant Biologists, Rockville, MD, 2000; pp. 158-1249.

[2]   Chinnusamy V, Schumaker K, Zhu JK. Molecular genetic perspectives on cross-talk and specificity in abiotic stress signalling in plants. J Exp Bot 2004; 55: 225-36.

[3]   Bartels D, Sunkar R. Drought and salt tolerance in plants. Crit Rev Plant Sci 2005; 21:1-36.

[4]   Rushton PJ, Somssich IE. Transcriptional control of plant genes responsive to pathogens. Curr Opin Plant Biol 1998; 1: 311-5.

[5]   Riechmann JL, Heard J, Martin G, Reuber L, Jiang C, Keddie J, *et al. Arabidopsis* transcription factors: genome-wide comparative analysis among eukaryotes. Science 2000; 290: 2105-10.

[6]   Seki M, Narusaka M, Abe H, Kasuga M, Yamaguchi- Shinozaki K, Carninci P, *et al.* Monitoring the expression pattern of 1300 Arabidopsis genes under drought and cold stresses by using a full-length cDNA Microarray. Plant Cell 2001; 13: 61-72.

[7]   Chen TH, Murata N. Enhancement of tolerance of abiotic stress by metabolic engineering of betaines and other compatible solutes. Curr Opin Plant Biol 2002; 5: 250-7.

[8]   Souer E, van Houwelingen A, Kloos D, Mol J, Koes R. The No Apical Meristem gene of petunia is required for pattern formation in embryos and flowers and is expressed at meristem and primordia boundaries. Cell 1996; 85: 159-70.

[9]   Aida M, Ishida T, Fukaki H, Fujisawa H, Tasaka M. Genes involved in organ separation in *Arabidopsis*: an analysis of the *cup-shaped cotyledon* mutant. Plant Cell 1997; 9: 841-57.

[10]   Nakashima K, Ito Y, Yamaguchi-Shinozaki K. Transcriptional regulatory networks in response to abiotic stresses in *Arabidopsis* and grasses. Plant Physiol 2009; 149: 88-95.

[11]   Olsen AN, Ernst HA, Leggio LL, Skriver K. NAC transcription factors: structurally distinct, functionally diverse. Trends Plant Sci 2005; 10: 79-87.

[12]   Tran LS, Nakashima K, Sakuma Y, Simpson SD, Fujita Y, Maruyama K, *et al.* Isolation and functional analysis of *Arabidopsis* stress-inducible NAC transcription factors that bind to a drought-responsive *cis*-element in the early responsive to dehydration stress 1 promoter. Plant Cell 2004; 16: 2481-98.

[13]   Hu HH, Dai MQ, Yao JL, Xiao BZ, Li XH, Zhang QF, *et al.* Overexpressing a NAM, ATAF, and CUC (NAC) transcription factor enhances drought resistance and salt tolerance in rice. Proc Nat Acad Sci USA 2006; 103: 12987-92.

[14]   Uauy C, Distelfeld A, Fahima T, Blechl A, Dubcovsky J. A NAC gene regulating senescence improves grain protein, Zinc, and Iron content in wheat. Science 2006; 314: 1298-1301.

[15]   Fang Y, You J, Xie K, Xie W, Xiong L. Systematic sequence analysis and identification of tissue-specific or stress-responsive genes of NAC transcription factor family in rice. Mol Genet Genomics 2008; 280: 535-46.

[16]   Gong W, Shen YP, Ma LG, Pan Y, Du YL, Wang DH, *et al.* Genome-wide ORFeome cloning and analysis of Arabidopsis transcription factor genes. Plant Physiol 2004; 135: 773-82.

[17]   Xiong Y, Liu T, Tian C, Sun S, Li J, Chen M. Transcription factors in rice: a genome-wide comparative analysis between monocots and eudicots. Plant Mol Biol 2005; 59:191-203.

[18]   Mochida K, Yoshida T, Sakurai T, Yamaguchi- Shinozaki K, Shinozaki K, Tran LSP. *In silico* analysis of transcription factor repertoire and prediction of stress responsive transcription factors in soybean. DNA Res 2009; PMID: 19884168.

[19]   Ooka H, Satoh K, Doi K, Nagata T, Otomo Y, Murakami K, *et al.* Comprehensive analysis of NAC family genes in *Oryza sativa* and *Arabidopsis thaliana*. DNA Res 2003; 20: 239-47.

[20]   Rushton PJ, Bokowiec MT, Han S, Zhang H, Chen X, Laudman TW, *et al.* Tobacco transcription factors: novel insights into transcriptional regulation in the Solanaceae. Plant Physiol 2008; 147: 280-95.

[21]   Sablowski RWM, Meyerowitz EM. A homolog of NO APICAL MERISTEM is an immediate target of the floral homeotic genes APETALA3/PISTILLATA. Cell 1998; 92: 93-103.

[22]   Xie Q, Frugis G, Colgan D, Chua NH. Arabidopsis NAC1 transduces auxin signal downstream of TIR1 to promote lateral root development. Genes Dev 2000; 14: 3024-36.

[23]   Takada S, Hibara KI, Ishida T, Tasaka M (2001) The *CUP-SHAPED COTYLEDON1* gene of *Arabidopsis* regulates shoot apical meristem formation. Development 128; 1127-35.

[24]   Vroemen CW, Mordhorst AP, Albrecht C, Kwaaitaal MACJ, de Vries SC. The CUP-SHAPED COTYLEDON3 gene is required for boundary and shoot meristem formation in Arabidopsis. Plant Cell 2003; 15: 1563-77.

[25]   Weir I, Lu J, Cook H, Causier B, Schwarz-Sommer Z, Davies B. CUPULIFORMIS establishes lateral organ boundaries in Antirrhinum. Development 2004; 131: 915-22.

[26]   Xie Q, Sanz-Burgos AP, Guo H, Garc´ıa JA, Guti´errez C. GRAB proteins, novel members of the NAC domain family, isolated by their interaction with a geminivirus protein. Plant Mol Biol 1999; 39: 647-56.

[27]   Ren T, Qu F, Morris TJ. HRT gene function requires interaction between a NAC protein and viral capsid protein to confer resistance to turnip crinkle virus. Plant Cell 2000; 12: 1917-25.

[28]   Collinge M, Boller T. Differential induction of two potato genes, Stprx2 and StNAC, in response to infection by *Phytophthora infestans* and to wounding. Plant Mol Biol 2001; 46: 521-9.

[29]   Hegedus D, Yu M, Baldwin D, Gruber M, Sharpe A, Parkin I, *et al.* Molecular characterization of *Brassica napus* NAC domain transcriptional activators induced in response to biotic and abiotic stress. Plant Mol Biol 2003; 53: 383-97.

[30]   Fujita M, Fujita Y, Maruyama K, Seki M, Hiratsu K, Ohme-Takagi M, *et al.* A dehydration-induced NAC protein, RD26, is involved in a novel ABA-dependent stress-signaling pathway. Plant J 2004; 39: 863-76.

[31]   Ernst HA, Olsen AN, Larsen S, Lo Leggio L. Structure of the conserved domain of ANAC, a member of the NAC family of transcription factors. EMBO Rep 2004; 5: 297-303.

[32]   Xie Q, Guo HS, Dallman G, Fang S, Weissman AM, Chua NH. *SINAT5* promotes ubiquitin-related degradation of NAC1 to attenuate auxin signals. Nature 2002; 419:167-70.

[33]   Tran LSP, Nakashima K, Sakuma Y, Osakabe Y, Qin F, Simpson SD, *et al.* Co-expression of the stressinducible zinc finger homeodomain ZFHD1 and NAC transcription factors enhances expression of the *ERD1* gene in *Arabidopsis*. Plant J 2007; 49: 46-63.

[34]   Puranik S, Bahadur RP, Srivastava PS, Prasad M. Molecular cloning and characterization of a membrane associated NAC family gene, SiNAC from foxtail millet [*Setaria italica* (L.) P. Beauv.]. Mol Biotech (2011) DOI: 10.1007/s12033-011-9385-7.

[35]   Duval M, Hsieh TF, Kim SY, Thomas TL. Molecular characterization of AtNAM: a member of the *Arabidopsis* NAC domain superfamily. Plant Mol Biol 2002; 50: 237-48.

[36]   Nakashima K, Tran LSP, Van Nguyen D, Fujita M, Maruyama K, Todaka D, *et al.* Functional analysis of a NAC-type transcription factor *OsNAC6* involved in abiotic and biotic stress-responsive gene expression in rice. Plant J 2007; 51: 617-30.

[37]   Jensen MK, Rung JH, Gregersen PL, Gjetting T, Fuglsang AT, Hansen M, *et al.* The HvNAC6 transcription factor: a positive regulator of penetration resistance in barley and *Arabidopsis*. Plant Mol Biol 2007; 65: 137-50.

[38]   Kikuchi K, Ueguchi-Tanaka M, Yoshida KT, Nagato Y, Matsusoka M, Hirano HY. Molecular analysis of the NAC gene family in rice. Mol Gen Genet 2000; 262: 1047-51;

[39]   Kim HS, Park BO, Yoo JH, Jung MS, Lee SM, Han HJ, *et al.* Identification of a calmodulin-binding NAC protein (CBNAC) as a transcriptional repressor in *Arabidopsis*. J Biol Chem 2007; 282: 36292-302.

[40]   Hao YJ, Song QX, Chen HW, Zou HF, Wei W, Kang XS, *et al.* Plant NAC-type transcription factor proteins contain a NARD domain for repression of transcriptional activation. Planta 2010; 232: 1033-43.

[41]   Kim YS, Kim SG, Park JE, Park HY, Lim MH, Chua NH *et al.* A membrane-bound NAC transcription factor regulates cell division in *Arabidopsis*. Plant Cell 2006; 18: 3132-44.

[42]   Kim YS, Kim SG, Kim YS, Seo PJ, Bae M, Yoon HK *et al.* Exploring membrane-associated NAC transcription factors in *Arabidopsis*: implications for membrane biology in genome regulation. Nucl Acids Res 2007; 35: 203-13.

[43]   Morishita T, Kojima Y, Maruta T, Nishizawa-Yokoi A, Yabuta Y, Shigeoka S. *Arabidopsis* NAC transcription factor, ANAC078, regulates flavonoid biosynthesis under high-light. Plant Cell Physiol 2009; 50: 2210-22.

[44]   Kim SG, Kim SY, Park CM. A membrane-associated NAC transcription factor regulates salt-responsive flowering *via FLOWERING LOCUS T* in *Arabidopsis*. Planta 2007; 226: 647-54.

[45]   Kim SG, Park CM. Gibberellic acid-mediated salt signaling in seed germination. Plant Sig Behav 2008; 3: 877-9.

[46]   Kim SG, Lee AK, Yoon HK, Park CM. A membrane bound NAC transcription factor NTL8 regulates gibberellic acid-mediated salt signaling in Arabidopsis seed germination. Plant J 2008; 55:77-88.

[47]   Jeong JS, Park YT, Jung H, Park SH, Kim JK. Rice NAC proteins act as homodimers and heterodimers. Plant Biotechnol Rep 2009; 3: 127-34.

[48]   Ogo Y, Kobayashi T, Itai RN, Nakanishi H, Kakei Y, Takahashi M, *et al.* A Novel NAC Transcription Factor, IDEF2, That Recognizes the Iron Deficiency-responsive Element 2 Regulates the Genes Involved in Iron Homeostasis in Plants. J Biol Chem 2008; 283: 13407-17.

[49]   Bu Q, Jiang H, Li C-B, Zhai Q, Zhang J, Wu X, *et al.* Role of the *Arabidopsis thaliana* NAC transcription factors ANAC019 and ANAC055 in regulating jasmonic acid-signaled defense responses. Cell Research 2008; 18: 756-67.

[50]   Yabuta Y, Morishita T, Kojima Y, Maruta T, Nishizawa-Yokoi A, Shigeoka S. Identification of recognition sequence of ANAC078 protein by the cyclic amplification and selection of targets technique. Plant Sigl Behav 2010; 5: 1-3.

[51]   John I, Hackett R, Cooper W, Drake R, Farrell A, Grierson D. Cloning and characterization of tomato leaf senescence-related cDNAs. Plant Mol Biol 1997; 33:641-51.

[52]   Hayama R, Izawa T, Shimamoto K. Isolation of rice genes possibly involved in the photoperiodic control of flowering by a fluorescent differential display method. Plant Cell Physiol 2002; 43: 494-504.

[53]   Ulm R, Baumann A, Oravecz A, Mate Z, Adam E, Oakeley EJ, *et al.* Genome-wide analysis of gene expression reveals function of the bZIP transcription factor HY5 in the UV-B response of Arabidopsis. Proc Natl Acad Sci USA 2004; 101: 1397-402.

[54]   Vandenabeele S, Vanderauwera S, Vuylsteke M, Rombauts S, Langebartels C, Seidlitz HK *et al.* Catalase deficiency drastically affects gene expression induced by high light in *Arabidopsis thaliana*. Plant J 2004; 39: 45-58.

[55]   Gechev TS, Gadjev IZ, Hille J. An extensive microarray analysis of AALtoxin- induced cell death in *Arabidopsis thaliana* brings new insights into the complexity of programmed cell death in plants. Cell Mol Life Sci 2004; 61: 1185-97.

[56]   He XJ, Mu RL, Cao WH, Zhang ZG, Zhang JS, Chen SY. *AtNAC2*, a transcription factor downstream of ethylene and auxin signaling pathways, is involved in salt stress response and lateral root development. Plant J 2005; 44:903-16.

[57]   Lu PL, Chen NZ, An R, Su Z, Qi BS, Ren F, *et al.* A novel drought-inducible gene, ATAF1, encodes a NAC family protein that negatively regulates the expression of stress-responsive genes in Arabidopsis. Plant Mol Biol 2007; 63: 289-305.

[58]   Ohnishi T, Sugahara S, Yamada T, Kikuchi K, Yoshiba Y, Hirano H-Y, *et al. OsNAC6*, a member of the NAC gene family, is induced by various stresses in rice. Genes Genet Sys 2005; 80: 135-9.

[59]   Hu H, You J, Fang Y, Zhu X, Qi Z, Xiong L. Characterization of transcription factor gene *SNAC2* conferring cold and salt tolerance in rice. Plant Mol Biol 2008; 67:169-81.

[60]   Xia N, Zhang G, Sun Y-F, Zhu L, Xu L-S, Chen X-M, *et al.* TaNAC8, a novel NAC transcription factor gene in wheat, responds to stripe rust pathogen infection and abiotic stresses, Physiol Mol Plant Pathol 2010; doi:10.1016/j.pmpp.2010.06.005.

[61]   Pinheiro GL, Marques CS, Costa MD, Reis PA, Alves MS, Carvalho CM, *et al.* Complete inventory of soybean NAC transcription factors: sequence conservation and expression analysis uncover their distinct roles in stress response. Gene 2009; 444:10-23.

[62]   Peng H, Yu X, Cheng H, Shi Q, Zhang H, Li J, *et al.* Cloning and characterization of a novel NAC family gene *CarNAC1* from chickpea (*Cicer arietinum* L.). Mol Biotechnol 2010; 44: 30-40.

[63]   Peng H, Cheng H-Y, Chen C, Yu X-W, Yang J-N, Gao W-R, *et al.* A NAC transcription factor gene of Chickpea (*Cicer arietinum*), CarNAC3, is involved in drought stress response and various developmental processes. J Plant Physiol 2009; 166: 1934-45.

[64]   Peng H, Cheng H-Y, Yu X-W, Shi Q-H, Zhang H, Li J-G, *et al.* Characterization of a chickpea (*Cicer arietinum* L.) NAC family gene, CarNAC5, which is both developmentally- and stress-regulated. Plant Physiol Biochem 2009; 47: 1037-45.

[65]   Ganesan G, Sankararamasubramanian HM, Narayanan JM, Sivaprakash KR, Parida A. Transcript level characterization of a cDNA encoding stress regulated NAC transcription factor in the mangrove plant *Avicennia marina*. Plant Physiol Biochem 2008; 46: 928-34.

[66]   Giordani T, Buti M, Natali L, Pugliesi C, Cattonaro F, Morgante M, *et al.* An analysis of sequence variability in eight genes putataively involved in drought response in sunflower (*Helianthus annus* L.). Theor Appl Genet 2011; 122: 1039-49.

[67]   Singla-Pareek SL, Reddy MK, Sopory SK. Genetic engineering of the glyoxalase pathway in tobacco leads to enhanced salinity tolerance. Proc Natl Acad Sci USA 2003; 100: 14672-7.

[68]   Yokotani N, Ichikawa T, Kondou Y, Matsui M, Hirochika H, Iwabuchi M, *et al.* Tolerance to various environmental stresses conferred by the salt-responsive rice gene *ONAC063* in transgenic Arabidopsis. Planta 2009; 229:1065-75.

[69]   Liu X, Hong L, Li X-Y, Yao Y, Hu B, Li L. Improved drought and salt tolerance in transfenic *Arabidopsis* overexpressing a NAC transcriptional factor from *Arachis hypogea*. Biosci Biotechnol Biochem 2011; 75: 443-50.

[70]   Jensen MK, Kjaersgaard T, Nielsen MM, Galberg P, Petersen K, O'shea C, *et al.*The *Arabidopsis thaliana* NAC transcription factor family: structure–function relationships and determinants of ANAC019 stress signalling. Biochem J 2010; 426: 183-96.

[71]   Jeong JS, Kim YS, Baek HK, Jung H, Ha S-H, Choi YD, *et al.* Root-specific expression of *OsNAC10* improves drought tolerance and grain yield in rice under field drought conditions. Plant Physiol 2010; 153: 185-97.

[72]   Zheng X, Chen B, Lu G, Han B. Overexpression of a NAC transcription factor enhances rice drought and salt tolerance. Biochem Biophys Res Commun 2009; 379: 985-9

[73]   Takasaki H, Maruyama K, Kidokoro S, Ito Y, Fujita Y, Shinozaki K, *et al.* The abiotic stress-responsive NAC-type transcription factor OsNAC5 regulates stress-inducible genes and stress tolerance in rice. Mol Genet Genomics 2010. DOI 10.1007/s00438-010-0557-0.

[74]   Balazadeh S, Siddiqui H, Allu AD, Matallana-Ramirez LP, Caldana C, *et al.* A gene regulatory network controlled by the NAC transcription factor ANAC092/AtNAC2/ORE1 during salt-promoted senescence. Plant J 2010; 62: 250-64.

[75]   Delessert C, Kazan K, Wilson IW, Straeten DVD, Manners J, Dennis ES, *et al.* The transcription factor ATAF2 represses the expression of pathogenesis-related genes in Arabidopsis. Plant J 2005; 43: 745-57.

[76]   Meng C, Cai C, Zhang T, Guo W. Characterization of six novel NAC genes and their responses to abiotic stresses in *Gossypium hirsutum* L. Plant Sci 2009; 176: 352-9.

[77]   Yang R, Deng C, Ouyang B, Ye Z. Molecular analysis of two salt-responsive NAC-family genes and their expression analysis in tomato. Mol Biol Rep 2010; DOI 10.1007/s11033-010-0177-0.

[78]   Fan J, Gao X, Yang Y-W, Deng W, Li Z-G. Molecular cloning and characterization of a NAC-like gene in "navel" orange fruit response to postharvest stresses. Plant Mol Biol Rep 2007; 25: 145-53.

[79]   Nogueira FTS, Schlo¨gl PS, Camargo SR, Fernandez JH, De Rosa Jr. VE, Pompermayer P, Arruda P. SsNAC23, a member of the NAC domain protein family, is associated with cold, herbivory and water stress in sugarcane. Plant Science 2005; 169: 93-106.

[80]   Xue G-P, Bower NI, McIntyre CL, Riding GA, Kazan K, Shorter R. *TaNAC69* from the NAC superfamily of transcription factors is up-regulated by abiotic stresses in wheat and recognises two consensus DNA-binding sequences. Funct Plant Biol 2006; 33: 43-57.

CHAPTER 10

Phytoremediation: A New Hope for the Environment

Sarvjeet Kukreja[*] and Umesh Goutam

Lovely Professional University, Jalandhar, Punjab, India

Abstract: A proportion of major environmental and human health problems' are imposed by contaminated soils and water. These natural resources are prone to contamination by both organic and inorganic contaminants. Heavy metals like Cu, Cd, Zn, As, Hg are the major inorganic contaminants. These contaminants are the result of various controlled and uncontrolled human activities like disposal of waste, mining *etc.* The major danger from these contaminants is their entry into human food chain because of possibility of certain plants to accumulate and translocate these contaminants to edible and harvested parts. Most of the conventional remedial technologies can be successful in certain specific situations, but, they are expensive and inhibit the soil fertility; thus subsequently causes negative impacts on the ecosystem. Different emerging phytoremediation technologies which imply the use of plants to remove or lower down the metal contamination may be used to combat the problem. This cost-effective plant-based approach to remediation takes advantage of the remarkable ability of plants to extract, sequester and detoxify pollutants. There are several types of phytoremediation *viz.*, Phytoextraction, Phytostabilization, Rhizofiltration and Phytodegradation/Phytovolatalization. Hyperaccumulators are the best candidates for phytoremediation process. In recent years, knowledge of the physiological and molecular mechanisms of phytoremediation began to emerge together with biological and engineering strategies designed to optimize and improve phytoremediation. Transgenic plants have been developed for metal uptake, tolerance and detoxification. Genetic engineering is surely a powerful tool allowing investigating, evaluating and improving the potential of phytoremediation.

Keywords: Heavy metals, phytoremediation, phytoextraction, phytostabilization.

INTRODUCTION

Earth's natural resources like air, ground water and soil face continuous threats due to the human, industrial, mining and military activities. The growth and development of any human society depends on the judicious and economic use of its available natural resources. Unfortunately, these resources have been subjected to maximum exploitation and thus are polluted and degraded by human interference due to urbanization and industrialization. Hence, a major environmental concern is the contamination of soil and water due to dispersal of industrial and urban wastes generated by human activities. Controlled and uncontrolled disposal of waste, accidental and process spillage, mining and smelting of metalliferous ores, sewage sludge application to agricultural soils are responsible for the migration of contaminants into non-contaminated sites as dust or leachate and contribute towards contamination of our ecosystem. As a result over recent decades an annual worldwide release of heavy metals reached 22,000 t (metric ton) for cadmium, 939,000 t for copper, 783,000 t for lead and 1,350,000 t for zinc [1]. A wide range of inorganic and organic compounds cause contamination, which includes heavy metals, combustible and putriscible substances, hazardous wastes, explosives and petroleum products. Major component of inorganic contaminates are heavy metals, which present a different problem than organic contaminants [2, 3]. Organic contaminants can be degraded by soil microorganisms, while metals like mercury, lead, cadmium, copper and zinc are immutable by all biochemical reactions and hence remain in ecosystem [4]. They would seep into surface water, ground water or even channel into food chain by crops growing in such a soil [5]. Although many metals are essential, they are toxic at higher concentrations. Thus mechanisms must exist both to satisfy the requirements of cellular metabolism on one hand and also to protect cells from the toxic effects of these heavy metals. High concentrations of heavy metals in soil hampers uptake of essential ions,

*Address correspondence to Sarvjeet Kukreja: Lovely Professional University, Jalandhar, Punjab, India; Phone: +91- 1824-404404, Fax: +91-1824-506111; E-mail: sarvjeetkukreja@gmail.com

Aakash Goyal and Priti Maheshwari (Eds)

chlorophyll biosynthesis, lipids and nucleic acid metabolism and thus have an ultimate effect on nutrition, respiration and photosynthesis. The main reason behind this toxicity is the generation of oxidative stress due to the formation of free radicals. Another reason why metals may be toxic is that they can replace essential metals in pigments or enzymes disrupting their function [6]. Thus, metals render the land unsuitable for plant growth and destroy the biodiversity.

Because of heavy metal accumulation in the food chain and their persistence in nature, it becomes necessary to remove toxic metals from waste water. Traditional techniques which are employed for removal of metals include chemical precipitation, ion exchange or electrochemical processes. Limitations of these methods are that they are neither economical nor effective especially when dealing with low concentrations of heavy metals and are thus responsible for causing further disturbance in the already damaged environment [7]. Various chemical, physical and biological techniques broadly characterized under *ex situ and in situ* methods have been employed for remedifying heavy metal contaminated soil. *Ex situ* method involves on or off-site treatment of contaminated soil and then returning the treated soil back to the original site. This method largely depends on excavation, detoxification and destruction of contaminant by either physical or chemical method. On the other hand, *in situ* method employs remediation without excavation of contaminated soil. This method involves destruction or transformation of contaminant, immobilization of contaminant to decrease its availability and separation of contaminant from soil. *In situ* techniques are preferred over *ex situ* techniques because of their reduced cost and impact on the ecosystem. Moreover, the off-site burial of contaminated soil used in *ex situ* method is not a better option because it is simply shifting the contaminated soil to some other place and is not eliminating the problem [8].

To overcome the limitations of traditional techniques, now research has focused on use of cost effective and innovative biological technique such as phytoremediation [8, 9]. It is an emerging technology that uses various plants to degrade, extract, contain, or immobilize contaminants from soil and ground water. Because of its elegance and extent of removal of contaminants from the polluted areas, the technique has received significant scientific attention [10].

CONCEPT OF PHYTOREMEDIATION

Phytoremediation can be defined as the engineered use of green plants, including grasses, forbs, and woody species, to remove, contain, or render harmless some environmental contaminants such as heavy metals, trace elements, organic compounds, and radioactive compounds accumulated in soil or water because of undesirable human interference. This definition includes all plant-influenced biological, chemical, and physical processes that aid in the uptake, sequestration, degradation, and metabolism of contaminants, either by plants or by the free-living organisms that constitute the plant's rhizosphere. Phytoremediation takes advantage of the unique and selective uptake capabilities of plant root systems, together with the translocation, bioaccumulation, and contaminant storage/degradation abilities of the entire plant body [11].

The idea of using metal accumulating plants to remove heavy metals and other compounds was first introduced in 1983, but the concept has actually been implemented for the past 300 years [12]. The generic term 'Phytoremediation' consists of the Greek prefix phyto (plant), attached to the Latin root remedium (to clean or restore) [13]. This technology can be applied to both organic and inorganic pollutants present in soil (solid substrate), water (liquid substrate) and the air [14-17]. The physico-chemical techniques for soil remediation render the land useless for plant growth as they remove all biological activities, including useful microbes such as nitrogen fixing bacteria, mycorrhiza, fungi, as well as fauna in the process of decontamination [18-20].

The main strategy adopted by terrestrial plants to combat heavy metal stress is to store excess metals in root cell walls and vacuole. This sequesters heavy metals both from root cytoplasm as well as from shoots and thus helps in minimizing the damage to photosynthetic apparatus. However, there are some plants species (hyperaccumulators) that have emerged on soil contaminated with heavy metals and thus are endemic to metalliferous soils. But, their small size and slow growth proved to be a limiting factor for speed of metal removal [21-26]. By definition, a hyperaccumulator must accumulate at least 1000 μg g^{-1} of Co, Cu, Cr, Pb, or Ni, or 10,000 µg g^{-1} (*i.e.* 1%) of Mn or Zn in the dry matter [27, 28]. Some plants tolerate and accumulate high concentrations of metal in their tissue but not at the level required to be called as hyperaccumulators. These plants are often called moderate metal-accumulators, or just moderate accumulators [29]. Many hyperaccumulators responsible for phytoremediation belong to Brassica family [25]. Several high-biomass, metal-accumulating species have been identified by various researchers [29, 30]. Actually, terrestrial plants employ three basic strategies for growing on contaminated metalliferous soils: a) Metal excluders: excludes metals by restricting their entry in above ground parts and keeping them in roots only by either altering membrane permeability and metal binding capacity of cell walls or by releasing chelating substances[31]; b) Metal indicators: accumulates metals in their aerial parts. They serve as the indicators of soil metal levels; c) Metal accumulators: concentrate metals in their aerial parts to concentrations much higher than those present in soil. Hyperaccumulators can absorb high levels of contaminants and concentrate in their roots, shoots or leaves [32-34].

The phytoremediation concept is based on the well-known ability of plants and their associated rhizosphere to concentrate and/or degrade highly dilute contaminants. Plant roots and rhizosphere microorganisms "sense" the immediate soil environment in which they are growing and have complex feedback mechanisms that permit them to adapt to changing conditions as they grow. Critical components of the rhizosphere, in addition to a variety of free-living microorganisms, include root exudates. These complex root secretions, which "feed" the microorganisms by providing carbohydrates, also contain natural chelating agents (citric, acetic, and other organic acids) that make the ions of both nutrients and contaminants more mobile in the soil. Root exudates may also include enzymes, such as nitroreductase, dehalogenases, and laccases. These enzymes have important natural functions, but they may also degrade organic contaminants that contain nitro groups (*e.g.*, TNT, other explosives) or halogenated compounds (*e.g.*, chlorinated hydrocarbons, many pesticides) [35]. In some plants growing in phosphorus-deficient soil, the root exudates contain large amounts of citric acid, in an attempt to mobilize and make available for uptake any phosphorus compounds present. Some rhizosphere microorganisms secrete plant hormones that increase root growth, and thereby the secretion of root exudates that contain metabolites they use as an energy source [35].

DIFFERENT TECHNOLOGIES OR PROCESSES OF PHYTOREMEDIATION

Phytoremediation processes can be classified based on the contaminant fate: degradation, extraction, containment, or a combination of these (Fig. **1**). Phytoremediation consists of a collection of four different plant-based technologies, each having a different mechanism of action for the remediation of metal-polluted soil, sediment, or water. These include: a) phytoextraction: the use of plants to remove metals from soils and to transport and concentrate them in above-ground biomass; b) rhizofiltration: involves the use of plants to clean various aquatic environments; c) phytostabilization: the use of plants to minimize metal mobility in contaminated soil through accumulation by roots or precipitation within the rhizosphere; and d) phytovolatilization: involves the use of plants to extract certain metals from soil and then release them into the atmosphere through volatilization (Fig. **1**).

PHYTOEXTRACTION

Removal of contaminants from metal polluted soil or ground water using phytoextraction is one of the best approaches amongst various techniques employed for phytoremediation which has received much scientific attention. It involves removal and isolation of contaminant from soil without disrupting structure and fertility of the soil. It is also referred as phytoaccumulation [36]. This technique uses metal accumulating plants as a means to remove toxic metals from soil [37]. Metal accumulating plants transport and concentrate metals from the soil into the harvestable parts of roots and above-ground shoots. As the plant absorb, concentrate and precipitate toxic metals and radionuclide from contaminated soils into the biomass, it is best suited for the remediation of diffusely polluted areas, where pollutants occur only at relatively low concentration and superficially.

The primary aim of phytoextraction is to reduce the metal concentration in contaminated soils to acceptable level within reasonable time duration. The ability of selected plants to grow and accumulate metals under the specific conditions (climatic and soil) decides the efficiency of phytoextraction. Various plants (Hyperaccumulator) which can accumulate different contaminants belonging to *Brassicaceae, Euphorbiaceae, Asteraceae, Lamiaceae,* or *Scrophulariaceae* families [38] are shown in Table **1**.

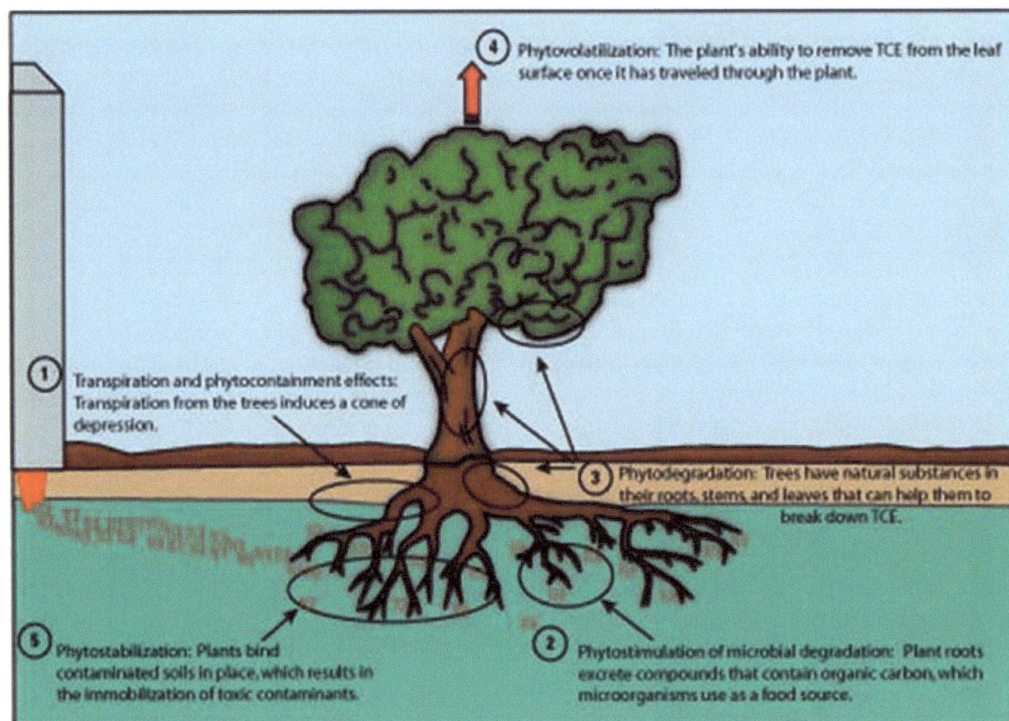

Figure 1: Different strategies employed for phytoremediation.

Table 1: Hyperaccumulator plants which shows accumulation of different heavy metals

Plant	Contaminant	References
Brassica juncea (Indian mustard)	Pb, Cr (VI), Cd, Cu, Ni, Zn, Sr, B, and Se	[37, 39, 40]
Thlaspi caerulescens (Alpine pennycress)	Ni and Zn	[41-43]
Thlaspi rotundifolium ssp. *Cepaeifolium, Lemna minor*	Pb	[39, 44]
Alyssum wulfenianum, Boxwood (*Buxaceae*) and cactus-like succulents (*Euphoribiaceae*)	Ni	[23, 38]
Indian mustard (*Brassica juncea*) and canola (*Brassica napus*), Kenaf (*Hibiscus cannabinus* L. cv. Indian) and tall fescue (*Festuca arundinacea* Schreb cv. Alta)	Se and B	[45]
Hybrid poplar trees	As and Cd	[46]
Sunflower	Cs (roots) and Sr (shoots)	[47]
Desert broom (*Baccharis sarothroids* Gray)	Pb, Cr, Cu, Ni, Zn and As	[48]
Water lettuce	Pb, Cr, Cu, Ni, Zn and Fe	[49]
Chrysopogon zizanioides	As	[50]

Generally, the metal phytoextraction process involves the following steps (1) cultivation of plants on the contaminated site: roots of established plants absorb metal elements from the soil and translocate as well as accumulate them to the above-ground shoots. If metal availability in the soil is not adequate for sufficient plant uptake, chelates or acidifying agents may be used to liberate them into the soil [51, 52] (2) removal of harvested metal-rich biomass: after sufficient plant growth and metal accumulation, the above-ground portions of the plant are harvested and removed, resulting in the permanent removal of metals from the site

(3) post harvest treatments and subsequent disposal of the biomass as a hazardous waste: incineration of harvested plant tissue dramatically reduces the volume of the material requiring disposal [29] and (4) eventual recuperation of metals from the metal-enriched biomass: valuable metals can be extracted from the metal-rich ash and serve as a source of revenue, thereby offsetting the expense of remediation [24, 53, 54] (Fig. **2**). In recent years, metal phytoextraction has received increasing attention, being in development to meet the growing market for phytoremediation products [55].

Phytoextraction should be viewed as a long-term remediation effort, requiring many cropping cycles to reduce metal concentrations to acceptable levels [24]. The time required for remediation ranges from 1-20 years and is dependent on various factors, *viz.*, the type and extent of metal contamination, the length of the growing season, and the efficiency of metal removal by plants [29, 56]. This technology is suitable for the remediation of large areas of land that are contaminated at shallow depths with low to moderate levels of metal-contaminants [29, 28].

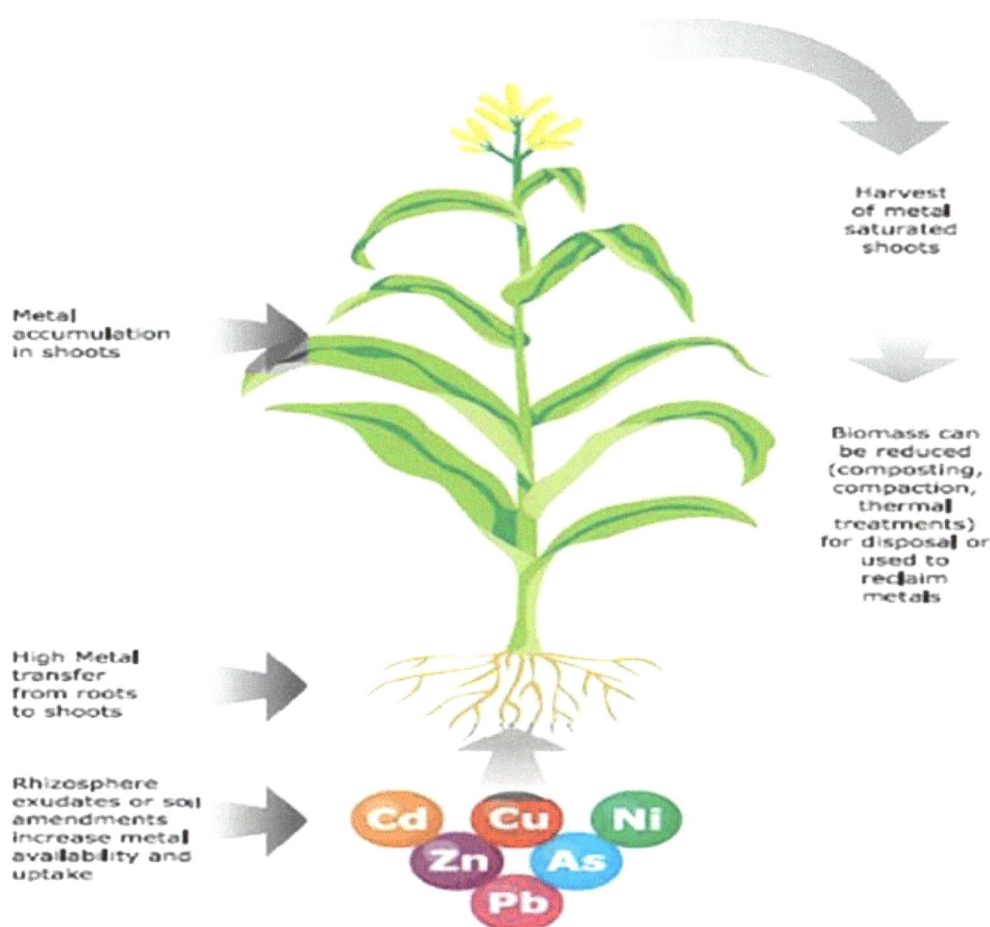

Figure 2: Schematic representation of the processes involved in phytoextraction of metals.

STRATEGIES OF PHYTOEXTRACTION

Two approaches have been used for remedial of contaminated soil using phytoextraction: a) the use of plants with exceptional, natural metal-accumulating capacity, the so-called *hyperaccumulators*, and b) the utilization of high-biomass crop plants, such as corn, barley, peas, oats, rice, and Indian mustard with a chemically enhanced method of phytoextraction [14, 57, 58]. So, the ideal plant for the phytoextraction process should not only be able to tolerate and accumulate high level of heavy metals in its harvestable parts but should also have a rapid growth rate and the potential to produce a high biomass in the field. As the ability of plants to accumulate metals is somewhat dependent on their capacity to tolerate high levels of metals in tissues, some mechanisms involved in metal accumulation by plants such as compartmentation in the vacuole and chelation in the cytoplasm have also been extensively studied. So there are two basic strategies of phytoextraction- 1) chelate-assisted phytoextraction also known as induced phytoextraction and, 2) long- term continuous or natural phytoextraction (Figs. **3** and **4**) and Table **2**.

INDUCED PHYTOEXTRACTION

This strategy employs the addition of artificial chelates to increase the mobility and uptake of metal contaminant (Fig. **3**). Within the plant cell heavy metal may trigger the production of oligopeptide ligands known as phytochelatins (PCs) and metallothioneins (MTs) [59]. These peptides bind and form stable complex with the heavy metal and thus neutralize the toxicity of the metal ion [60]. Glutathione serves as building blocks for the synthesis of PCs, resulting in a peptide with structure Gly-(γ-Glu-Cys-)n; {where, n varies from 2-11}. A number of plant species have been reported to produce phytochelating ligands in the presence of heavy metals [51]. Metallothioneins (MTs), are small gene encoded, Cys-rich polypeptides and are functionally equivalent to PCs [60]. Chelators have been isolated from plants that are strongly involved in the uptake of heavy metals and their detoxification. Application of chelating agents like ethylenediamine tetraacetic acid (EDTA) to Pb contaminated soils significantly increases the amount of bioavailable lead in the soil on one hand and on the other hand, facilitates a greater accumulation in plants [61, 62]. According to a study, the addition of chelates to a lead contaminated soil (total soil Pb 2500 mg kg^{-1}) increased shoot lead concentration of *Zea mays* (corn) and *Pisum sativum* (pea) from less than 500 mg kg^{-1} to more than 10,000 mg kg^{-1}. This was achieved by adding synthetic chelate EDTA to the soil, similar results using citric acid to enhance uranium uptake have been documented. These results indicate that both the transport of Pb in the xylem and its translocation from roots to shoots is enhanced or facilitated by chelators. The order of effectiveness of chelates in increasing Pb desorption from the soil was EDTA > Hydroxyethylethylene-diaminetriacetic acid (HEDTA) > Diethylenetriaminepentaacetic acid (DTPA) > Ethylenediamine di(o-hyroxyphenylacetic acid) EDDHA [61]. Metal accumulation efficiency is reported to be directly related to the affinity of the applied chelate for the metal. Thus, it can be inferred that synthetic chelates having high affinity for metal of interest (EDTA for lead, EGTA for cadmium, citrate for uranium) should be used as they increases the efficiency of phytoextraction process. Chelate assisted strategy consists of two basic processes- release of bound metals into the soil solution combined with transport of metals to the harvestable shoot.

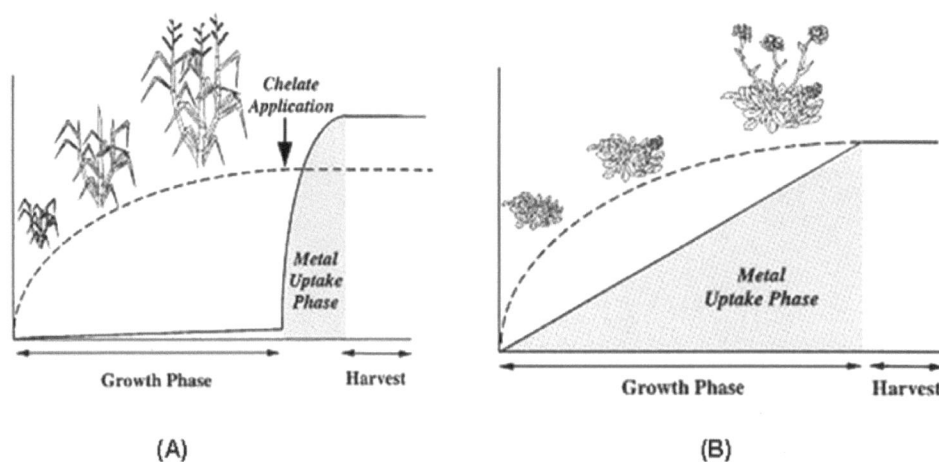

Figure 3: Process of (A)chelate-assisted and (B) natural metal extraction by plants. —— line represents metal concentration in shoot biomass; ----- line represents shoot biomass.

CONTINUOUS OR NATURAL PHYTOEXTRACTION

"Natural" phytoextraction (Fig. **3**) is usually conducted by planting (or transplanting) selected plant species in the contaminated soil. These plants are grown under normal farming conditions (fertilized and irrigated as necessary) until they reach their maximum size. The aboveground parts of the plants containing the contaminants are then harvested and disposed of appropriately. The plants are highly specialized, occur naturally, and can tolerate very elevated concentrations of metals that would be toxic to other plants. Typically, these plants are small, have a small and shallow root system, and grow relatively slowly. This type of metal uptake is facilitated by the use of hyperaccumulating plants that grow on soils rich in heavy metals. These plants can accumulate >1% of shoot dry biomass as zinc, nickel, manganese or selenium. Thus, as compared to induced phytoextraction, this method is based on the genetic and physiological capacity of specialized plants to accumulate, translocate and resist high amounts of metals.

Table 2: Main characteristics of the two strategies of phytoextraction of metals from soils

Chemically Assisted (Induced) Phytoextraction	Natural Phytoextraction
Plants are normally metal excluders	Plants naturally hyperaccumulate metals
Fast growing, high biomass plants	Slow growing, low biomass production
Synthetic chelators and organic acids are used to enhance metal uptake	Natural ability to extract high amount of metals from soils
Chemical amendments increase the metal transfer from roots to shoots	Efficient translocation of metals from roots to shoots
Low tolerance to metals; the increase in absorption leads to plant death	High tolerance; survival with high concentrations of metals in tissues

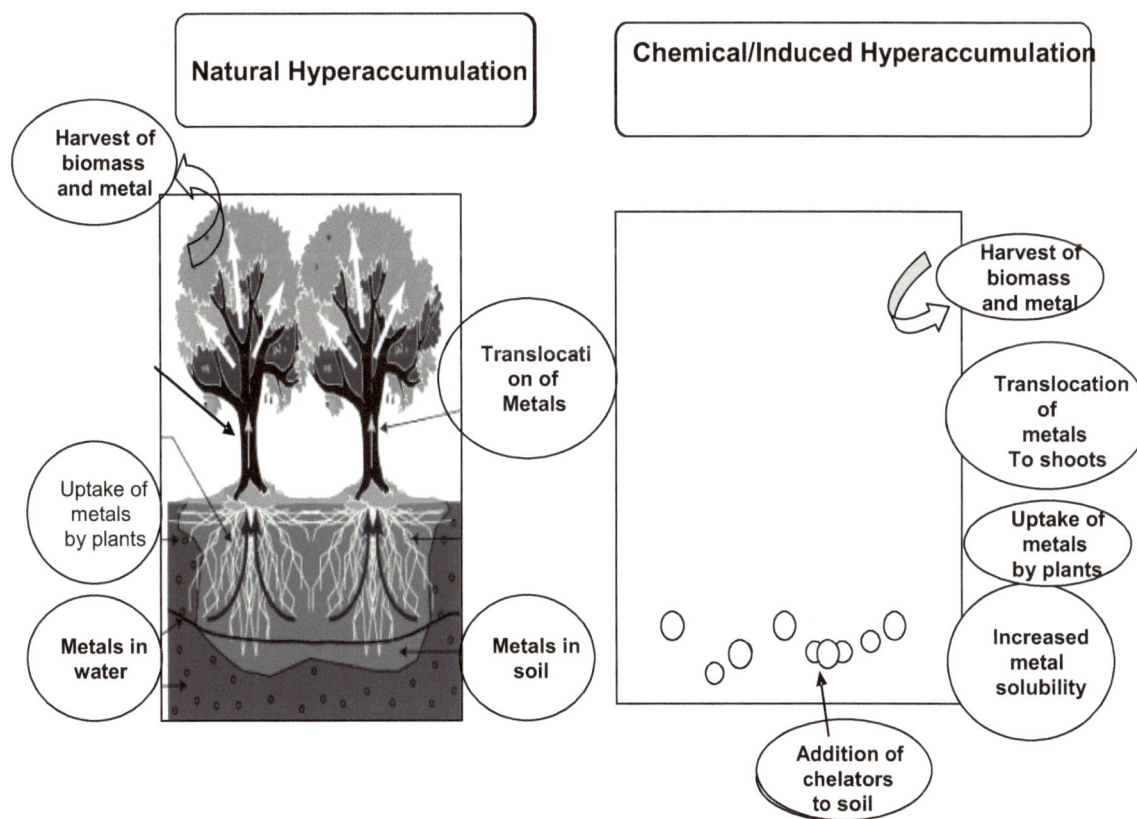

Figure 4: Comparison of natural and induced phytoextraction.

MECHANISM OF PHYTOEXTRACTION

Following mobilization, a metal has to be captured by root cells. Metals are first bound by the cell wall followed by their uptake across the plasma membrane facilitated through secondary transporters such as channel proteins and/or H^+- coupled carrier proteins [63].

Once inside the plant, most metals usually form carbonate, sulphate or phosphate precipitates and are thus immobilized in apoplastic (extracellular) and symplastic (intracellular) compartments [64]. Apoplastic pathway is relatively unregulated, because water and dissolved substance can flow and diffuse without having to cross a membrane. The cell walls of the endodermal cell layer act as a barrier for apoplastic diffusion into the vascular system. In general, solutes have to be taken up into the root symplasm before they can enter the xylem [65]. Subsequent to metal uptake into the root symplasm, three processes govern the movement of metals from the root into the xylem: sequestration of metals inside root cells, symplastic transport into the stele and release into the xylem mediated *via* membrane transport proteins which functions as specific or generic metal ion carriers or channels [66, 67]. After heavy metals have entered the root they are either stored in the root or translocated to the shoots. Metal ions can be actively transported across the tonoplast as free ions or as metal–chelate complexes [68, 69]. The vacuole is an important component of the metal ion storage where they are often chelated either by organic acid or phytochelatins. Insoluble precipitates may form under certain conditions. Precipitation, compartmentalization and chelating are the most likely major events that take place in resisting the damaging effects of metals (Fig. **5**) [25]. Transporters mediate uptake into the symplast, and distribution with in the leaf occurs *via* the apoplast or the symplast [70]. Plants transpire water to move nutrients from the soil solution to leaves and stems, where photosynthesis occurs. Willows, hybrid poplar are also good phytoremediators, because they take up and process large volumes of soil water. For example, data show that a single willow tree, on a hot summer day, can transpire more than 19,000 litres of water [71].

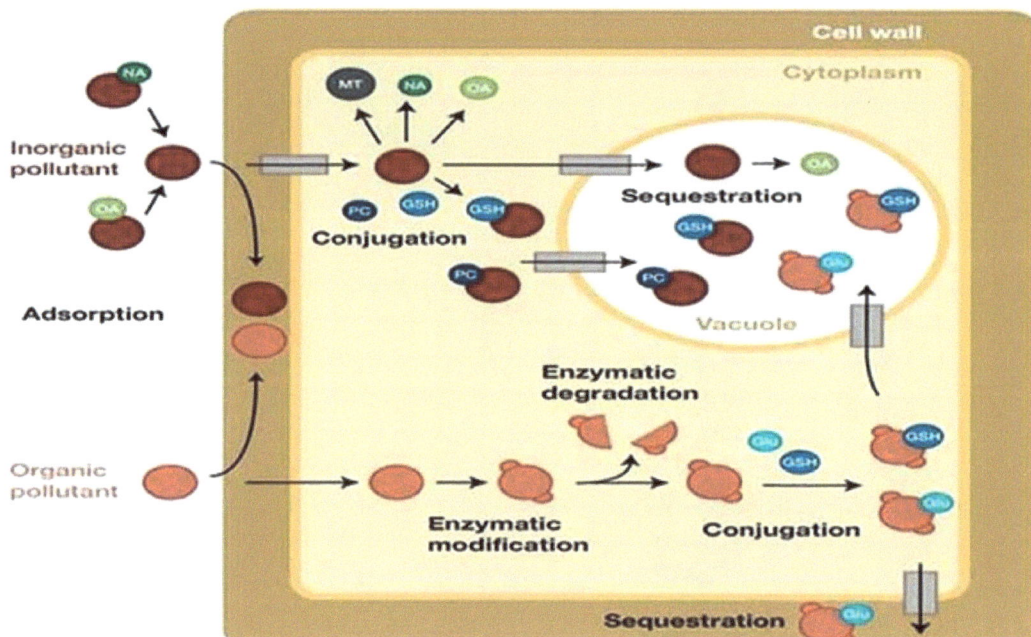

Figure 5: Tolerance mechanisms for inorganic and organic pollutants in plant cells.

Detoxification generally involves conjugation followed by active sequestration in the vacuole and apoplast, where the pollutant can do the least harm. Chelators shown are GSH: glutathione, Glu: glucose, MT:

metallothioneins, NA: nicotianamine, OA: organic acids, PC: phytochelatins. Active transporters are shown as boxes with arrows. (Adapted from Smits, 2005 [135]).

STRATEGIES TO ENHANCE PHYTOEXTRACTION PROCESS

Increasing Metal Availability in Soil

A major factor influencing the efficiency of phytoextraction is the ability of plants to absorb large quantities of metal in a short period of time. Hyperaccumulators accumulate appreciable quantities of metal in their tissue regardless of the concentration of metal in the soil [72]. This property is unlike moderate accumulators now being used for phytoextraction. The quantity of absorbed metal by a moderate accumulator is a reflection of the concentration of the particular metal in the soil. Although the total soil metal content may be high, it is the fraction that is readily available in the soil solution that determines the efficiency of metal absorption by plant roots.

Soil environment contains metals in either of the following forms: a) free metal ions and soluble metal complexes in soil solutions; b) metal ions occupying ion exchangeable sites and specifically adsorbed on inorganic soil constituents; c) organically bound metals; d) precipitated or insoluble compounds, especially of oxides, carbonates and hydroxides; and e) metals in the structure of silicate minerals. Contamination resulting from human activities results in types of fractions from (a-d), however fraction (e) is the indicator of background or indigenous soil concentrations [73]. Out of these five kinds of fractions, only a and b types of fractions are readily available to plants. This necessitates for manipulation of soil environment for effective phytoremediation.

To enhance the speed and quantity of metal removal by plants, various chemicals for example, acidifying agents [41, 53, 54, 74], fertilizer salts [52, 75] and chelating materials [51, 76] are added to the soil. These chemicals functions by either liberating or displacing metal from the solid phase of the soil or by making precipitated metal species more soluble and thus increase the amount of bioavailable metal in the soil solution.

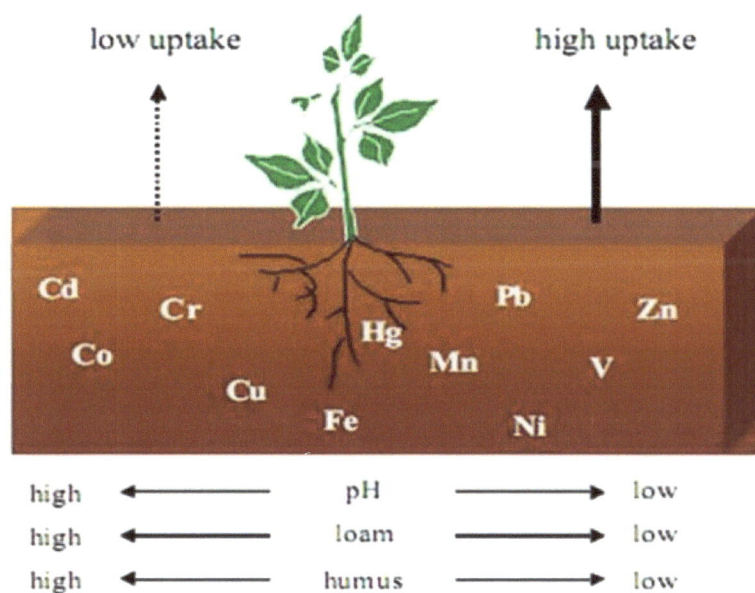

Figure 6: Factors affecting phytoextraction.

Soil pH is a major factor influencing the availability of elements in the soil for plant uptake [77]. Under acidic conditions, H^+ ions displace metal cations from the cation exchange complex (CEC) of soil components and cause metals to be released from sesquioxides and variable-charged clays to which they have been chemisorbed *i.e.* specific adsorption [78]. The retention of metals to soil organic matter is also weaker at low pH, resulting in more available metal in the soil solution for root absorption. Many metal cations are more soluble and available in the soil solution at low pH (below 5.5) including Cd, Cu, Hg, Ni, Pb, and Zn (Fig. **6**) [54, 78]. Phytoextraction process is enhanced when metal availability to plant roots is facilitated through the addition of acidifying agents to the soil [37, 41, 54].The addition of chelating materials to soil, such as EDTA, HEDTA, and EDDHA, is the most effective means of liberating labile metal-contaminants in to the soil solution. Chelates complex the free metal ion in solution, allowing further dissolution of the sorbed or precipitated phases until equilibrium is reached between the complexed metal, free metal, and insoluble metal fraction [79]. Chelates are used to enhance the phytoextraction of a number of metal contaminants including Cd, Cu, Ni, Pb, and Zn [51, 70, 80].

ROLE OF ROOT EXUDATES IN METAL PHYTOEXTRACTION

Bioavailability of metal to plants is increased by secreting phytosidrophores into the rhizosphere to chelate and solubilize metals that are soil bound [81]. The exudates in rhizosphere of the plants have been clearly demonstrated to be associated with increase of metals uptake from soil and their translocation to shoots [37, 82-85]. Low molecular-weight organic acids are probably the most important exudates in natural phytoextraction systems. They influence the acquisition of metals by either forming complexes with metal ions or decreasing the pH around the roots and altering soil characteristics. Despite the fact that metals uptake may be increased due to decreasing pH [41], it is clear that the complexing capacity of organic acids, rather than their capacity to decrease pH, is the main factor related to mobilization of metals in soil and their accumulation in plants [86-89]. Indirect effects of root exudates on microbial activity, rhizosphere physical properties and root growth dynamics may also influence ion solubility and uptake [77, 90]. For instance, microorganisms have been shown to mobilize Zn for hyperaccumulation by *Thlaspi caerulescens via* dissolution of Zn from the non-labile phase in soil [91].

RHIZOFILTRATION

It is defined as the use of plants, both terrestrial and aquatic; to absorb, concentrate, and precipitate contaminants from polluted aqueous sources with low contaminant concentration in their roots. In other words, it can also be defined as the adsorption or precipitation onto plant roots, or absorption into the roots of contaminants that are in solution surrounding the root zone, due to biotic or abiotic processes [36]. There is no involvement of shoots in this process, so the plants responsible for this mechanism should have rapidly growing roots in order to remove toxic metals from water and soil for an extended period of time (Fig. **7**).

Plant uptake, concentration, and translocation might occur, depending on the contaminant. Some metals are precipitated by the exudates released from plant roots. Rhizofiltration first results in contaminant containment, in which the contaminants are immobilized or accumulated on or within the plant. Contaminants are then removed by physically removing the plant [36].

Rhizofiltration is similar to phytoextraction, but the plants are used primarily to clean contaminated ground water rather than soil. The plants to be used for cleanup are raised in greenhouses with their roots in water (hydroponic culture) rather than in soil. Once the plants are acclimatized and develop a large root system, contaminated water is collected from a waste site and brought to the plants where it is substituted for their water source. The plants are then transferred to the contaminated area where their roots take up contaminants along with the water. Harvesting of roots is done when they become saturated with the contaminants [30, 37, 92, 93].

Rhizofiltration occurs within the root zone in water. For rhizofiltration to occur, the water must come into contact with the roots. This contact zone can be maximized by designing the engineered systems by matching the depth of the unit to the depth of the roots. Groundwater may be extracted from any depth and piped to an engineered hydroponic system for *ex situ* treatment [36].

For *in situ* technologies, such as natural water bodies, the depth of the roots might not be the same as the depth of the water body. The water must be adequately circulated in such cases to ensure complete treatment, which is likely to become more difficult as the depth of the water increases [36].

Different mechanisms are employed by plants for the removal of different metals by roots. For example, in the case of lead (Pb), the fastest means of metal removal by the roots is passive surface absorption which includes chelation, ion exchange and specific adsorption. Other biological process which are relatively slow but can also help in metal removal includes intracellular uptake, vacuolar deposition and translocation to the shoots. Root mediated precipitation of lead in the form of lead phosphate is the slowest process [94].

This mechanism may offer cost advantage in water treatment because of the ability of plants to remove upto 60% of their dry weight as toxic metals, thus markedly reducing the generation and disposal cost of the hazardous or radioactive residue. It can also be a particularly cost-competitive technology in the treatment of surface of ground water containing relatively low levels of toxic metals, where various conventional techniques (precipitation or flocculation ; ion exchange; reverse osmosis and microfiltration) sometimes fails to deliver.

Rhizofiltration can partially treat industrial discharge, agricultural runoff, or acid mine drainage. It can be used for lead, cadmium, copper, nickel, zinc and chromium, which are primarily retained with in the roots [36, 95]. The advantages of rhizofiltration include its ability to be used as *in situ* or *ex situ* applications and species other than hyperaccumulators can also be used. Plants like sunflower, Indian mustard, tobacco, rye, spinach and corn have been studied for their ability to remove lead from effluent, with sunflower having the greatest ability. Indian mustard has proven to be effective in removing a wide concentration range of lead (4 –500 mg/l) [96]. The technology has been tested in the field with uranium (U) contaminated water at concentrations of 21-874 µg/l; the treated U concentration reported by Dushenkov was < 20 µg/l before discharge into the environment [97].

PHYTOSTABILIZATION

Phytostabilization, also known as phytorestoration, involves the use of plants to eliminate the bioavailability of metals in soils. It is defined as (1) immobilization of a contaminant in soil through absorption and accumulation by roots, adsorption onto roots, or precipitation within the root zone of plants, and (2) the use of plants and plant roots to prevent contaminant migration *via* wind and water erosion, leaching, and soil dispersion (Fig. **7**).

This particular technique provides hydraulic control, which suppresses the vertical migration of contaminants in to groundwater; and physically and chemically immobilizes contaminants by root sorption and by chemical fixation with various soil amendments [25, 37, 92, 98, 99]. This technique is actually a modified version of the in-place inactivation method in which the function of plants is secondary to the role of soil amendments. Unlike other phytoremediative techniques, the goal of phytostabilization is not to remove metal contaminants from a site, but rather to stabilize them and reduce the risk to human health and the environment.

Phytostabilization occurs through the chemistry and microbiology of the root-zone, and/or alteration of the soil environment or contaminant chemistry. Plant root exudates or production of carbon-dioxide may be responsible for change in soil pH. Phytostabilization can change metal solubility and mobility or impact the dissociation of organic compounds. The plant affected soil environment can convert metals from a soluble to an insoluble oxidation state [37]. Phytostabilization can occur through sorption, precipitation, complexation, or metal valence reduction [100].

Soils polluted with heavy metals usually lack established vegetation cover due to toxic effects of pollutants and other physical disturbances. Barren soils being more prone to erosion and leaching are responsible for spreading of pollutants in the environment. Re-vegetation with metal tolerant plant species is the most feasible solution to this problem. Re-vegetation aids in reducing the mobility of metals and in turn reduces the risk of leaching of metals in ground water or airborne spread. This technique is very effective when rapid immobilization is needed to preserve ground and surface water and disposal of biomass is not required. However the major disadvantage is that, the contaminant remains in soil as it is, and therefore requires regular monitoring.

Agrostis tenuis cv. Gogian (for acid, Pb and Zn wastes), *Festuca rubra* cv. Merlin (for calcareous Pb/Zn wastes) and *Agrostis tenuis* cv. Parys (for copper wastes), all from grass family, have been widely used in USA and UK to provide thick cover on contaminated sites [94].

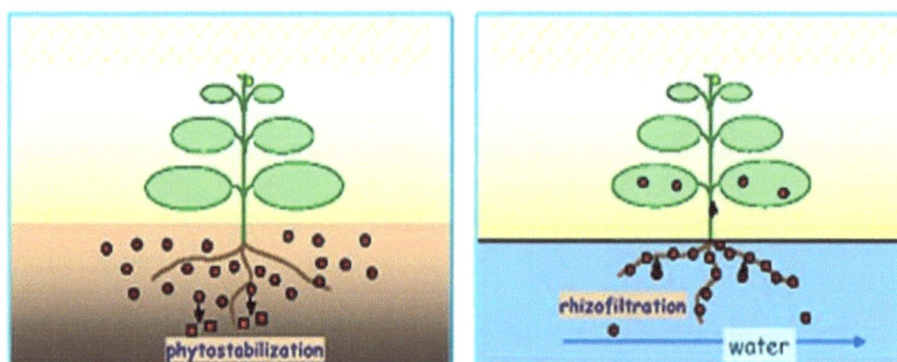

Figure 7: Difference between phytostabilization and rhizofiltration.

PHYTOVOLATILIZATION

Phytovolatilization involves the use of growing trees and other plants to take up water soluble organic and inorganic contaminants from the soil and release them into the atmosphere during transpiration of water by leaves. Contaminants usually become modified in their path of translocation through vascular tissues from roots to leaves. From leaves, they volatilize into the atmosphere at comparatively low concentrations [101].

Phytovolatilization has been primarily used for the removal of mercury where the mercuric ion is transformed into less toxic elemental mercury. However, there are chances of precipitation of the released mercury and hence, its re-deposition back into the environment [6]. According to a study by Gary Banuelos of USDS's Agricultural Research Service, some plants grow in high selenium media produce volatile selenium in the form of dimethylselenide and dimethyldiselenide [102]. The idea of selenium detoxification in plants was derived from the same mechanism of detoxification observed in case of animals. Selenium in animals is detoxified by releasing it as volatile dimethyl selenide from lungs. Dogs when injected with sodium tellurite were found to release the odour of dimethyl telluride in their breath. Similarly, odour of garlic in plants accumulating selenium may be the indicator of release of volatile selenium compounds. Both selenium accumulator and non-accumulator species have been reported to volatize selenium [103]. For example, *Astragalus racemosus* (selenium accumulator) have been reported to release dimethyl diselenide as a volatile form of selenium [104]. However, dimethyl selenide, a different volatile form of selenium have been reported to be released from a selenium non-accumulator, alfalfa [105].

PHYTODEGRADATION

This technique involves the use of plants' metabolic processes to degrade organic contaminants so that they can be converted into non-toxic forms. The contaminant can be transformed, broken-down, stabilized or is converted from non-volatile to volatile form [95]. Certain plant based enzymes like dehalogenases, oxygenases, nitrilase, reductases and peroxidases are used for this technique. These enzymes are responsible for degradation and transformation of ammunition wastes (TNT, dinitromonoaminotoluene and mononitrodiaminotoluene), chlorinated solvents such as trichloroethylene and other herbicides [106].

RHIZODEGRADATION

Rhizodegradation is the breakdown of organics in the soil through microbial activity of the root zone (rhizosphere) and is a much slower process than phytodegradation. Yeast, fungi, bacteria and other microorganisms consume and digest organic substances like fuels and solvents. Rhizodegradation is also known as plant-assisted degradation, plant-assisted bioremediation, plant-aided *in situ* biodegradation, and enhanced rhizosphere biodegradation. Root-zone biodegradation is the mechanism for implementing rhizodegradation. Root exudates are compounds produced by plants and released from plant roots. They include sugars, amino acids, organic acids, fatty acids, sterols, growth factors, nucleotides, flavanones, enzymes, and other compounds [107, 108]. The microbial populations and activity in the rhizosphere can be increased due to the presence of these exudates, and can result in increased organic contaminant biodegradation in the soil. Additionally, the rhizosphere substantially increases the surface area where active microbial degradation can be stimulated. Degradation of the exudates can lead to co-metabolism of contaminants in the rhizosphere.

Plant roots can affect soil conditions by increasing soil aeration and moderating soil moisture content, thereby creating conditions more favorable for biodegradation by indigenous microorganisms. Thus, increased biodegradation could occur even in the absence of root exudates. All phytoremediation technologies are not exclusive and may be used simultaneously, but the metal extraction depends on its bio available fraction in soil (Fig. **8**).

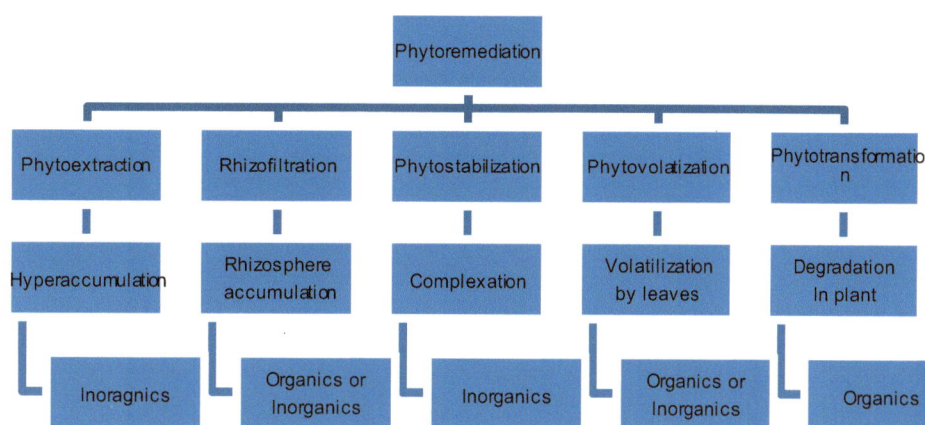

Figure 8: Summary of various mechanisms adopted by different phytoremediation processes including the type of contaminant being removed by the respective process.

ADVANTAGES AND DISADVANTAGES OF PHYTOREMEDIATION

Advantages

1) It is a cost effective technique as compared to other conventional methods of remediation.

2) Reduces both the spread of contaminants as well as destruction of soil characteristics due to *in situ* application.

3) Produces less waste.

4) No additional carbon-dioxide is released into atmosphere after burning of harvested biomass.

5) Biomass can be further used for heat and energy production.

Disadvantages

1) As compared to conventional techniques, it is a relatively slow process: needs years to remedify soil from pollutants. Time to remediate soil from contaminants depends on: type and number of plants being used; type and amount of contaminants present; size and depth of polluted area and type of soil and conditions present [109].

2) Toxicity of pollutants to the plants.

3) Risks posed to consumers of plants.

4) Dependent on season.

5) Dependent on root system of remediating plants.

APPLICATION OF GENETIC ENGINEERING TO IMPROVE PHYTOREMEDIATION

Plants have an inherent capability of removing organic and inorganic contaminants from the environment by means of phytoremediation. But remedying the soil by this method takes years because of slow rate of growth of plants and in addition is dependent on the total harvestable biomass of a plant. This results in slowing down the pace of removal of contaminants. Therefore, identification of fast growing and hyperaccumulating genotypes became indispensable (Fig. **9**).

By the application of genetic engineering, plants have been modified with respect to their phenotypic and genotypic characters. Biological functions of plants have also been successfully altered through modification of primary and secondary metabolism. This helped in understanding and improving the phytoremediation capacities of the plants [110].

Metal-hyperaccumulating plants and microbes with unique abilities to tolerate, accumulate and detoxify metals, represent an important reservoir of unique genes controlling these traits [111]. This provides an idea that such genes can be utilized effectively by genetic engineering by transferring them to fast-growing plants species for introgression of traits for improved phytoremediation [112]. Genetic studies have revealed that adaptive metal tolerance is a trait which is controlled by some major genes and a few minor modifier genes. The collective effects of these genes results in improved hyperaccumulation of contaminants by plants. A genetic analysis of copper tolerance with copper-tolerant and susceptible lines of *Mimulus guttatus* showed that a modifier gene that is active only in the presence of the tolerance gene is responsible for difference in Cu-tolerance in this species [113]. This and many more studies have revealed that the desired traits of phytoremediation may be improved by identifying the genes governing the expression of some particular protein, metal chelators or transport proteins and introgression of these genes in suitable plants. This can also be influenced by over expression of a particular gene. Therefore, it can be concluded from various published reports that genetic engineering can be successfully applied for modification of physico-chemical and molecular mechanisms of plants' for heavy metal uptake and tolerance/resistance by introduction of bacterial gene or mutant cells on the basis of desired phenotype in plant genome for enhanced metal uptake.

Recombinant DNA technology is one of the promising approaches for manipulation of plant's character and has been used to enhance phytoremediation by making modification in plants characters. This technique utilizes the concept of selectively choosing the desired trait to be introduced into the plants cells *via* introduction of genes from other living organisms, by preparing the gene construct accompanying promoters and regulators for the gene of interest.

CHANGING THE OXIDATION STATE OF HEAVY METALS

Introduction of genes into the plants, governing the expression of enzymes responsible for catalyzing the reactions which either results in changing the oxidation state of the metal or convert them into the less toxic form. For instance, bacterial *merA* and *merB* genes encoding mercuric oxide reductase and organomercurial lyase enzymes, respectively, and enzyme that methylate selenium into dimethylselenate [114, 115]. Both *merA* and *merB* coded enzymes are encoded on the mer operon and require sulfhydryl-bound substrates. All these enzymes results in the production of volatile form of heavy metals leading to remediation by phytovolatilization. MerB catalyzes the protonolysis of the carbon-mercury bond, resulting in the formation of a reduced carbon compound and inorganic ionic mercury. MerA catalyzes the conversion of Hg (II) to

Hg (0).The gene *merA* was also artificially synthesized using codon usage. The new gene *merApe9* thus produced was responsible for releasing free mercury which was both less toxic as well as volatile as compared to Hg^{2+}.

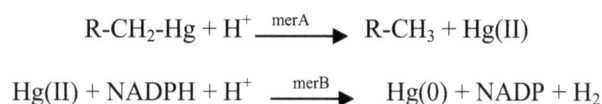

$$R\text{-}CH_2\text{-}Hg + H^+ \xrightarrow{\text{merA}} R\text{-}CH_3 + Hg(II)$$

$$Hg(II) + NADPH + H^+ \xrightarrow{\text{merB}} Hg(0) + NADP + H_2$$

Transgenic plants carrying *merA* and *merB* genes can grow on medium containing 50-folds higher methylmercury concentration as compared to wild type plants. The study has great significance as methylmercury is highly toxic and is found to occur in mercury polluted wetlands and coastal sediments throughout the world [94].

USE OF PHYTOCHELATINS

One method for enhanced hyperaccumulation is the expression of proteins, peptides or metal chelators secreted either into soil or plant cells. Phytochelatins (PCs) are small metal-binding peptides found in plants. The principal classes of metal chelators include phytochelatins, metallothioneins, organic acids and amino acids. Iso-PCs, a series of PC-like homologous chelating peptides are reported with varying terminal amino acids and have a C-terminal modified residue other than glycine. The PC and iso-PC molecules form complexes with heavy metals like Cd. In addition to PC-Cd complex other PC-metal-complexes include Ag, Cu and As. PC reactivate metal poisoned plant enzymes such as nitrate reductase up to 1000-fold better than chelators such as glutathione (GSH) or citrate, showing again the extraordinary sequestering potential of these peptides [116].

It has been reported that there are genes controlling the synthesis of peptides that sequester metals, like phytochelatins *e.g.*, the *Arabidopsis* cad1 gene [117]. An allelic series of cad1, cadmium-sensitive mutants of *Arabidopsis thaliana*, was isolated. These mutants were sensitive to cadmium to different extents and were deficient in their ability to form cadmium-peptide complexes as detected by gel-filtration chromatography. Each mutant was deficient in its ability to accumulate phytochelatins (PCs) as detected by high-performance liquid chromatography and the amount of PCs accumulated by each mutant correlated with its degree of sensitivity to cadmium. The mutants had wild-type levels of glutathione, the substrate for PC biosynthesis, and *in vitro* assays demonstrated that each of the mutants was deficient in PC synthase activity. These results demonstrate conclusively the importance of PCs for cadmium tolerance in plants [118].

Arsenic is a metal which contaminates thousands of sites throughout the world and has adverse effects on human health. Genetic engineering has been used for the production of arsenic tolerant and hyperaccumulating transgenic plants (*Arabidopsis thaliana*). Transgenic plants were made by the introduction of two different bacterial genes, arsC and γ-ECS, responsible for encoding arsenate reductase (ArsC) and γ-glutamylcysteine synthetase (γ-ECS), respectively, under the influence of specific regulatory sequences. ArsC catalyzes a reaction which converts arsenate (ASO_4^{3-}) to arsenite (ASO_3^{3-}), while γ-ECS converts amino acids, glutamine and cysteine into γ-glutamylcysteine, which is further used for the synthesis of organic thiols (RS) including glutathione (GSH) and phytochelatins (PCs). Phytochelatins are able to bind with arsenite and not arsenate. This led to increased degree of tolerance against hyperaccumulation of arsenic [94].

GENES FOR PLANT METAL TRANSPORTERS

Another promising approach could be the identification of metal transporter proteins and introduction of protein coding gene. This would help in enhanced ability of metal ions to enter plant cells. These are generally proteins that are localized in plasma membrane and either have an affinity for metal ions, or which create favorable energy conditions to allow metals to enter the cell. Several plant metal transporters are identified and reported till date (Table **3**).

Table 3: Various plant metal transporter proteins in plants along with their source and function

Gene	Function	Source	Reference
IRT1	Encodes protein that regulates uptake of iron and other metals	Arabidopsis	[119]
MRP1	Encodes an Mg-ATPase transporter	Arabidopsis	[120]
Nramp family (AtNramp1;OsNramp1: class1)and (AtNramp2-5; OsNramp2:class2)	AtNramp3 involved in Cd^{2+} uptake. Disruption of gene enhanced Cd tolerance whereas its over-expression led to Cd hypersensitivity	Rice and Arabidopsis	[121]
YCF1	MgATP-energized vacuolar transporter responsible for sequestration of compounds after their S-conjugation with glutathione	*S. cerevisiae*	[122]
AtHMA3	P1B ATPase, allows vacuolar storage of Cd/Zn/Co/Pb	Arabidopsis	[123]

METAL ACCUMULATING PROTEINS

A protein named NtCBP4 has been reported to modulate plant tolerance to heavy metals. Several independent transgenic lines expressing NtCBP4 had higher than normal levels of NtCBP4, exhibiting improved tolerance to Ni and hypersensitivity to Pb, which is associated with reduced Ni accumulation and enhanced Pb accumulation, respectively. This was the first report of a plant protein (probably involved in metal uptake across the plasma membrane) that modulates plant tolerance and accumulation of Pb. This gene could be useful for improving phytoremediation strategies [124].

Metallothioneins (metal-binding proteins that confer heavy metal tolerance and accumulation) have also been used to produce transgenic plants. Transgenic plants thus produced showed enhanced tolerance to high metal concentrations but on the other hand no enhancement in metal uptake was reported. To enhance higher plant metal sequestration, the yeast metallothionein CUP1 was introduced into tobacco plants, and the cup1 gene expression and Cu and Cd phytoextraction were determined [125]. Over-expression of copper inducible MT *cup 1* also enhanced Cu tolerance in plants [126].

TRANSGENIC PLANTS FOR METAL UPTAKE, TOLERANCE AND DETOXIFICATION

The genetic and biochemical basis is becoming an interesting target for genetic engineering. A fundamental understanding of both uptake and translocation processes in normal plants and hyperaccumulators, regulatory control of these activities, and the use of tissue-specific promotors offers great promise that the use of molecular biology tools can be effective to develop efficient and economic phytoremediation plants for soil metals [127]. Various transgenic plants thus produced for the above said purpose are compiled in the form of Table **4**.

Table 4: Transgenic Plants for Metal uptake, Tolerance and Detoxification

Gene/Enzyme	Transgenic Plant	Improved Trait	References
Glutamylcystine synthetase	*Populus augustifolia, Nicotiana tabacum & Silene cucubalis*	Enhanced metal accumulation	[118]
gsh1	Brassica juncea	γ-ECS transgenic seedlings showed increased tolerance to cadmium and had higher concentrations of PCs, γ - GluCys, glutathione, and total nonprotein thiols	[128]
Selenocysteine methyl transferase (SMT)	Arabidopsis, Indian mustard	Enhanced tolerance to Se, volatized Se faster	[129]
Cystathione-gamma-synthase (CGS)	Indian mustard	Enhanced tolerance to selenite, volatize Se faster	[130]
merA	*Arabidopsis thaliana,* yellow poplar	Enhanced resistance to $HgCl_2$ accompanied with volatilization	[131, 132]

Table 4: cont….

merB	*Arabidopsis thaliana*	Enhanced resistance to methyl mercury	[132, 133]
ACC deaminase	Tomato	Enhanced metal (Cd, Cu, Ni, Mn, Pb and Zn) accumulation and tolerance	[134]
*CSb(Citrate synthase)**	Tobacco, papaya, rice and corn	Tolerance to aluminium toxicity	[94]

* Citrate synthase : synthesizes high level of citrate and secreted into the soil by transgenic plants through their roots. Citrate binds with available aluminium, rendering it incapable of entering the roots.

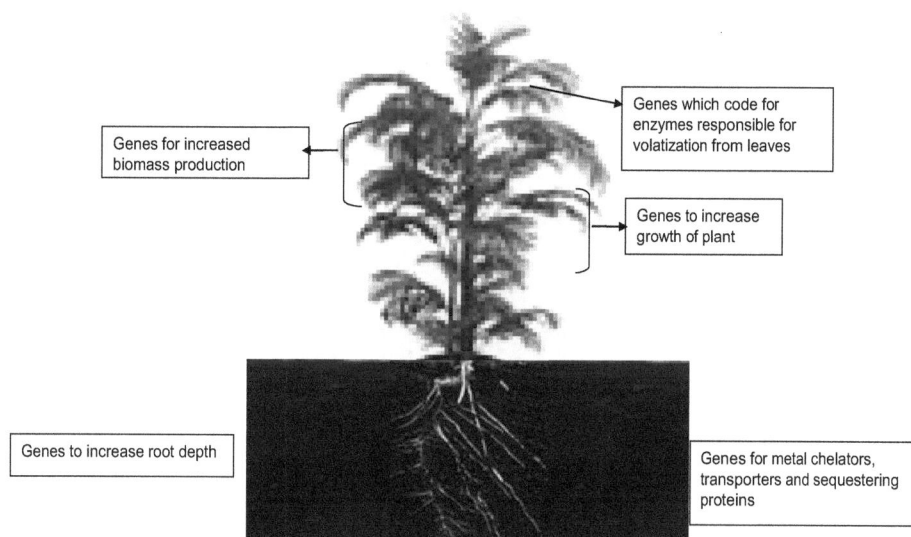

Figure 9: Introduction of various genes to improve phytoremediation efficiency of a plant.

FUTURE PROSPECTS

Although plants show some ability to reduce the hazards of organic pollutants, the greatest progress in phytoremediation has been made with metals. Phytoremediative technologies which are soil-focused are suitable for large areas that have been contaminated with low to moderate levels of contaminants. Sites which are heavily contaminated can not be cleaned through phytoremediative means because the harsh conditions will not support plant growth. The depth of soil which can be cleaned or stabilized is restricted to the root zone of the plants being used. Depending on the plant, this depth can range from a few inches to several meters. Phytoremediation should be viewed as a long-term remediation solution because many cropping cycles may be needed over several years to reduce metals to acceptable regulatory levels. This new remediation technology is competitive with, and may be superior to existing conventional technologies at sites where phytoremediation is applicable. Phytoremediation is not the solution for all hazardous waste problems but is rather a tool that can be used, possibly in conjunction with other clean-up methods, to remediate polluted environments. Although, the use of genetic engineering for modifying the biochemical mechanisms of plants, in addition to enhancing some special traits in metal hyperaccumulators is the most convincing and promising approach for phytoremediation. A better understanding of the biochemical processes involved in plant heavy metal uptake, transport, accumulation and resistance will help to enhance or improve phytoremediation of genetically engineered plants. In the case of lack of the information about known phytoremediation genes, the objective can be accomplished by the use of somatic and sexual hybridization succeeded by extensive screening and back crossing of the progeny. There calls for an extensive need to explore the genetic content available in the plant species and genotypes. The information can thus be further used to genetically engineer the desired plants, to improve the phytoremediation efficiency. This further requires a multidisciplinary approach, including areas like plant biology, agronomy, soil science, agriculture engineering and genetic engineering. Bioinformatics can prove to be an important tool in predicting metal binding sites for phytoremediation.

ABBREVIATIONS USED

T: Metric ton.

PCs: Phytochelatins

MTs: Metallothioneins

EDTA: Ethylenediamine tetraacetic acid

HEDTA: Hydroxyethylethylene-diaminetriacetic acid

DTPA: Diethylenetriaminepentaacetic acid

EDDHA: Ethylenediamine di(o-hyroxyphenylacetic acid)

CEC: Cation exchange complex

REFERENCES

[1]   Singh SP, Ghosh M. A comparative study on effect of cadmium, chromium and lead on seed germination of weed and accumulator plant species. Ind J of Environ Protect 2003; 23(5): 513-8.

[2]   Adriano DC. Trace elements in the terrestrial environment. Springer-Verlag New York. 1986; pp. 533.

[3]   Alloway BJ – In: Alloway B J. Heavy Metals in Soils. Blackie, Glasgow. 1990.

[4]   Kramer U, Chardonnens AN. The use of transgenic plants in the bioremediation of soils contaminated with trace elements. Appl Microbiol Biotechnol 2001; 55: 661-72.

[5]   Lin CF, Lo SS, Lin HY Lee YC. Stabilization of cadmium contaminated soils using synthesized zeolite. J Hazard Mat 1998; 60: 217-26.

[6]   Henry JR. In: NNEMS Report. An Overview of Phytoremediation of Lead and Mercury. Washington 2000; pp. 3-9.

[7]   Klimmek S, Stan HJ, Wilke A, Bunke G, Buchholz R. Comparative analysis of the biosorption of cadmium, lead, nickel and zinc by algae. Environ Sci Tech 2001; 35: 4283-8.

[8]   Ghosh M, Singh SP. A review on phytoremediation of heavy metals and utilization of its by products. Appl Ecol Environ Res 2005; 3(1): 1-18.

[9]   Lone MI, He ZL, Stoffella PJ, Yang XE. Phytoremediation of heavy metal polluted soils and water: Progress and perspectives. J Zhejiang Univ Sci B 2008; 9(3): 210-20.

[10]   Cappuana M. Heavy metals and woody plants- biotechnologies for phytoremediation. J Biogeosci Forest 2011; 4: 7-15.

[11]   Hinchman RR, Cristina NM. Providing the baseline science and data for real-life phytoremediation applications. In: Seattle WA, Ed. Proceedings of the 2nd Intl. Conference on Phytoremediation, Partnering for Success 1997; Chapter 1.5.

[12]   Henry JR. In: NNEMS Report. An Overview of Phytoremediation of Lead and Mercury. Washington, DC. 2000; pp. 9-12.

[13]   Cunningham SD, Huang JW, Chen J, Berti WR.: In: Abstracts of Papers of the American Chemical Society 1996; 212: pp. 87.

[14]   Salt DE, Smith RD, Raskin I. Phytoremediation. Annu Rev Plant Physiol Plant Mol Biol 1998; 49: 643-68.

[15]   Raskin I, Kumar PBAN, Dushenkov S, Salt D. Bioconcentration of heavy metals by plants. Curr Opin Biotechnol 1994; 5: 285-90.

[16]   Weisman D, Alkio M, Carmona AC. Transcriptional responses to polycyclic aromatic hydrocarbon-induced stress in *Arabidopsis thaliana* reveals the involvement of hormone and defense signaling pathways. BMC Plant Biol 2010; 10: 59-71.

[17]   James CA, Strand SE. Phytoremediation of small organic contaminants using transgenic plants. Curr Opin Biotechnol 2009; 20(2): 237-41.

[18]   Burns RG, Rogers S, McGhee I.: In: Naidu R, Kookana RS, Oliver DP, Rogers S, McLaughlin MJ, Eds. Contaminants and the Soil Environment in the Australia Pacific Region London, Kluwer Academic Publishers. 1996; pp. 361-410.

[19]   Brooks RR, Chambers MF, Nicks LJ, Robinson BH. Phytomining. Trends in Plant Sci 1998; 1: 359-62.

[20]   Cunningham SD, Shann JR, Crowley D, Anderson TA. In: Krueger EL, Anderson TA, Coats JP, Eds. Phytoremediation of Soil and Water Contaminants. American Chemical Society, Washington 1997.

[21]   Baker AJM, Brooks RR. Terrestrial higher plants which hyperaccumulate metallic elements. A review of their distribution, ecology and phytochemistry. Biorecovery 1989; 1: 81-126.

[22]   Baker AJM, Reeves RD, McGrath SP. In: Hinchee RE, Olfenbuttel RF, Eds. *In situ* decontamination of heavy metal polluted soils using crops of metal accumulating plants- A feasible study: *In situ* bioreclmation. Butterworth-Hinemann Boston 1991; pp. 539-44.

[23]   Reeves RD, Brooks RR. Hyperaccumulation of lead and zinc by two metallophytes from mining areas of Central Europe. Environ Pollut Ser A.1983; 31: 277-85.

[24]   Comis D. Green remediation: Using plants to clean the soil. J Soil Water Conserv 1996; 51(3): pp. 184-7.

[25]   Cunningham SD, Berti WR, Huang JW. Phytoremediation of contaminated soils. Trends Biotech 1995; 13: pp. 393-7.

[26]   Ebbs SD, Lasat MM, Brady DJ, Cornish J, Gordon R, Kochian LV. Phytoextraction of cadmium and zinc from a contaminated soil. J Environ Qual 1997; 26: 1424-30.

[27]   Reeves RD, Baker AJM. In: Raskin I, Ensley BD, Ed. Phytoremediation of toxic metals- using plants to clean-up the environment. In: Metal-accumulating plants. John Wiley & Sons, Inc., New York 2000; pp. 193-230.

[28]   Watanabe ME. Phytoremediation on the brink of commercialization. Environ Sci Technol 1997; 31(4):182A-6A.

[29]   Kumar PBAN, Dushenkov V, Phytoextraction: The use of plants to remove heavy metals from soils. Environ Sci Technol 1995; 29:1232-8.

[30]   Dushenkov V, Kumar PBAN, Motto H, Raskin I. Rhizofiltration: the use of plants to remove heavy metals from aqueous streams. Environ Sci Technol 1995; 29:1239-45.

[31]   Cunningham SD. Current Topics in Plant Biochemistry, Physiology, and Molecular Biology. In: Proceedings/Abstracts of the Fourteenth Annual Symposium, Columbia 1995; pp. 47-48.

[32]   Baker AJM, Walker PL. In: Shaw AJ, Ed. Heavy Metal Tolerance in Plants: Evolutionary Aspects. Boca Raton, CRC Press. 1990; pp. 155–77.

[33]   Cunningham SD, Ow DW. Promises and prospects of phytoremediation. Plant Physiol 1996; 110: 715-19.

[34]   Baker AJM, McGrath SP, Sidoli CMD, Reeves RD. The possibility of *in situ* heavy metal decontamination of polluted soils using crops of metal-accumulating plants. Resour Conserv Recycl 1994; 11: 41-9.

[35]   Hinchman RR, Negri MC, Gatliff EG. Phytoremediation: Using green plants to clean up contaminated soil, groundwater, and wastewater. Argonne National LaboratoryApplied Natural Sciences, Inc. 1998.

[36]   United States Protection Agency Reports, Introduction to Phytoremediation. – EPA 600/R-99/107, 2000.

[37]   Salt DE M, Blaylock PBA, Nanda K, Dushenkov V, Ensley BDI, Raskin I. Phytoremediation: A novel strategy for the removal of toxic metals from the environment using plants. Biotechnol 1995; 13: 468-74.

[38]   Baker AJM. Metal Hyperaccumulation by Plants: Our Present Knowledge of the Ecophysiological Phenomenon. In: Interdisciplinary Plant Group, Eds. Will Plants Have a Role in Bioremediation? Proceedings/Abstracts of the Fourteenth Annual Symposium Current Topics in Plant Biochemistry, Physiology, and Molecular Biology; University of Missouri: Columbia 1995.

[39]   Nanda Kumar PBA., Dushenkov V, Motto H, Raskin I. Phytoextraction: The use of plants to remove heavy metals from soils. Environ Sci Technol 1995; 29(5):1232-8.

[40]   Raskin I, Nanda Kumar PBA, Dushenkov S, Blaylock MJ, Salt D. Phytoremediation – Using plants to clean up soils and waters contaminated with toxic metals. Emerging technologies in hazardous waste management VI. In: Proceedings/Abstract of ACS Industrial & Engineering Chemistry Division Special Symposium, Atlanta, GA 1994; 1:19.

[41]   Brown SL, Chaney RL, Angle JS, Baker AJM. Phytoremediation potential of *Thlaspi caerulescens* and *Bladder campion* for Zinc and Cadmium contaminated soil. J Environ Qual 1994; 23: 1151-7.

[42]   Milner JM, Kochian LV. Investigating heavy-metal Hyperaccumulation using *Thlaspi caerulescens* as a model system. Ann Bot 2008; 102: 3-13.

[43]   Liu G, Chai T, Sun T. Sheng Wu Gong Cheng Xue Bao Heavy metal absorption, transportation and accumulation mechanisms in hyperaccumulator *Thlaspi caerulescens*. 2010; 26(5):561-8.

[44]   Leblebici Z, Aksoy A. Growth and lead accumulation capacity of *Lemna minor* and *Spirodela polyrhiza* (Lemnaceae): Interactions with nutrient enrichment. Water Air Soil Pollut 2011; 214(1-4):175-84.

[45]   Bañuelos, GS, Ajwa HA, Mackey B, Wu LL, Cook C, Akohoue S, Zambrzuski S. Evaluation of different plant species used for phytoremediation of high soil selenium. J Environ Qual 1997b; 26: 639-46.

[46]   Pierzynski GM, Schnoor JL, Banks MK, Tracy JC, Licht LA, Erickson LE. Vegetative remediation at superfund sites, mining and its environment impact. Royal Soc Chem Issues in Environ Sci Technol 1994; 1: pp. 49-69.

[47]   Adler T. Botanical Cleanup Crews. Sci. News 1996; 150: 42-3.

[48]   Haque N, Peralta JR, Jones GL, Gill TE, Gardea JL. Screening the phytoremediation potential of desert broom (*Baccharis sarothroides* Gray) growing on mine tailings in Arizona, USA. Environ Pollut 2008; 153(2): 362-8.

[49]   Lu Q, He ZL, Graetz DA, Stoffella PJ, Yang X. Uptake and distribution of metals by water lettuce (*Pistia stratiotes* L.). Environ Sci Pollut Res Int 2011.

[50]   Datta R, Quispe MA, Sarkar D. Greenhouse study on the phytoremediation potential of vetiver grass, *Chrysopogon zizanioides* L., in arsenic-contaminated soils. Bull Environ Contam Toxicol 2011; 86(1):124-8.

[51]   Huang, JW, Chen J, Berti WR, Cunningham SD. Phytoremediation of lead contaminated soils-Role of synthetic chelates in lead phytoextraction. Environ Sci Technol 1997a; 31: 800-6.

[52]   Lasat MM, Fuhrmann M, Ebbs SD, Cornish JE, Kochian LV. Phytoremediation of a radiocesium-contaminated soil: Evaluation of cesium-137 bioaccumulation in the shoots of three plant species. J Environ Qual 1998; 27:163-9.

[53]   Cunningham SD, Ow DW. Promises and prospects of phytoremediation. Plant Physiol 1996; 110: 715-9.

[54]   Blaylock MJ, Huang JW In: Raskin I, Ensley BD, Ed. Phytoextraction of metals. Phytoremediation of toxic metals - using plants to clean-up the environment. John Wiley & Sons, Inc., New York 2000; pp. 53-70.

[55]   Glass DJ. In: Raskin I, Ensley BD, Ed. Economic potential of phytoremediation. Phytoremediation of toxic metals - using plants to clean-up the environment. John Wiley & Sons, Inc., New York 2000; pp. 15-32.

[56]   Abou-Shanab R, Ghanem N, Ghanem K, Al-Kolaibe A. Phytoremediation potential of crop and wild plants for multi-metal contaminated soils. Res J Agric Biol Sci 2007; 3(5): pp. 370-6.

[57]   Lombi E, Zhao FJ, Dunham SJ, McGrath SP. Phytoremediation of heavy-metal contaminated soils: natural hyperaccumulation versus chemically enhanced phytoextraction. J Environ Qual 2001; 30: pp.1919-26.

[58]   Chen Y, Xiangdong L, Shen Z. Leaching and uptake of heavy metals by ten different species of plants during an EDTA-assisted phytoextraction process. Chemosphere 2004; 57: pp.187-96.

[59]   Cobbett CS. Phytochelatins and their role in Heavy Metal Detoxification. Plant Physiol 2000; 123: 825-32.

[60]   Grill E, Winnacker L, Zenk HM. Phytochelatins, the heavy-metal-binding peptides of plants, are synthesized from glutathione by a specific glutamylcysteine dipeptidyl transpeptidase (Phytochelatin Synthase). Proc Natl Acad Sci 1987; 86: 6838-42.

[61]   Rauser WE. Structure and function of metal chelators produced by plants: The case for organic acids, amino acids, phytin, and metallothioneins. Cell Biochem and Biophys 1999; 31: 19-48.

[62]   Vashegyi I, Cseh E, Lévai L, Fodor F. Chelator-enhanced lead accumulation in *Agropyron elongatum cv. Szarvasi-1* in hydroponic culture. Int J Phytoremed 2011; 13: 302-15.

[63]   Hirsch RE, Bryan DL, Edgar PS, Michael RS. A role for the AKT1 potassium channel in plant nutrition. Sci 1998; 280: 918-21.

[64]   Raskin I, Smith RD, Salt DE. Phytoremediation of metals: Using plants to remove pollutants from the environment. Curr Opin Biotechnol 1997; 8(2): 221-6.

[65]   Tester M, Leigh RA. Partitioning of nutrient transport processes in roots. J Exp Bot 2001; 52: 445–57.

[66]   Gaymard F. Identification and disruption of a plant shaker-like outward channel involved in K+ release into the xylem sap. Cell 1998; 94: 647–55.

[67]   Bubb JM, Lester JN. The impact of heavy metals on lowland rivers and the implications for man and the environment. Sci Total Environ 1991; 100: 207–33.

[68]   Cataldo DA, Wildung RE.: Soil and plant factors influencing the accumulation of heavy metals by plants. Environ and Health perspective 1978; 27: 149-59.

[69]   Dierberg FE, DeBusk TA, Goule NA Jr. In: Reddy KB, Smith WH, Ed. Aquatic Plants for Water Treatment and Resource Recovery. Florida, Magnolia Publishing Inc. 1987; pp. 497–504.

[70]   Karley AJ, Leigh RA, Sanders D. Where do all the ions go? The cellular basis of differential ion accumulation in leaf cells. Trends Plant Sci 2000; 5: 465-70.

[71]   Hinchman R, Negri C. Phytoremediation becoming quite "Poplar". Haz Waste Consult 1997; 15(3): 1-16.

[72]   Baker AJM. Accumulators and excluders - strategies in the response of plants to heavy metals. J Plant Nutr 1981; 3(1-4): 643-54.

[73]   Ramos L, Hernandez LM, Gonzalez JM, 1994. Sequential fractionation of copper, cadmium and zinc in soils from or near Donana National Park. J Environ Qual 23: 50-7.

[74]   Huang, JW, Blaylock MJ, Kapulnik Y, Ensley BD. Phytoremediation of Uranium-contaminated soils: Role of organic acids in triggering uranium hyperaccumulation in plants. Environ Sci Technol 1998; 32: 2004-8.

[75]   Lasat MM, Norvell WA, Kochian LV. Potential for phytoextraction of ^{137}Cs from a contaminated soil. Plant Soil 1997; 195: 99-106.

[76]   Blaylock MJ, Salt DE, Dushenkov S, Zakharova O, Gussman C, Kapulnik Y, *et al.* Enhanced accumulation of Pb in Indian Mustard by soil-applied chelating agents. Environ Sci Technol 1997; 31: 860-5.

[77]   Marschner H. Mineral nutrition of higher plants. 2nd ed. Academic Press: New York 1995.

[78]   McBride MB. Environmental chemistry of soils. 1st ed. Oxford University Press: New York 1994.

[79]   Norvell WA. In: Mortvedt JJ, Ed. Reactions of metal chelates in soil and nutrient solution. Micronutrients in agriculture. SSSA, Madison, WI 1991; pp. 187-228.

[80]   Huang JW, Chen J, Cunningham SD. Phytoextraction of lead from contaminated soils. In: Kruger EL, Ed. ACS symposium series 664 Washington, DC, American Chemical Society 1997b; pp. 283-98.

[81]   Kinnersely AM. The role of phytochelates in plant growth and productivity. Plant Growth Regul 1993; 12: 207–17.

[82]   Mench M, Martin E. Mobilization of cadmium and other metals from two soils by root exudates of *Zea mays* L., *Nicotiana tabacum* L. and *Nicotiana rustica* L. Plant and Soil 1991; 132: pp.187-96.

[83]   Krishnamurti GSR, Cieslinki G, Huang PM, Vanpees KCJ. Kinetics of cadmium release from soils as influenced by organic acids: implication in cadmium availability. J Environ Qual 1997; 26: pp. 271-7.

[84]   Lin Q, Chen YX, Chen HM, Yu YL, Luo YM, Wong MH. Chemical behavior of Cd in rice rhizosphere. Chemosphere 2003; 50: pp.755-61.

[85]   Wenzel WW, Unterbrunner R, Sommer P, Sacco P. Chelate-assisted phytoextraction using canola (*Brassica napus* L.) in outdoors pot and lysimeter experiments. Plant and Soil 2003; 249: pp.83-96.

[86]   Beranl MP, McGrath SP, Miller AJ, Baker AJM. Comparison of the chemical changes in the rhizosphere of the nickel hyperaccumulator *Alyssum murale* with the non-accumulator *Raphanus sativus*. Plant and Soil 1994; 164: pp. 251-9.

[87]   McGrath SP, Shen ZG, Zhao FJ. Heavy metal uptake and chemical changes in the rhizosphere of *Thlaspi caerulescens* and *Thlaspi ocholeucum* grown in contaminated soils. Plant and Soil 1997; 188: pp.153-9.

[88]   Gupta SK, Herren T, Wenger K, Krebs R, Hari T. In: Terry N, Banuelos G, Ed. *In situ* gentle remediation measures for heavy metal-polluted soils. Boca Raton, Lewis Publishers 2000; pp. 303-22.

[89]   Quartacci MF, Baker AJM, Navari-Izzo F. Nitrilotriacetate- and citric acid-assisted phytoextraction of cadmium by Indian mustard (*Brassica juncea* (L.) Czernj, Brassicaceae). Chemosphere 2005; 59: pp.1249-55.

[90]   Walker TS, Bals HP, Grotewold E, Vivanco JM. Root exudation and rhizosphere biology. Plant Physiol 2003; 132: pp. 44- 51.

[91]   Whiting SN, De Souza MP, Terry N. Rhizosphere bacteria mobilize Zn for hyperaccumulation by *Thlaspi caerulescens*. Environ Sci Tech 2001; 35: pp.3144-50.

[92]   Flathman PE, Lanza GR. Phytoremediation: current views on an emerging green technology. J Soil Contam 1998; 7(4): 415-32.

[93]   Zhu YL, Zayed AM, Quian JH, de Souza M, Terry N. Phytoaccumulation of trace elements by wetland plants: II. Water Hyacinth. J Environ Qual 1999; 28: 339-44.

[94]   Gupta PK. Biotechnology and Genomics. Ist Ed. Rastogi Publications, Meerut, India 2005-06; pp. 686-94.

[95]   Chaudhary TM, Hayes WJ, Khan AG, Khoo CS. Phytoremediation - focusing on accumulator plants that remediate metal-contaminated soils. Aust J Ecotoxicol 1998; 4: 37-51.

[96]   Raskin I, Ensley BD. Phytoremediation of Toxic Metals: Using Plants to Clean Up the Environment. John Wiley & Sons, Inc. New York 2000; pp. 53-70.

[97]   Dushenkov S, Vasudev D, Kapolnik Y, Gleba D, Fleisher D, Ting KC, Ensley B. Removal of Uranium from Water Using Terrestrial Plants. Environ Sci Technol 1997; 31(12): 3468-76.

[98]   Berti WR, Cunningham SD. In: Raskin, I, Ed. Phytoremediation of Toxic Metals: Using Plants to Clean Up the Environment. John Wiley and Sons, Inc. New York 2000; pp 71- 88.

[99]   Schnoor JL. In: Raskin I, Ensley BD, Ed. Phytostabilization of metals using hybrid poplar trees. John Wiley & Sons, Inc, New York 2000; pp. 133-50.

[100] EPA (Environmental Protection Agency). Electrokinetic laboratory and field processes applicable to radioactive and hazardous mixed waste in soil and groundwater. EPA 402/R-97/006. Washington 1997.

[101] Mueller B, Rock S, Gowswami D, Ensley D. Phytoremediation Decision Tree. Interstate Technology and Regulatory Cooperation Work Group 1999; pp. 1-36.

[102] Bañuelos GS. Phytoextraction of selenium from soils irrigated with selenium-laden effluent. Plant and Soil 2000; 224(2): 251-8.

[103] Lewis BG, Johnson CM, Broyer TC. Volatile selenium in higher plants: The production of dimethyl selenide in cabbage leaves by enzymic cleavage of Se-methyl selenomethionine selenonium salt. Plant Soil 1974; 40:107–18.

[104] Evans CS, Asher CJ, Johnson CM. Isolation of dimethyl diselenide and other volatile selenium compounds from *Astragalus racemosus* (Pursh.). Aust J Biol Sci 1968; 21: 13–20.

[105] Lewis BG, Johnson CM, Delwiche CC. Release of volatile selenium compounds by plants: Collection procedures and preliminary observations. J Agric Food Chem 1966; 14: 638–40.

[106] Black H. Absorbing possibilities: Phytoremediation. Environ Health Prespect 1995; 103(12): 1106-8.

[107] Shimp JF, Tracy JC, Davis LC, Lee E, Huang W, Erickson LE, Schnoor JL. Beneficial effects of plants in the remediation of soil and groundwater contaminated with organic materials. Crit Rev Environ Sci Technol 1993; 23: 41-77.

[108] Schnoor JL, Licht LA, McCutcheon SC, Wolfe NL, Carreira LH. Phytoremediation of organic and nutrient contaminants. Environ Sci Technol 1995a; 29: 318A-23A.

[109] Chaney RL, Li YM, Angle JS, Baker AJM, Reeves RD, Brown SL, Homer FA, Malik M, Chin M. In : Terry N, Banelos G, Ed. Phytoremediation of contaminated soil and water. Lewis Publishers, Boca Raton, FL. 2000; pp. 129–58.

[110] Davison J. Risk mitigation of genetically modified bacteria and plants designed for bioremediation. J Ind Microbiol Biotechnol 2005; 32: 639-50.

[111] Danika L, LeDuc Norman T. Phytoremediation of toxic trace elements in soil and water; J Ind Microbiol Biotechnol 2005; 32: 514-20.

[112] De Souza MP, Pilon-Smits EAH, Lytle CM, Hwang S, Tai J, Honma TSU, Yeh L, Terry N. Rate-limiting steps in selenium assimilation and volatilization by Indian mustard. Plant Physiol 1998; 117: 1487-94.

[113] Smith SE, McNair MR. Hypostatic modifiers cause variation in degree of copper tolerance in *Mimulus guttatus*. Heredity 1998; 80: 760-8.

[114] Rugh CL. Mercury detoxification with transgenic plants and other biotechnological breakthrough for phytoremediation *in vitro* cell development. Biol Plant 2001; 37: 321-5.

[115] Hansen D, Duda P, Zayed AM, Terry N. Selenium removal by constructed wetlands: role of biological volatilization. Environ Sci Technol 1998; 32: 591-7.

[116] Shah K, Nongkynrih J. Metal hyperaccumulation and bioremediation. Biol Plant 2007; 51(4): 618-34.

[117] Howden R, Goldsbrough PB, Andersen CR, Cobbett CS. Cadmium- sensitive, cad1 mutants of *Arabidopsis thaliana* are phytochelatin deficient. Plant Physiol 1995; 107(4): 1059-66.

[118] Fulekar MH, Singh A, Bhaduri AM. Genetic engineering strategies for enhancing phytoremediation of heavy metals. Afr J Biotechnol 2009; 8 (4): pp. 529-35.

[119] Eide D, Broderius M, Fett JM, Guerinot ML. A Novel Iron-Regulated Metal Transporter from Plants Identified by Functional Expression in Yeast. Proc Natl Acad Sci 1996; 93(11): 5624-8.

[120] Lu YP, Li ZS, Rea PA. AtMRP1 gene of Arabidopsis encodes a glutathione S-conjugate pump: isolation and functional definition of a plant ATP-binding cassette transporter gene. Proc Natl Acad Sci USA 1997; 94: 8243-8.

[121] Curie C, Panaviene Z, Loulerguech C, Delaporta SL, Briat JF, Walker EL. Maize yellow stripe encodes a membrane protein directly involved in Fe (III) uptake. Nature 2001; pp. 409.

[122] Tommasini R, Vogt E, Fromenteau M, Hortensteiner S, Matile P, Amrhein N, Martinoia E. An ABC transporter of *Arabidopsis thaliana* has both glutathione conjugate and chlorophyll catabolite transport activity. Plant J 1998; 13: 773-80.

[123] Morel M, Couuzet J, Gravot A, Auroy P, Leonhardt N, Vavasseur A, Richaud P. AtHMA3, a P_{1B}- ATPase allowing Cd/Zn/Co/Pb vacuolar storage in Arabidopsis. Plant Physiol 2009; 149: 894-04.

[124] Alkorta I, Hernandez-Allica J, Becerril JM, Amezaga I, Albizu I, Garbisu I. Recent findings on the phytoremediation of soils contaminated with environmentally toxic heavy metals and metalloids such as zinc, cadmium, lead and arsenic. Environ Sci Biotechnol 2004; 3: 71-90.

[125] Thomas JC, Davies EC, Malick FK *et al.* Yeast metallothionein in transgenic tobacco promotes copper uptake from contaminated soils. Biotechnol Prog 2003; 19: 273-80.

[126] Hamer DH. Metallothioneins. Ann Rev Biochem 1986; 55: 913-51.

[127] Chaney RL, Malik M, Li YM, Brown SL, Angle JS Baker AJM. Phytoremediation of soil metals. Curr Opin Biotechnol 1997; 8: 279-84.

[128] Ow DW. Heavy metal tolerance genes-prospective tools for bioremediation. Res Conserv Recycling 1996; 18: 135-49.

[129] Lee J, Bae H, Jeong J, Lee JY, Yang YY, Hwang I, Martinoia E, Lee Y. Functional expression of a bacterial heavy metal transporter in Arabidopsis enhances resistance to and decrease uptake of heavy metals. Plant Physiol 2003a; 133: 589-96.

[130] Van Huysen T, Abdel-Ghany S, Hale KL, LeDuc D, Terry N, Pilon-Smits EA. Overexpression of cystathionine-gamma-synthase enhances selenium volatilization in *Brassica juncea*. Planta 2003; 218: 71-8.

[131] Meagher RB. Phytoremediation of toxic elemental and organic pollutants. Curr Opin Plant Biol 2000; 3: 153-62.

[132] Ruiz ON, Daniell H. Genetic engineering to enhance mercury phytoremediation. Curr Opin Biotechnol 2009; 20(2): 213-19.

[133] Bizily SP, Kim T, Kandasamy MK, Meagher RB. Subcellular targeting of methyl mercury lyase enhances its specific activity for organic mercury detoxification in plants. Plant Physiol 2003; 131: 463-71.

[134] Grichko VP, Filby B, Glick BR. Increased ability of transgenic plants expressing the enzyme ACC deaminase to accumulate Cd, Co, Cu, Ni, Pb and Zn. J Biotechnol 2000; 81: 45-53.

[135] Smits EP. Phytoremediation. Annu Rev Plant Biol 2005; 56: 15-29.

INDEX

2,4-dichlorophenoxyacetic acid, 65
Alkaloids, 10
Amplification, 126
Amplified fragment length polymorphism, (AFLP) 65, 131
Anti-Bacterial, 86
Antibiotics, 4, 99
Antimicrobial, 86
Aromatic plants, 5
Bacillus, 100
Bacteria, 100
Benzyl amino purine, 65
Biocontrol agent, 101
Biofertilizers, 39
Biological control, 99
Biomedicines, 5, 4, 7, 12
Biotransformation, 5, 12, 21
Biological nitrogen fixation, 52
Biomedicines, 4
Bio-PCR, 124
Biopesticides, 39
Carbohydrates, 50
cDNA microarray, 120
Chitinase, 99, 104, 106, 107, 113, 114
Chitinolytic enzymes, 106
CMV, 130
cocaine, 5, 4, 10
Coumarins, 9
Crop, 60
Crop Improvement, 24
Crystallographic Studies, 86
Cumulative difference, 96
Cytoplasmic male sterilit, 67
DAF, 131
Deoxyribonuclease, 109
DICER, 26
Diphtheria, 86
Disease, 101
Diseases, 5
DNA, 31, 118
Dot-blot hybridization, 119
Double-stranded RNA, 35
Early nodulins, 52
ELISA, 130
Environment, 69
Euphorbia, 87
Exopolysaccharides, 52
Fingerprinting, 126
Flavonoids, 49
Gateway system, 33
Gene silencing, 32

Genetic engineering, 13
Genetic improvement, 29
Genetic manipulation, 12
Genic male sterility, 67
Genomics, 3, 24, 146, 148, 169
Genotypic, 68
GFP, 130
Glycosides, 11
Hybridization based, 119
In vitro, 7, 12, 20, 21, 58, 111
Indole acetic acid, 52, 109
Induced systemic resistance, 109
Infection, 43
Investigations, 62
Jatropha Gossypifolia, 86
Kilogram, 96
Late nodulins, 52
Least standard deviation, 96
Leguminous, 39
Macrocyclic diterpenoids, 86
Macrocyclic natural products, 95
Male sterility, 67
Male sterility, 67, 69, 71, 72, 77, 80, 82
McFarland standards, 89
Medicago truncatula, 51
Medicinal plants, 13
Medicine, 4, 15, 17
Megaplasmid, 48
Mentha arvensis, 10
Meperidine, 4
Meristematic activity, 46
Metabolic engineering, 5, 4, 13, 36, 72, 82, 145
Metabolome, 16
metal transporter proteins, 163
Metroxylon sagu, 60
Microbiological, 39
Micrograms, 96
Micropropagation, 60
Microarray, 116, 119, 120, 129, 132, 142, 147
Microfibrils, 45
Microparticle bombardment, 34
Micropropagation, 4, 5, 12, 60, 63, 64, 65
Microspores, 67, 72
Milligram, 96
Mimosoideae,, 39
Molecular Approaches, 116
Morphine, 4, 10
Mueller- Hinton agar, 89
Multiplex PCR, 121
M urashige and Skoog medium, 64
Mutation, 72
Mycobacterium tuberculosis, 86
NAC, 138
NAC domain proteins, 137, 138, 139, 143

N-acetylglucosamine, 48

N-acyl homoserine lactone, 105

Naphthoquinones, 9

Nauclea latifolia, 87

Neisseria, 87

Nested PCR, 121

Nicotine, 5, 10, 11

Nitrogen fertilizers, 40

Nitrogen Fixation, 3, 39, 46, 56

Nitrogenase, 40

Nitrogen-fixing symbiotic microbes, 40

Nod factors, 48

nodABC genes, 51

NodD gene, 51

Nodulation (NOD) factors, 51

Nodulation genes, 39, 49, 51, 58

Nodule primordium, 48

Nodules, 40

Nodulins, 50, 51, 52

Nucleic Acid Sequence- Based Amplification, 124

Ocimum sanctum, 7

Papilionoideae, 39

Parthenocarpic fruits, 65

Pathogen, 5, 99, 102, 126, 129, 130, 147

Pathogen quantification, 129

PCR-restriction fragment length polymorphism, 116

Peperomia galioides, 86

Pest control, 31

Phagocytosis, 48

pHELLSGATE, 33

Phenolics, 5, 107, 111

Phenotypic basis of male sterility, 68

Phenylpropanoid, 5, 9, 20

Phenylpropanoids, 8, 12, 21

Phytoalexins, 7, 18

Phytochelatins, 163

Phytochemicals, 5

Phytodegradation, 149

Phytoextraction, 149, 153, 155, 158, 167, 168, 169, 170

Phytopharmaceuticals, 12

Phytoremediation, 4, 149,

Phytostabilization, 149, 159, 169

Phytovolatalization, 149

Plant diseases, 99

Plant growth promoting rhizobacteria, 108

Plant pathogens, 99

Plasmalemma, 48

Pollination, 70

Pollination control systems, 67

Polygalacturonase, 48

Polyphenolics, 9

Primordium, 44

Pyrethrins, 5

Quinones, 9, 20

Random amplified polymorphic DNA, 126
Reactive oxygen species, 137
Real-time PCR, 121
Regulatory proteins, 74
Resveratrol, 6
Reverse-transcriptase PCR, 121
Rf genes, 77
Rhizobia, 40
Rhizobial pathway, 48
Rhizobium, 39
Rhizobium Genes, 48
Rhizobium-legume symbiosis, 39, 48, 56
Rhizofiltration, 149, 158, 159, 167
Rhizopines, 49
Rhizosphere, 40, 99
RISC complex, 24, 26, 27, 28
RNA interference, 2, 14, 24, 25, 26, 27, 28, 34, 35, 36, 37, 38, 73
RNA silencing, 24
RNA-dependent RNA polymerases, 26
RNAi, 3, 4, 24, 73, 76
RNAi discovery, 25
Root exudates, 158
Root nodulation, 40
Rosa damacena, 7
Sago palm, 60
Sago palm micropropagation, 60
Sago palm propagation, 64
Salmonella paratyphi-A & B, 88
Salmonella typhi, 11, 87, 88
Salmonella typhimurium, 86
Schizophyllan, 6
Secondary metabolites, 4, 105, 131
Selective ablation, 71
Serratia, 3, 99,
Shigella flexneri, 86, 88
Shikimate pathway, 7, 8
Shikonin, 5, 9, 12, 13, 14, 21
Siderophores, 106
Simple sequence repeats, 128
siRNA, 26, 27, 28, 32, 34, 35
Site-directed mutagenesis, 48
Sorghum, 69
Sporogenous tissues, 67
Stamens, 67
Staphyllococcus aureus, 11
Starch yield of sago palm, 62
Starter Culture, 89
Suspension cultures, 12
Symbiosis, 39, 51, 57
Systemic resistance, 107
Tannins, 9
Terpene, 5
Terpenes, 5, 10, 16
Terpenes, 10

Terpenoids, 6, 7, 9
thiosulfinates, 11
TLC plate, 93
Transactivation Domains, 139
transcription factors, 15, 137, 138, 139, 140, 142, 144, 145, 146, 147, 148
Transcription regulators, 138
Transcriptional gene silencing, 32
Transcriptomics, 16
Transgene silencing, 34
Transgenic crops, 68
Transgenic male sterility, 69
Vinblastine, 5, 10
Vincristine, 5, 10
Virus-induced gene silencing, 31
Withania somnifera, 4, 14, 22
Xanthohumol, 8, 20
X-Ray Diffraction Crystallography, 86
Zingiber officinalis, 4

www.ingramcontent.com/pod-product-compliance
Lightning Source LLC
Chambersburg PA
CBHW041705210326
41598CB00007B/537